Lecture Notes in Computer Science   11674

More information about this series at http://www.springer.com/series/7407

Emmanuel Filiot · Raphaël Jungers ·
Igor Potapov (Eds.)

Reachability Problems

13th International Conference, RP 2019
Brussels, Belgium, September 11–13, 2019
Proceedings

 Springer

Editors
Emmanuel Filiot
Université Libre de Bruxelles
Brussels, Belgium

Raphaël Jungers
Université Catholique de Louvain
Louvain-la-Neuve, Belgium

Igor Potapov
University of Liverpool
Liverpool, UK

ISSN 0302-9743   ISSN 1611-3349   (electronic)
Lecture Notes in Computer Science
ISBN 978-3-030-30805-6   ISBN 978-3-030-30806-3   (eBook)
https://doi.org/10.1007/978-3-030-30806-3

LNCS Sublibrary: SL1 – Theoretical Computer Science and General Issues

This Springer imprint is published by the registered company Springer Nature Switzerland AG
The registered company address is: Gewerbestrasse 11, 6330 Cham, Switzerland

Preface

This volume contains the papers presented at RP 2019, the 13th International Conference on Reachability Problems, organized in Brussels during September 11–13, 2019 by the Université Catholique de Louvain (UCL), Louvain-La-Neuve, Belgium and Université libre de Bruxelles (ULB), Brussels, Belgium. Previous events in the series were located at: Aix-Marseille University (2018), Royal Holloway, University of London (2017), Aalborg University (2016), the University of Warsaw (2015), the University of Oxford (2014), Uppsala University (2013), the University of Bordeaux (2012), the University of Genoa (2011), Masaryk University Brno (2010), École Polytechnique (2009), the University of Liverpool (2008), and Turku University (2007).

The aim of the conference is to bring together scholars from diverse fields with a shared interest in reachability problems, and to promote the exploration of new approaches for the modeling and analysis of computational processes by combining mathematical, algorithmic, and computational techniques. Topics of interest include (but are not limited to): reachability for infinite state systems; rewriting systems; reachability analysis in counter/timed/cellular/communicating automata; Petri nets; computational game theory, computational aspects of semi-groups, groups, and rings; reachability in dynamical and hybrid systems; frontiers between decidable and undecidable reachability problems; complexity and decidability aspects; predictability in iterative maps; and new computational paradigms.

We are very grateful to our invited speakers:

- Thomas A. Henzinger (IST Austria): "Temporal Logics for Multi-Agent Systems"
- Vladimir Protasov (HSE Moscow): "Primitivity and Synchronizing Automata: A Functional Analytic Approach"
- Slawomir Lasota (University of Warsaw): "The Reachability Problem for Petri Nets is Not Elementary"
- Sriram Sankaranarayanan (University of Colorado Boulder): "Reaching Out Towards Fully Verified Autonomous Systems"
- Jean-François Raskin (ULB Brussels): "Tutorial on Game Graphs for Reactive Synthesis"

The conference received 26 paper submissions, from which 2 papers were withdrawn. Each submission was carefully reviewed by three Program Committee (PC) members. Based on these reviews, the PC decided to accept 14 papers, in addition to the 4 invited talks and 1 invited tutorial. The members of the PC and the list of external reviewers can be found on the next pages. We are grateful for the high quality work produced by the PC and the external reviewers. Overall this volume contains 14 contributed papers and 3 papers from invited speakers which cover their talks.

The conference also provided the opportunity to other young and established researchers to present work in progress or work already published elsewhere. This year,

the PC selected 16 high-quality presentations on various reachability aspects in theoretical computer science.

List of accepted talk-only submissions:

- Amaury Pouly, Nathanaël Fijalkow, James Worrell, Joël Ouaknine, and Joao Sousa
  Pinto: On the Decidability of Reachability in Linear Time-Invariant Systems
- Bernadette Charron-Bost and Patrick Lambein-Monette: Randomization and
  Quantization for Average Consensus
- Cas Widdershoven and Stefan Kiefer: Efficient Analysis of Unambiguous Automata
  Using Matrix Semigroup Techniques
- Emilie Charlier, Célia Cisternino, and Adeline Massuir: State Complexity of the
  Multiples of the Thue-Morse Set
- Florent Delgrange, Thomas Brihaye, Youssouf Oualhadj, and Mickael Randour:
  Life is Random, Time is Not: Markov Decision Processes with Window Objectives
- Frederik M. Bønneland, Peter Gjøl Jensen, Kim Guldstrand Larsen, Marco Muniz,
  and Jiri Srba: Partial Order Reduction for Reachability Games
- Léo Exibard, Emmanuel Filiot, and Pierre-Alain Reynier: Synthesis of Data Word
  Transducers
- Marie van den Bogaard: The Complexity of Subgame Perfect Equilibria in Quan-
  titative Reachability Games
- Mehran Hosseini, Joël Ouaknine, and James Worrell: Termination of Affine Loops
  over the Integers
- Nikolaos Athanasopoulos and Raphaël Jungers: Path-Complete Reachability for
  Switching Systems
- Pablo Barceló, Chih-Duo Hong, Xuan-Bach Le, Anthony Widjaja Lin, and Reino
  Niskanen: Monadic Decomposability of Regular Relations
- Patricia Bouyer, Thomas Brihaye, Mickael Randour, Cédric Rivière, and Pierre
  Vandenhove: Reachability in Stochastic Hybrid Systems
- Pierre Ohlmann, Nathanal Fijalkow, and Pawe l Gawrychowski: The Complexity of
  Mean Payoff Games Using Universal Graphs
- Stéphane Le Roux, Arno Pauly, and Mickael Randour: Extending Finite-Memory
  Determinacy by Boolean Combination of Winning Conditions
- Thomas Colcombet, Joël Ouaknine, Pavel Semukhin, and James Worrell: On
  Reachability Problems for Low-Dimensional Matrix Semigroups
- Tobias Winkler, Sebastian Junges, Guillermo Pérez, and Joost-Pieter Katoen: On
  the Complexity of Reachability in Parametric Markov Decision Processes

Overall, the conference program consisted of 4 invited talks, 1 invited tutorial, 14
presentations of contributed papers, and 16 informal presentations in the area of
reachability problems stretching from results on fundamental questions in mathematics
and computer science to efficient solutions of practical problems.

A special thanks to the team behind the EasyChair system and the *Lecture Notes in
Computer Science* team at Springer, who together made the production of this volume
possible in time for the conference. Finally, we thank all the authors and invited
speakers for their high-quality contributions, and the participants for making RP 2019 a
success. We are also very grateful to Alfred Hofmann for the continuous support of the

event in the last decade, to F.R.S.-FNRS and Springer for their financial sponsorship, and the following people for their help in organizing RP 2019:

- The Formal Methods and Verification Team of ULB, in particular Nicolas Mazzocchi
- The ICTEAM Institute at UCLouvain, in particular Pascale Premereur, Marie-Christine Joveneau, Ludovic Taffin, Guillaume Berger, and Franois Gonze.

September 2019

Emmanuel Filiot
Igor Potapov
Raphaël Jungers

Organization

Program Committee

S. Akshay	IIT Bombay, India
Nikolaos Athanasopoulos	Queen's University Belfast, UK
Christel Baier	TU Dresden, Germany
Paul Bell	Liverpool John Moores University, UK
Nathalie Bertrand	Inria, France
Bernard Boigelot	University of Liege, Belgium
Ahmed Bouajjani	IRIF, University Paris Diderot, France
Alessandro D'Innocenzo	University of L'Aquila, Italy
Nathanaël Fijalkow	CNRS, LaBRI, University of Bordeaux, France
Emmanuel Filiot	Université Libre de Bruxelles, Belgium
Gilles Geeraerts	Université libre de Bruxelles, Belgium
Matthew Hague	Royal Holloway University of London, UK
Vesa Halava	University of Turku, Finland
Raphaël Jungers	Université Catholique de Louvain (UCL), Belgium
Martin Lange	University of Kassel, Germany
Fribourg Laurent	LSV, CNRS, ENS Paris-Saclay, France
Axel Legay	Université Catholique de Louvain (UCL), Belgium
Guillermo Perez	University of Antwerp, Belgium
Igor Potapov	University of Liverpool, UK
Pavithra Prabhakar	Kansas State University, USA
Maria Prandini	Politecnico di Milano, Italy
Alexander Rabinovich	Tel Aviv University, Israel
Mickael Randour	F.R.S.-FNRS, UMONS - Université de Mons, Belgium
Pierre-Alain Reynier	Aix-Marseille Université, France
Shinnosuke Seki	The University of Electro-Communications, Japan
Patrick Totzke	University of Liverpool, UK

Additional Reviewers

Almagor, Shaull	Delgrange, Florent	Niskanen, Reino
Ausiello, Giorgio	Fujiyoshi, Akio	Phawade, Ramchandra
Berwanger, Dietmar	Given-Wilson, Thomas	Rahimi Afzal, Zahra
Blondin, Michael	Hundeshagen, Norbert	Rossi, Massimiliano
Bollig, Benedikt	Lal, Ratan	Schnoebelen, Philippe
Bradfield, Julian	Mazowiecki, Filip	Taghian Dinani, Soudabeh
Bruse, Florian D.	Narayan Kumar, K.	Zimmermann, Martin

Abstracts of Invited Talks

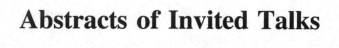

Bidding Games on Markov Decision Processes

Guy Avni[1], Thomas A. Henzinger[1], Rasmus Ibsen-Jensen[2],
and Petr Novotný[3]

[1] IST Austria
[2] University of Liverpool
[3] Masaryk University

Abstract. In two-player games on graphs, the players move a token through a graph to produce an infinite path, which determines the qualitative winner or quantitative payoff of the game. In bidding games, in each turn, we hold an auction between the two players to determine which player moves the token. Bidding games have largely been studied with concrete bidding mechanisms that are variants of a first-price auction: in each turn both players simultaneously submit bids, the higher bidder moves the token, and pays his bid to the lower bidder in richman bidding, to the bank in poorman bidding, and in taxman bidding, the bid is split between the other player and the bank according to a predefined constant factor. Bidding games are deterministic games. They have an intriguing connection with a fragment of stochastic games called random-turn games. We study, for the first time, a combination of bidding games with probabilistic behavior; namely, we study bidding games that are played on Markov decision processes, where the players bid for the right to choose the next action, which determines the probability distribution according to which the next vertex is chosen. We study parity and mean-payoff bidding games on MDPs and extend results from the deterministic bidding setting to the probabilistic one.

This research was supported in part by the Austrian Science Fund (FWF) under grants S11402-N23 (RiSE/SHiNE), Z211-N23 (Wittgenstein Award), and M 2369N33 (Meitner fellowship), and the Czech Science Foundation grant no. GJ19-15134Y.

The Reachability Problem for Petri Nets is Not Elementary

Slawomir Lasota

University of Warsaw
sl@mimuw.edu.pl

Abstract. The central algorithmic problem for Petri nets (or vector addition systems) is reachability: whether from the given initial configuration there exists a sequence of valid execution steps that reaches the given final configuration. The complexity of the problem has remained unsettled since the 1960s, and it is one of the most prominent open questions in the theory of verification. Decidability was proved by Mayr in his seminal STOC 1981 work, and the currently best published upper bound is non-primitive recursive Ackermannian of Leroux and Schmitz from LICS this year. We establish a non-elementary lower bound: the reachability problem needs a tower of exponentials of time and space. Until this work, the best lower bound has been exponential space, due to Lipton in 1976. The new lower bound is a major breakthrough for several reasons. Firstly, it shows that the reachability problem is much harder than the coverability (i.e., state reachability) problem, which is also ubiquitous but has been known to be complete for exponential space since the late 1970s. Secondly, it implies that a plethora of problems from formal languages, logic, concurrent systems, process calculi, and other areas, that are known to admit reductions from the Petri nets reachability problem, are also not elementary. Thirdly, it makes obsolete the currently best lower bounds for the reachability problems for two key extensions of Petri nets: with branching and with a pushdown stack.

Primitivity and Synchronizing Automata: A Functional Analytic Approach

Vladimir Yu. Protasov[1,2]

[1] University of L'Aquila, Italy
[2] Moscow State University, Russia
v-protassov@yandex.ru

Abstract. We give a survey of a function-analytic approach in the study of primitivity of matrix families and of synchronizing automata. Then we define the m-synchronising automata and prove that the existence of a reset m-tuple of a deterministic automata with n states can be decided in less than $mn^2\left(\log_2 n + \frac{m+4}{2}\right)$ operations. We study whether the functional-analytic approach can be extended to m-primitivity and to m-synchronising automata. Several open problems and conjectures concerning the length of m-reset tuples, m-primitive products, and finding those objects algorithmically are formulated.

Keywords: Nonnegative matrix · Primitive semigroups · Synchrinizing automata · Functional equation · Contraction · Affine operator

The research is supported by FRBR grants 17-01-00809 and 19-04-01227.

Game Graphs for Reactive Synthesis

Jean-François Raskin

Université libre de Bruxelles, U.L.B., Belgium
jraskin@ulb.ac.be

Abstract. In this talk, I will recall how to use infinite duration games played on graphs to formalize and solve the reactive synthesis problem. I will review the main results concerning the two-player zero-sum graph games including algorithms to solve them for ω-regular objectives. Then I will show some limitations of this framework and hint at solutions, based on non-zero games, to overcome them.

Reaching Out Towards Fully Verified Autonomous Systems

Sriram Sankaranarayanan[1], Souradeep Dutta[1], and Sergio Mover[2]

[1] University of Colorado, Boulder, USA
{srirams,souradeep.dutta}@colorado.edu
[2] Ecole Polytechnique, Institut Polytechnique de Paris, Palaiseau, France
smover@lix.polytechnique.fr

Abstract. Autonomous systems such as "self-driving" vehicles and closed-loop medical devices increasingly rely on learning-enabled components such as neural networks to perform safety critical perception and control tasks. As a result, the problem of verifying that these systems operate correctly is of the utmost importance. We will briefly examine the role of neural networks in the design and implementation of autonomous systems, and how various verification approaches can contribute towards engineering verified autonomous systems. In doing so, we examine promising initial solutions that have been proposed over the past three years and the big challenges that remain to be tackled.

Keywords: Formal verification · Autonomous systems · Constraint solvers

Contents

Bidding Games on Markov Decision Processes 1
Guy Avni, Thomas A. Henzinger, Rasmus Ibsen-Jensen,
and Petr Novotný

Primitivity and Synchronizing Automata: A Functional Analytic Approach . . . 13
Vladimir Yu. Protasov

Reaching Out Towards Fully Verified Autonomous Systems 22
Sriram Sankaranarayanan, Souradeep Dutta, and Sergio Mover

On the m-eternal Domination Number of Cactus Graphs 33
Václav Blažej, Jan Matyáš Křišt'an, and Tomáš Valla

On Relevant Equilibria in Reachability Games 48
Thomas Brihaye, Véronique Bruyère, Aline Goeminne,
and Nathan Thomasset

Partial Solvers for Generalized Parity Games 63
Véronique Bruyère, Guillermo A. Pérez, Jean-François Raskin,
and Clément Tamines

Reachability in Augmented Interval Markov Chains 79
Ventsislav Chonev

On Solving Word Equations Using SAT 93
Joel D. Day, Thorsten Ehlers, Mitja Kulczynski, Florin Manea,
Dirk Nowotka, and Danny Bøgsted Poulsen

Parameterised Verification of Publish/Subscribe Networks
with Exception Handling 107
Giorgio Delzanno

Cellular Automata for the Self-stabilisation of Colourings and Tilings 121
Nazim Fatès, Irène Marcovici, and Siamak Taati

On the Termination Problem for Counter Machines
with Incrementing Errors 137
Christopher Hampson

Reachability Problems on Partially Lossy Queue Automata 149
Chris Köcher

On the Computation of the Minimal Coverability Set of Petri Nets   164
    Pierre-Alain Reynier and Frédéric Servais

Deciding Reachability for Piecewise Constant Derivative Systems
on Orientable Manifolds.   178
    Andrei Sandler and Olga Tveretina

Coverability Is Undecidable in One-Dimensional Pushdown Vector
Addition Systems with Resets.   193
    Sylvain Schmitz and Georg Zetzsche

Synthesis of Structurally Restricted b-bounded Petri Nets:
Complexity Results   202
    Ronny Tredup

Reachability of Five Gossip Protocols   218
    Hans van Ditmarsch, Malvin Gattinger, Ioannis Kokkinis,
    and Louwe B. Kuijer

Author Index   233

Bidding Games on Markov Decision Processes

Guy Avni[1(✉)], Thomas A. Henzinger[1], Rasmus Ibsen-Jensen[2],
and Petr Novotný[3]

[1] IST Austria, Klosterneuburg, Austria
guy.avni@ist.ac.at
[2] University of Liverpool, Liverpool, UK
[3] Masaryk University, Brno, Czechia

Abstract. In two-player games on graphs, the players move a token through a graph to produce an infinite path, which determines the qualitative winner or quantitative payoff of the game. In *bidding games*, in each turn, we hold an auction between the two players to determine which player moves the token. Bidding games have largely been studied with concrete bidding mechanisms that are variants of a first-price auction: in each turn both players simultaneously submit bids, the higher bidder moves the token, and pays his bid to the lower bidder in *Richman bidding*, to the bank in *poorman bidding*, and in *taxman bidding*, the bid is split between the other player and the bank according to a predefined constant factor. Bidding games are deterministic games. They have an intriguing connection with a fragment of stochastic games called *random-turn* games. We study, for the first time, a combination of bidding games with probabilistic behavior; namely, we study bidding games that are played on Markov decision processes, where the players bid for the right to choose the next action, which determines the probability distribution according to which the next vertex is chosen. We study parity and mean-payoff bidding games on MDPs and extend results from the deterministic bidding setting to the probabilistic one.

1 Introduction

Two-player infinite-duration games on graphs are a central class of games in formal verification [2], where they are used, for example, to solve the problem of reactive synthesis [12], and they have deep connections to foundations of logic [14]. A graph game proceeds by placing a token on a vertex in the graph, which the players move throughout the graph to produce an infinite path ("play") π. The game is zero-sum and π determines the winner or payoff.

A graph game is equipped with a set of rules, which we call the "mode of moving", that determine how the token is moved in each turn. The simplest

This research was supported in part by the Austrian Science Fund (FWF) under grants S11402-N23 (RiSE/SHiNE), Z211-N23 (Wittgenstein Award), and M 2369-N33 (Meitner fellowship), and the Czech Science Foundation grant no. GJ19-15134Y.

© Springer Nature Switzerland AG 2019
E. Filiot et al. (Eds.): RP 2019, LNCS 11674, pp. 1–12, 2019.
https://doi.org/10.1007/978-3-030-30806-3_1

mode of moving is *turn based* in which the vertices are partitioned between the two players, and when the token is placed on a vertex v, the player who owns v decides to which neighbor of v it proceeds to. Turn-based games are used to model antagonistic behavior and are appropriate in worst-case analysis. On the other hand, probabilistic transitions conveniently model lack of information and are appropriate for average-case analysis. In *Markov chains*, the token proceeds from each vertex according to a probability distribution on neighboring vertices. A *Markov decision process* (MDP, for short) is associated with a set of actions Γ, and each vertex v is associated with a probability distribution $\delta(v, \gamma)$ on neighboring vertices, for each action $\gamma \in \Gamma$. Thus, an MDP can be thought of as a 1.5-player game in which, assuming the token is placed on a vertex v, the single player chooses an action γ, and *Nature* chooses the vertex to move the token to according to the distribution $\delta(v, \gamma)$. *Stochastic games*, a.k.a. 2.5-player games, combine turn-based games and probabilistic transitions [7]. The vertices in a stochastic game are partitioned between two players and a Nature player. Whenever the token is placed on a vertex that is controlled by a player, we proceed as in turn-based games, and whenever it is placed on a vertex that is controlled by Nature, we proceed randomly as in Markov chains.

Bidding is another mode of moving. In bidding games, both players have budgets and an auction is held in each turn to determine which player moves the token. Bidding games where introduced in [9, 10], where several concrete bidding rules were defined. In *Richman* bidding (named after David Richman), each player has a budget, and before each turn, the players submit bids simultaneously, where a bid is legal if it does not exceed the available budget. The player who bids higher wins the bidding, pays the bid to the other player, and moves the token. A second bidding rule called *poorman* bidding in [9], is similar except that the winner of the bidding pays the "bank" rather than the other player. Thus, the bid is deducted from his budget and the money is lost. A third bidding rule called *taxman* in [9], spans the spectrum between poorman and Richman bidding. Taxman bidding is parameterized by a constant $\tau \in [0, 1]$: the winner of a bidding pays portion τ of his bid to the other player and portion $1 - \tau$ to the bank. Taxman bidding with $\tau = 1$ coincides with Richman bidding and taxman bidding with $\tau = 0$ coincides with poorman bidding.

We study for the first time, a combination of the bidding and probabilistic modes of moving by studying bidding games that are played on MDPs; namely, the bidding game is played on an MDP, and in each turn we hold a bidding to determine which player chooses an action. One motivation for the study of bidding games on MDPs is practical; the extension expands the modelling capabilities of bidding games. A second motivation is theoretical and aims at a better understanding of a curious connection between bidding games and stochastic games, which we describe below.

Up to now, we have only discussed modes of moving the token. A second classification for graph games is according to the players' objectives. The simplest objective is *reachability*, where Player 1 wins iff an infinite play visits a designated target vertex. Bidding reachability games were studied in [9, 10], and these are the only objectives studied there. A central quantity in bidding games is the *initial ratio* of the players' budgets. The central question that was studied in [9] regards the existence of a necessary and sufficient initial ratio to guarantee winning the game. Formally, assuming that, for $i \in \{1,2\}$, Player i's initial budget is B_i, we say that Player 1's initial ratio is $B_1/(B_1 + B_2)$. The *threshold ratio* in a vertex v, denoted $\texttt{Thresh}(v)$, is such that if Player 1's initial ratio exceeds $\texttt{Thresh}(v)$, he can guarantee winning the game, and if his initial ratio is less than $\texttt{Thresh}(v)$, Player 2 can guarantee winning the game[1]. Existence of threshold ratios in reachability games for all three bidding mechanisms was shown in [9].

Moreover, the following probabilistic connection was shown for reachability games with Richman-bidding and only for this bidding rule. *Random-turn* games are a fragment of stochastic games. A random-turn game is parameterized by $p \in [0,1]$. In each turn, the player who moves is determined according to a (possibly) biased coin toss: with probability p, Player 1 chooses how to move the token, and Player 2 chooses with probability $1 - p$. Consider a reachability Richman-bidding game $\mathcal{G}$. We construct a "uniform" random-turn game on top of $\mathcal{G}$, denoted $\texttt{RT}^{0.5}(\mathcal{G})$, in which we toss an unbiased coin in each turn. The objective of Player 1 remains reaching his target vertex. It is well known that each vertex in $\texttt{RT}^{0.5}(\mathcal{G})$ has a *value*, which is, informally, the probability of reaching the target when both players play optimally, and which we denote by $val(\texttt{RT}^{0.5}(\mathcal{G}), v)$. The probabilistic connection that is observed in [10] is the following: For every vertex v in the reachability Richman-bidding game $\mathcal{G}$, the threshold ratio in v equals $1 - val(\texttt{RT}(\mathcal{G}), v)$. We note that such a connection is not known and is unlikely to exist in reachability games with neither poorman nor taxman bidding. Indeed, very simple poorman games have irrational threshold ratios [4]. Random-turn games have been extensively studied in their own right, mostly with unbiased coin tosses, since the seminal paper [11].

Infinite-duration bidding games were studied with Richman- [3], poorman- [4], and taxman-bidding [5]. The most interesting results in these papers regards an extended probabilistic connection for *mean-payoff* bidding games. Mean-payoff games are quantitative games; an infinite play is associated with a payoff that is Player 1's reward and Player 2's cost. Accordingly, we refer to the players in a mean-payoff game as Max and Min, respectively. Consider a strongly-connected mean-payoff taxman-bidding game $\mathcal{G}$ with taxman parameter $\tau \in [0,1]$ and initial ratio $r \in (0,1)$. The probabilistic connection is the following: the value of $\mathcal{G}$ w.r.t. τ and r, namely the optimal payoff Max can guarantee assuming his budget exceeds r, equals the *value* of the mean-payoff random-turn game $\texttt{RT}^{F(\tau,r)}(\mathcal{G})$ for $F(\tau,r) = \frac{r+\tau(1-r)}{1+\tau}$, where the value of $\texttt{RT}^{F(\tau,r)}(\mathcal{G})$ is the expected payoff when both players play optimally. Specifically, for Richman-bidding, the value does not depend on the initial ratio and equals the value of $\texttt{RT}^{0.5}(\mathcal{G})$. For

---

[1] When the initial ratio is exactly $\texttt{Thresh}(v)$, the winner depends on the mechanism with which ties are broken. Our results do not depend on a specific tie-breaking mechanism.Tie-breaking mechanisms are particularly important in discrete-bidding games [1].

poorman bidding, the value of $\mathcal{G}$ equals the value of $\mathrm{RT}^r(\mathcal{G})$. We highlight the point that bidding games are deterministic. One way to understand the probabilistic connection is as a "derandomization"; namely, Max has a deterministic bidding strategy in $\mathcal{G}$ that ensures a behavior that mimics the probabilistic behavior of $\mathrm{RT}^{F(\tau,r)}(\mathcal{G})$.

For qualitative objectives, we show existence of surely-winning threshold ratios in Richman-bidding reachability games. We then focus on strongly-connected games and show that in a strongly-connected parity taxman-bidding game, one of the players wins almost-surely with any positive initial budget. For mean-payoff objectives, we extend the probabilistic connection for strongly-connected mean-payoff taxman-bidding games from the deterministic setting to the probabilistic one. Namely, we show that the optimal expected payoff in a taxman-bidding game $\mathcal{G}$ w.r.t. τ and r equals the value of $\mathrm{RT}^{F(\tau,r)}(\mathcal{G})$. The proof is constructive and we show an optimal bidding strategy for the two players.

2   Preliminaries

A *Markov decision process* (MDP, for short) is $\mathcal{M} = \langle V, \Gamma, \delta \rangle$, where V is a set of vertices, Γ is a set of actions, and $\delta : V \times \Gamma \to [0,1]^V$ is a probabilistic transition function, where for every $v \in V$ and $\gamma \in \Gamma$, we have $\sum_{u \in V} \delta(v,\gamma)(u) = 1$. We say that an MDP $\mathcal{M}$ is *strongly-connected* if from every two vertices v and u, both players have a strategy that forces the game from v to u with probability 1. We focus on strongly-connected MDPs, where the initial position of the token is not crucial and we sometimes omit it.

We study bidding games that are played on MDPs. The game proceeds as follows. Initially, a token is placed on some vertex and the players start with budgets, which are real numbers. Suppose the token is placed on $v \in V$ in the beginning of a turn. We hold a bidding in which both players simultaneously submit bids, where a bid is legal if it does not exceed the available budget. The player who bids higher wins the bidding and chooses an action $\gamma \in \Gamma$, and the next position of the token is chosen at random according to the distribution $\delta(v,\gamma)$. The bidding rules that we consider differ in the update to the players' budget, and specifically, in how the winning bid is distributed.

Definition 1. *Suppose the players budgets are B_1 and B_2 and Player 1 wins the bidding with a bid of b. The budgets in the next turn are obtained as follows.*

- **Richman bidding**: *Player 1 pays Player 2, thus $B_1' = B_1 - b$ and $B_2' = B_2 + b$.*
- **Poorman bidding**: *Player 1 pays the bank, thus $B_1' = B_1 - b$ and $B_2' = B_2$.*
- **Taxman bidding** *with parameter $\tau \in [0,1]$: Player 1 pays portion τ to Player 2 and portion $(1 - \tau)$ to the bank, thus $B_1' = B_1 - b$ and $B_2' = B_2 + b \cdot \tau$.*

Note that fixing the taxman parameter to $\tau = 1$ gives Richman bidding and fixing $\tau = 0$ gives poorman bidding.

A finite *play* of a bidding game is in $(V \times \Gamma \times \mathbb{R} \times \{1,2\})^* \cdot V$. A strategy is a function that takes a finite player and prescribes a bid as well as an action to perform upon winning the bid. Two strategies f_1 and f_2 for the two players and an initial vertex v_0 give rise to a distribution over plays of length $n \in \mathbb{N}$, which we denote by $Dist_n(v_0, f_1, f_2)$ and define inductively. For $n = 0$, the probability of the play v_0 is 1. Consider a finite play π that visits $n-1$ vertices. For $i \in \{1, 2\}$, let $\langle b_i, \gamma_i \rangle = f_i(\pi)$. If $b_i > b_{3-i}$, then Player i wins the bidding, and the next action to be played is γ_i. For $u \in V$, the probability of the n-lengthed play $\pi \cdot \langle v, \gamma_i, b_i, i \rangle \cdot u$ is $\Pr[\pi] \cdot \delta(v, \gamma_i)(u)$. The issue of draws, i.e., the case in which $b_i = b_{3-i}$, needs to be handled with a tie-breaking mechanism, and our results are not affected by which mechanism is used. The extension of the distribution $D_n(v_0, f_1, f_2)$ to infinite paths is standard.

Random-Turn Games. *Stochastic games* generalize MDPs; while an MDP can be thought of as a player playing against Nature, in a stochastic game, a player is playing against a second adversarial player as well as against Nature. We consider a fragment of stochastic games called *random-turn* games, which are similar to bidding games except that, in each turn, rather than bidding, the player who chooses an action is selected according to some fixed probability. Formally, let $\mathcal{G} = \langle V, \Gamma, \delta, w \rangle$ be a mean-payoff bidding game and $p \in (0, 1)$, then the random-turn game that is associated with $\mathcal{G}$ and p is $\mathrm{RT}^p(\mathcal{G}) = \langle V \cup (V \times \{1, 2\}), \Gamma, \delta', w \rangle$, where vertices in V are controlled by Nature and model coin tosses and a vertex $\langle v, i \rangle$, for $i \in \{1, 2\}$, models the case that Player i is chosen to play. Thus, for every $v \in V$ and $\gamma \in \Gamma$, we have $\delta'(v, \gamma)(\langle v, 1 \rangle) = p$ and $\delta'(v, \gamma)(\langle v, 2 \rangle) = 1 - p$. Also, Player i controls every vertex in $V \times \{i\}$, and we have $\delta'(\langle v, i \rangle, \gamma) = \delta(v, \gamma)$. Finally, it is technically convenient to assume that vertices in $V \times \{1, 2\}$ do not contribute to the energy of a play.

3 Qualitative Bidding Games on MDPs

In this section we study infinite-duration games with qualitative objectives. We adapt the concept of *surely winning* to bidding games played on MDPs.

Definition 2. *Let $\mathcal{G}$ be a game that is played on an MDP $\langle V, \Gamma, \delta \rangle$, let $O \subseteq V^\omega$ be an objective for Player 1, and let $v \in V$. The surely-winning threshold ratio in v, denoted $\mathbf{Thresh}(v)$, is such that*

- *If Player 1's initial ratio exceeds $\mathbf{Thresh}(v)$, then Player 1 has a strategy such that no matter how Player 2 plays, the resulting play is in O.*
- *If Player 2's initial ratio exceeds $1 - \mathbf{Thresh}(v)$, then he has a strategy such that no matter how Player 1 plays, the resulting play is not in O.*

In reachability games, Player 1 has a target vertex and an infinite play is winning for him iff it visits the target. We show existence of surely-winning threshold ratios in reachability Richman-bidding games.

Theorem 1. *Let $\mathcal{G}$ be a reachability Richman-bidding game. Surely-winning threshold ratios exist in $\mathcal{G}$ and can be found using a linear reduction to a stochastic reachability game.*

Proof. Recall that the random-turn game $\text{RT}^{0.5}(\mathcal{G})$ is a stochastic game that models the following process: in each turn, we toss a fair coin, and if it turns "heads" Player 1 determines the next action and otherwise Player 2 determines the next action. The action gives rise to a probability distribution with which the following vertex is chosen. We construct $\mathcal{G}'$ similarly, only that we replace the last probabilistic choice with a deterministic choice of Player 2. Formally, the vertices of $\mathcal{G}'$ are $V \cup (V \times \{1, 2\}) \cup (V \times \Gamma)$. The transition function δ' restricted to V is the same as in $\text{RT}^{0.5}(\mathcal{G})$, namely, for every action, we proceed from $v \in V$ to $\langle v, i \rangle$, for $i \in \{1, 2\}$, with probability 0.5. The vertex $\langle v, i \rangle$ is controlled by Player i. A vertex $u \in V$ is a neighbor of $\langle v, 2 \rangle$ iff there exists $\gamma \in \Gamma$ with $\delta(v, \gamma)(u) > 0$. The neighbors of $\langle v, 1 \rangle$ are $\{v\} \times \Gamma$, where moving to $\langle v, \gamma \rangle$ models Player 1 choosing the action γ at v. Each vertex $\langle v, \gamma \rangle$ is controlled by Player 2 and a vertex $u \in V$ is a neighbor of $\langle v, \gamma \rangle$ iff $\delta(v, \gamma)(u) > 0$.

Let $v \in V$. The *value* of v in $\mathcal{G}'$, denoted $val(\mathcal{G}', v)$ is the probability of reaching the target when both players play optimally. We claim that the surely-winning threshold ratio in v equals $1 - val(\mathcal{G}', v)$. Note that when $val(\mathcal{G}', v) = 0$, no matter how Player 1 plays, there is no path from v to t, thus Player 1 cannot win and we have $\texttt{Thresh}(v) = 1$. Suppose $val(\mathcal{G}', v) = 1$ and we claim that $\texttt{Thresh}(v) = 0$. We follow the construction in the deterministic setting [3,9]. Let $n = |V|$. It is not hard to show that if Player 1 wins n biddings in a row, he wins the game. Suppose Player 1's initial ratio is $\epsilon > 0$. He follows a strategy that guarantees that he either wins n biddings in a row or, if he loses, his budget increases by a constant that depends on ϵ and n. Thus, by repeatedly playing according to this strategy, he either wins the game or increases his budget arbitrarily close to 1, where he can force n bidding wins. The proof for vertices with $val(\mathcal{G}', v) \in (0, 1)$ is similar only that Player 1's strategy maintains the invariant that his budget exceeds $1 - val(\mathcal{G}', v)$ and his *surplus*, namely the difference between his budget and $1 - val(\mathcal{G}', v)$, increases every time he loses a bidding. The proof for Player 2 is dual.   □

Theorem 1 shows a reduction from the problem of finding threshold ratios to the problem of solving a stochastic reachability game. The complexity of the later is known to be in NP and coNP [7], thus we obtain the following corollary.

Corollary 1. *The problem of deciding, given a reachability Richman-bidding game on an MDP $\mathcal{G}$ and a vertex v in $\mathcal{G}$, whether the surely-winning threshold ratio is at least 0.5, is in NP and coNP.*

The solution to strongly-connected games is the key in the deterministic setting. We show that Player 1 almost-surely wins reachability games that are played on strongly-connected MDPs.

Proposition 1. *Let $\mathcal{G} = \langle V, \Gamma, \delta, w \rangle$ be a strongly-connected taxman-bidding game with taxman parameter τ. For every positive initial budget, initial vertex $v \in V$, and target vertex $u \in V$, Player i has a strategy that guarantees that u is reached from v with probability 1.*

Proof. Let f_i be a strategy for Player i in the MDP $\langle V, \Gamma, \delta \rangle$ that guarantees that u is reached from v with probability 1. Let $\epsilon > 0$ be an initial budget or Player i in the bidding game $\mathcal{G}$. It is shown in [5] that, for every $n \in \mathbb{N}$, there is a bidding strategy that guarantees that Player i eventually wins n biddings in a row. Intuitively, Player i splits his budget into n exponentially increasing parts $\epsilon_1, \ldots, \epsilon_n$ such that if Player i loses the j-th bidding, for $1 \leq j \leq n$, his budget increases by a constant factor. By repeatedly following such a strategy, Player i's ratio approaches 1, which guarantees n consecutive wins. Player i splits his budget into infinitely many parts $\epsilon_1, \epsilon_2, \ldots$, and, for $n \geq 1$, he plays as if his budget is ϵ_n until he wins n consecutive biddings. Upon winning a bidding, he chooses actions according to f_i. Thus, Player i essentially follows f_i for growing sequences thereby ensuring visiting u with a probability that approaches 1.   □

Consider a strongly-connected parity taxman-bidding game $\mathcal{G}$ in which the highest parity index is odd. A corollary of the above proposition is that Player 1 almost-surely wins in $\mathcal{G}$ with any positive initial budget. Indeed, in $\text{RT}^p(\mathcal{G})$, by repeatedly playing according to a strategy f_i that forces a visit to the vertex v with the highest parity index, Player 1 forces infinitely many visits to v with probability 1. A bidding strategy proceeds as in the proof of the proposition above and forces increasingly longer sequences of bidding winnings, which in turn implies following f_i for increasingly longer sequences.

Theorem 2. *Let $\mathcal{G}$ be a strongly-connected parity game. If the maximal parity index in $\mathcal{G}$ is odd, then Player 1 almost-surely wins in $\mathcal{G}$ with any positive initial budget, and if the maximal parity index in $\mathcal{G}$ is even, Player 2 almost-surely wins in $\mathcal{G}$ with any positive initial budget.*

4   Mean-Payoff Bidding Games on Strongly-Connected MDPs

Mean-payoff bidding games are played on a *weighted* MDP $\langle V, \Gamma, \delta, w \rangle$, where $\langle V, \Gamma, \delta \rangle$ is an MDP and $w : V \to \mathbb{Q}$ is a weight function. The *energy* of a finite play π, denoted $E(\pi)$, refers to the accumulated weights, thus $E(\pi) = \sum_{1 \leq i \leq n} w(v_i)$. Consider two strategies f_1 and f_2, and an initial vertex v_0. The *payoff* w.r.t f_1, f_2, and v_0, is $\text{MP}(v_0, f_1, f_2) = \liminf_{n \to \infty} \mathbb{E}_{\pi \sim Dist_n(v_0, f_1, f_2)}[E(\pi)/n]$. A mean-payoff game is a zero-sum game. The payoff is Player 1's reward and Player 2's cost. Accordingly, we refer to Player 1 as Max and Player 2 as Min.

We focus on strongly-connected mean-payoff games. Since the mean-payoff objective is prefix independent, Proposition 1 implies that the optimal payoff from each vertex in a strongly-connected game is the same.

Definition 3 (Mean-payoff values). *Consider a strongly-connected mean-payoff taxman-bidding game $\mathcal{G} = \langle V, \Gamma, \delta, w \rangle$, a ratio $r \in (0,1)$, and a taxman parameter $\tau \in [0,1]$. We say that $c \in \mathbb{R}$ is the value of $\mathcal{G}$ w.r.t. r and τ, denoted $MP^{\tau, r}(\mathcal{G})$, if for every $\epsilon > 0$,*

– when Max's initial ratio is $r + \epsilon$, he can guarantee an expected payoff of at least c, and
– when Max's initial ratio is $r - \epsilon$, Min can guarantee an expected payoff of at most c.

We describe an optimal bidding strategy for Max in $\mathcal{G}$ w.r.t. τ and r. The construction consists of two components. The first component assigns an "importance" to each vertex, which we call the *strength* of a vertex and denote by $\mathrm{St}^p(v)$, for every $v \in V$. Intuitively, if $\mathrm{St}^p(v) > \mathrm{St}^p(u)$, then it is more important to move in v than it is in u. The second ingredient is a "normalization scheme" for the strengths, which consists of a sequence $(r_x)_{x \geq 1}$ and associating normalization factors $(\beta_x)_{x \geq 1}$, where $\beta_x, r_x \in [0, 1]$. Max keeps track of a position on the sequence, where he maintains the invariant that when the position is x, his ratio exceeds r_x. One property of the sequence is that the invariant implies that position $x = 1$ is never reached. Assuming the token is placed on $v \in V$ and the position on the sequence is x, Max's bid is roughly $\beta_x \cdot \mathrm{St}^p(v)$. The outcome of the bidding determines the next position on the sequence, where winning means that we proceed up on the sequence and losing means that we proceed down on the sequence. A normalization scheme for Richman bidding was devised in [3], for poorman bidding in [4], and we use a unified normalization scheme that was devised in [5] for taxman bidding.

We start with assigning importance to vertices. Our definition relies on a solution to random-turn games.

Definition 4 (Values). *For a strongly-connected mean-payoff bidding game $\mathcal{G}$ and $p \in (0, 1)$, the* mean-payoff value *of $RT^p(\mathcal{G})$, denoted $MP(RT^p(\mathcal{G}))$, is the maximal expected payoff that Max guarantee from every vertex.*

A *positional strategy* is a strategy that always chooses the same action in a vertex. It is well known that there exist optimal positional strategies for both players in stochastic mean-payoff games. For some $p \in (0, 1)$, consider two optimal positional strategies f and g in $RT^p(\mathcal{G})$, for Min and Max, respectively. For a vertex $v \in V$, let $\gamma^+(v), \gamma^-(v) \in \Gamma$ denote the actions that f and g prescribe, thus $\gamma^+(v) = f(\langle v, 1 \rangle)$ and $\gamma^-(v) = g(\langle v, 2 \rangle)$.

The *potential* of v, denoted $\mathrm{Po}^p(v)$, is a known concept in probabilistic models and was originally used in the context of the strategy iteration algorithm for MDPs [8]. We use the potential to define the *strength* of v, denoted $\mathrm{St}^p(v)$, which intuitively measures how much the expected potentials of the neighbors of v differ. The potential and strengths of v are functions that satisfy the following:

$$\mathrm{Po}^p(v) = p \cdot \sum_{u \in V} \delta(v, \gamma^+(v))(u) \cdot \mathrm{Po}^p(u) + (1-p) \cdot \sum_{u \in V} \delta(v, \gamma^-(v))(u) \cdot \mathrm{Po}^p(u) - \mathrm{MP}(\mathrm{RT}^p(\mathcal{G})) \text{ and}$$

$$\mathrm{St}^p(v) = p(1-p) \Big(\sum_{u \in V} \delta(v, \gamma^+(v))(u) \cdot \mathrm{Po}^p(u) - \sum_{u \in V} \delta(v, \gamma^-(v))(u) \cdot \mathrm{Po}^p(u) \Big)$$

The existence of the potential and thus the strength is known to be guaranteed [13].

Consider a finite path $\eta = \langle v_1, \gamma_1 \rangle, \ldots, \langle v_{n-1}, \gamma_{n-1} \rangle, v_n$. Consider a partition of $\{1, \ldots, n-1\}$ to $W(\eta) \cup L(\eta)$ such that $i \in W(\eta)$ iff $\gamma_i = \gamma^+(v_i)$. Intuitively, we think of η as a play and the indices in $W(\eta)$ are the ones that Max wins whereas the ones in $L(\eta)$ represent the ones in which he loses. The probability of η is $\prod_{1 \leq i < n} \delta(v_i, \gamma_i(v_i))(v_{i+1})$. The energy of η, denoted $E(\eta)$, is $\sum_{1 \leq i < n} w(v_i)$. We define a random variable Ψ_n over paths of length n. Let η be such a path that ends in a vertex v, then

$$\Psi_n^p(\eta) = \mathrm{Po}^p(v) + E(\eta) - \sum_{i \in W(\eta)} \mathrm{St}^p(v_i)/p + \sum_{i \in L(\eta)} \mathrm{St}^p(v_i)/(1-p) - (n-1) \cdot \mathrm{MP}(\mathrm{RT}^p(\mathcal{G})).$$

Lemma 1. *For every game $\mathcal{G}$, $p \in [0,1]$, and $n \in \mathbb{N}$, we have $\mathbb{E}[\Psi_n^p - \Psi_{n+1}^p] \geq 0$. Thus, $\mathbb{E}[\Psi_n] \geq \mathbb{E}[\Psi_1] \geq \min_v \mathrm{Po}^p(v)$.*

Proof. Let $\eta = \langle v_1, \gamma_1 \rangle, \ldots, \langle v_{n-1}, \gamma_{n-1} \rangle, v_n$ and $\gamma \in \Gamma$. We show that $\mathbb{E}[\Psi_n(\eta) - \Psi_{n+1}(\eta')] \leq 0$, where η' is obtained from η by extending it with a last vertex that is chosen according to the distribution $\delta(v_n, \gamma)$. We prove for the case of $\gamma = \gamma^+(v_n)$. Since Max wins the last bidding, we have $W(\eta) = W(\eta') \cup \{n\}$ and $I(\eta) = I(\eta')$. In addition, we have $E(\eta) + w(v_n) = E(\eta')$. Thus,

$$\mathbb{E}[\Psi_n(\eta) - \Psi_{n+1}(\eta')] =$$

$$= \mathrm{Po}^p(v_n) - \left(\sum_{u \in V} \mathrm{Po}^p(u) \cdot \delta(v_n, \gamma^+(v_n))(u) + w(v_n) - \mathrm{St}^p(v_n)/p - \mathrm{MP}(\mathrm{RT}^p(\mathcal{G})) \right) =$$

$$= \mathrm{Po}^p(v_n) - \left((1-p) \sum_{u \in V} \mathrm{Po}^p(u) \cdot \delta(v_n, \gamma^-(v_n))(u) + \right.$$

$$+ p \sum_{u \in V} \mathrm{Po}^p(u) \cdot \delta(v_n, \gamma^+(v_n))(u) + w(v_n) - \mathrm{MP}(\mathrm{RT}^p(\mathcal{G}))) =$$

$$= \mathrm{Po}^p(v_n) - \mathrm{Po}^p(v_n) = 0$$

The proof for the case that $\gamma \neq \gamma^+(v_n)$ is similar. Since we define $\gamma^-(v)$ to be the action that minimizes $\min_a \sum_{u \in V} \delta(v_n, a)(u) \cdot \mathrm{Po}^p(u)$, we get $\mathbb{E}[\Psi_n(\eta) - \Psi_{n+1}(\eta')] \geq 0$. $\qquad \square$

We continue to describe the properties of a normalization scheme as well as show its existence.

Lemma 2. *[5] Let $S \subseteq \mathbb{Q}_{\geq 0}$, a ratio $r \in (0,1)$, and a taxman parameter $\tau \in [0,1]$. For every $K > \frac{\tau r^2 + r(1-r)}{\tau(1-r)^2 + r(1-r)}$ there exist sequences $(r_x)_{x \geq 1}$ and $(\beta_x)_{x \geq 1}$ with the following properties.*

1. *For each position $x \in \mathbb{R}_{\geq 1}$ and $s \in S$, we have $\beta_x \cdot s \cdot r \cdot (r-1) < r_x$.*
2. *For every $s \in S \setminus \{0\}$ and $1 \leq x < 1 + rs$, we have $\beta_x \cdot s \cdot r \cdot (r-1) > 1 - r_x$.*
3. *The ratios tend to r from above, thus for every $x \in \mathbb{R}_{\geq 1}$, we have $r_x \geq r$, and $\lim_{x \to \infty} r_x = r$.*

4. We have

$$\frac{r_x - \beta_x \cdot s \cdot r \cdot (r-1)}{1 - (1 - \tau) \cdot \beta_x \cdot s \cdot r \cdot (r-1)} \geq r_{x+(1-r)\cdot K \cdot s} \; and$$

$$\frac{r_x + \tau \cdot \beta_x \cdot s \cdot r \cdot (r-1)}{1 - (1 - \tau) \cdot \beta_x \cdot s \cdot r \cdot (r-1)} \geq r_{x-s\cdot r}$$

We combine the two ingredients to obtain the following.

Theorem 3. *Let $\mathcal{G}$ be a strongly-connected mean-payoff taxman-bidding game, $r \in (0,1)$ an initial ratio, and $\tau \in [0,1]$ a taxman constant. Then, the mean-payoff value of $\mathcal{G}$ w.r.t. r and τ equals the value of the random-turn game $RT^{F(\tau,r)}(\mathcal{G})$ in which Max is chosen to move with probability $F(\tau,r)$ and Min with probability $1 - F(\tau,r)$, where $F(\tau,r) = \frac{r+\tau(1-r)}{1+\tau}$.*

Proof. Since the definition of payoff favors Min, it suffices to show an optimal strategy for Max. Let $\mathcal{G}$ such that $RT^{F(\tau,r)}(\mathcal{G}) = 0$. For $\epsilon > 0$, we describe a strategy for Max that guarantees a payoff that is greater than $-\epsilon$, assuming his initial ratio is strictly greater than r. Following [5], we consider a slight change of parameters; we choose $K > \frac{\tau r^2 + r(1-r)}{\tau(1-r)^2 + r(1-r)}$, and define $\nu = r$, $\mu = K \cdot (1-r)$, and $p = \nu/(\nu+\mu)$, where we choose K such that $MP(RT^p(\mathcal{G})) > -\epsilon$, where this is possible due to the continuity of the mean-payoff value due to changes in the probabilities in the game structure [6,15]. We find potentials and strengths w.r.t. p and find a sequence $(r_x)_{x \geq 1}$ as in Lemma 2, where we set $S = \{St^p(v) : v \in V\}$.

Max maintains a position on the sequence. Recall that Max's ratio strictly exceeds r and that Point 3 implies that the sequence tends from above to r, thus Max can choose an initial position x_0 such that his initial ratio exceeds r_{x_0}. Whenever the token reaches a vertex v and the position on the sequence is x, Max bids $St^p(v) \cdot r(1-r)\beta_x$, and chooses the action $\gamma^+(v)$ upon winning. If Max wins the bidding, the next position on the sequence is $x + \mu St^p(v)$, and if he loses a bidding, the next position is $x - \nu \cdot St^p(v)$. Note that Point 4 implies the invariant that whenever the position is x, Max's ratio exceeds r_x; indeed, the first part of the point takes care of winning a bidding, and the second part of losing a bidding. The invariant together with Point 1 implies that Max has sufficient funds for bidding. Suppose the current position is x following a play π, then $x = x_0 + \mu \sum_{i \in W(\pi)} St^p(v) - \nu \sum_{i \in L(\pi)} St^p(v)$. Point 2 implies that $x > 1$; indeed, consider a position that is close to 1, i.e., a position such that if Min wins a bidding, the next position is $x \leq 1$, then Point 2 states that Max's bid is greater than Min's ratio, thus he necessarily wins the bidding and the next position is farther from 1. Rearranging, dividing by $\mu \cdot \nu$, and multiplying by (-1), we obtain $\sum_{i \in L(\pi)} St^p(v)/\mu - \sum_{i \in W(\pi)} St^p(v)/\nu = (x_0 - x)/(\mu \cdot \nu) < (x_0 - 1)/(\mu \cdot \nu)$, where recall that x_0 is a constant.

Let $n \in \mathbb{N}$. We adapt the notation in Lemma 1 from paths to plays in the straightforward manner. The lemma implies that $\mathbb{E}[\Psi_n] \geq c$, for some constant $c \in \mathbb{Q}$. On the other hand, recall that, for a play π of length n that ends in a vertex v, we have

$$\Psi_n(\pi) = \text{Po}^p(v) + E(\pi) - \sum_{i \in W(\pi)} \text{St}^p(v)/\nu + \sum_{i \in L(\pi)} \text{St}^p(v)/\mu - (n-1)\text{MP}(\text{RT}^p(\mathcal{G})).$$

For every vertex v, we have $\text{Po}^p(v) \leq \max_u \text{Po}^p(u)$. Also, as in the above, we have $\mathbb{E}[\sum_{i \in W(\pi)} \text{St}^p(v)/\nu - \sum_{i \in L(\pi)} \text{St}^p(v)/\mu]$ is bounded from above by a constant. Combining, we have that $\mathbb{E}[E(\pi)] \geq c' + (n-1) \cdot \text{MP}(\text{RT}^p(\mathcal{G}))$. We divide both sides by n and tend it to infinity, thus the constant c' vanishes, and we get a payoff that exceeds $-\epsilon$, as required. □

Theorem 3 shows a reduction from the problem of finding the value of a mean-payoff taxman-bidding game on a strongly-connected MDP to the problem of solving a stochastic mean-payoff game. The complexity of the later is known to be in NP and coNP, thus we obtain the following corollary.

Corollary 2. *The problem of deciding, given a mean-payoff taxman-bidding game $\mathcal{G}$ that is played on a strongly-connected MDP, an initial ratio r, a taxman parameter τ, and a value $k \in \mathbb{Q}$, whether $\text{MP}^{\tau,r}(\mathcal{G}) \geq k$, is in NP and coNP.*

5 Discussion

We study qualitative and mean-payoff bidding games on MDPs. For qualitative objectives, we show existence of surely-winning threshold ratios in reachability bidding games, and we study almost-surely winning in strongly-connected parity bidding games. For mean-payoff objectives, we extend the probabilistic connection from the deterministic setting to the probabilistic one. A problem that we leave open is a quantitative solution to reachability bidding games that are played on MDPs; namely, given an MDP with a target vertex t, an initial vertex v, and a probability p, find a necessary and sufficient budget with which Player 1 can guarantee that t is reached from v with probability at least p. We expect that a solution to this problem will imply a solution to parity and mean-payoff bidding games on general graphs.

References

1. Aghajohari, M., Avni, G., Henzinger, T.A.: Determinacy in discrete-bidding infinite-duration games. In: Proceedings 30th CONCUR (2019)
2. Apt, K.R., Grädel, E.: Lectures in Game Theory for Computer Scientists. Cambridge University Press, Cambridge (2011)
3. Avni, G., Henzinger, T.A., Chonev, V.: Infinite-duration bidding games. J. ACM **66**(4), 31:1–31:29 (2019)
4. Avni, G., Henzinger, T.A., Ibsen-Jensen, R.: Infinite-duration poorman-bidding games. In: Christodoulou, G., Harks, T. (eds.) WINE 2018. LNCS, vol. 11316, pp. 21–36. Springer, Cham (2018). https://doi.org/10.1007/978-3-030-04612-5_2
5. Avni, G., Henzinger, T.A., Žikelić, Đ.: Bidding mechanisms in graph games. In: Proceedings of the 44th MFCS (2019)

6. Chatterjee, K.: Robustness of structurally equivalent concurrent parity games. In: Birkedal, L. (ed.) FoSSaCS 2012. LNCS, vol. 7213, pp. 270–285. Springer, Heidelberg (2012). https://doi.org/10.1007/978-3-642-28729-9_18
7. Condon, A.: The complexity of stochastic games. Inf. Comput. **96**(2), 203–224 (1992)
8. Howard, A.R.: Dynamic Programming and Markov Processes. MIT Press, Cambridge (1960)
9. Lazarus, A.J., Loeb, D.E., Propp, J.G., Stromquist, W.R., Ullman, D.H.: Combinatorial games under auction play. Games Econ. Behav. **27**(2), 229–264 (1999)
10. Lazarus, A.J., Loeb, D.E., Propp, J.G., Ullman, D.: Richman games. Games No Chance **29**, 439–449 (1996)
11. Peres, Y., Schramm, O., Sheffield, S., Wilson, D.B.: Tug-of-war and the infinity laplacian. J. Amer. Math. Soc. **22**, 167–210 (2009)
12. Pnueli, A., Rosner, R.: On the synthesis of a reactive module. In: Proceedings of the 16th POPL, pp. 179–190 (1989)
13. Puterman, M.L.: Markov Decision Processes: Discrete Stochastic Dynamic Programming. Wiley, New York (2005)
14. Rabin, M.O.: Decidability of second order theories and automata on infinite trees. Trans. AMS **141**, 1–35 (1969)
15. Solan, E.: Continuity of the value of competitive Markov decision processes. J. Theor. Probab. **16**, 831–845 (2003)

Primitivity and Synchronizing Automata: A Functional Analytic Approach

Vladimir Yu. Protasov[1,2(✉)]

[1] University of L'Aquila, L'Aquila, Italy
v-protassov@yandex.ru
[2] Moscow State University, Moscow, Russia

Abstract. We give a survey of a function-analytic approach in the study of primitivity of matrix families and of synchronizing automata. Then we define the m-synchronising automata and prove that the existence of a reset m-tuple of a deterministic automata with n states can be decided in less than $mn^2(\log_2 n + \frac{m+4}{2})$ operations. We study whether the functional-analytic approach can be extended to m-primitivity and to m-synchronising automata. Several open problems and conjectures concerning the length of m-reset tuples, m-primitive products and finding those objects algorithmically are formulated.

Keywords: Nonnegative matrix · Primitive semigroups ·
Synchrinizing automata · Functional equation · Contraction ·
Affine operator

1 Introduction

A multiplicative semigroup of nonnegative matrices is called primitive if it possesses at least one strictly positive matrix. Such semigroups were introduced relatively recently and have been intensively studied in the literature due to applications to Markov chains, linear dynamical systems, graph theory, etc. Their relation to synchronizing automata are especially important. There are rather surprising links between primitive semigroups and functional equations with the contraction of the argument. Those equations are usually applied to generate fractals and self-similar tilings. The theory of those equations can produce short and clear proofs of some known results on primitivity. For example, the characterization of primitive families, the theorem of existence of a common invariant affine subspace for matrices of non-synchronizing automata, etc. A new approach is also useful to study Hurwitz primitive (or m-primitive) semigroups. We discuss the characterization theorem for Hurwitz primitivity, which looks very similar to usual primitivity in spite of the totally different proofs. This leads to the concept of Hurwitz-synchronizing automata and reset m-tuples. We prove

The research is supported by FRBR grants 17-01-00809 and 19-04-01227.

E. Filiot et al. (Eds.): RP 2019, LNCS 11674, pp. 13–21, 2019.
https://doi.org/10.1007/978-3-030-30806-3_2

polynomial decidability of the existence of reset m-tuples and formulate several open problems.

Throughout the paper we denote the vectors by bold letters: $\boldsymbol{x} = (x_1, \ldots, x_n) \in \mathbb{R}^n$. The vectors $\boldsymbol{e}_i$, $i = 1, \ldots, n$ denote the canonical basis in $\mathbb{R}^n$ (all but one components of $\boldsymbol{e}_i$ are zeros, the ith component is one). The norms of vectors and of matrices is always Euclidean. By norm of affine operator we mean the norm of its linear part. The spectral radius (the maximal modulus of eigenvalues) of a matrix A is denoted by $\rho(A)$. By convex body we mean a convex compact set with a nonempty interior, $\mathrm{co}\, M$ denotes the convex hull of M. The *support* of a non-negative vector (matrix) is the set of positions of its strictly positive entries.

2   Contraction Operators and Reachability Theorems

Many facts on reachability in graphs and in automata can be formulated in terms of contraction operators on convex domains. The following results proved in [17] implies at least two important results on reachability. Let $\mathcal{B}$ be an arbitrary family of affine operators acting in $\mathbb{R}^d$. This family is called *contractive* if for every $\varepsilon > 0$, there exists a product Π of operators from $\mathcal{B}$ (with repetitions permitted) such that $\|\Pi\| < \varepsilon$. Clearly, this is equivalent to the existence of a product with the spectral radius (maximal modulus of eigenvalues) smaller than one.

Theorem 1. *[17] Let $G \subset \mathbb{R}^n$ be a convex body and $\mathcal{B}$ be a family of affine operators respecting this body, i.e., $B\, G \subset G$ for all $B \in \mathcal{B}$. Then $\mathcal{B}$ is contractive unless all operators of $\mathcal{B}$ possess a common invariant affine subspace of some dimension q, $0 \le q \le d - 1$, that intersects G.*

In the next section we show how to prove this fact using tools of functional analysis. Now let us demonstrate two of its corollaries from reachability problems. The first one deals with synchronizing automata and was presented in 2016 by Berlinkov and Szykula:

Theorem 2. *[3] If an automaton is not synchronizing, then its matrices possess a proper invariant common linear subspace.*

Let us show how this fact can be deduced from Theorem 1.

Proof. To every matrix A of the automaton we associate the corresponding affine operator $A|_V$ on the affine hyperplane $V = \{\boldsymbol{x} \in \mathbb{R}^n \mid \sum_{i=1}^n x_i = 1\}$. For an arbitrary product Π of matrices of the automaton, we have either $\|\Pi|_V\| = 0$ if Π has a positive row, of $\|\Pi|_V\| \ge 1$ otherwise. If a product with positive row exists, then the automaton is synchronizing. Otherwise, the set of matrices of the automaton is not contractive on V. Applying Theorem 1 to the simplex $\Delta = \mathrm{co}\,\{\boldsymbol{e}_1, \ldots, \boldsymbol{e}_n\} \subset V$ we conclude the existence of a common invariant affine subspace in V. Its linear span with the origin is the desired invariant linear subspace. $\qquad\square$

The next fact that can be derived from Theorem 1 is the criterion of primitivity of matrix family proved in 2013 by Protasov and Voinov [16]. Let $\mathcal{A}$ be an arbitrary irreducible family of nonnegative $n \times n$ matrices. Irreducibility means that there is no coordinate subspace of $\mathbb{R}^n$, i.e., subspace spanned by several vectors of the canonical basis, which is invariant for every matrix from $\mathcal{A}$. The family $\mathcal{A}$ is called *primitive*, if there exists a strictly positive product of matrices from $\mathcal{A}$.

The concept of primitivity of matrix families was introduced in [16] and has been studied in the literature due to many applications, see the bibliograpghy in [2,4,11]. The importance of this property is explained by the fact that if the family $\mathcal{A}$ is finite and all its matrices have neither zero rows nor zero columns, then the primitivity of $\mathcal{A}$ implies that almost all long products of matrices from $\mathcal{A}$ are strictly positive.

Theorem 3. *[16] Let $\mathcal{A}$ be an irreducible family of non-negative matrices. Suppose all matrices of $\mathcal{A}$ have neither zero rows nor zero columns; then $\mathcal{A}$ is not primitive if and only if there exists a partition of the set $\Omega = \{1, \ldots, n\}$ to $r \geq 2$ nonempty subsets $\{\Omega_k\}_{k=1}^r$, on which all the matrices from $\mathcal{A}$ act as permutations.*

This means that for every $A \in \mathcal{A}$, there is a permutation σ of the set $\{1, \ldots, r\}$ such that for each $i \in \Omega_k$, the support of the vector Ae_i is contained in $\Omega_{\sigma(k)}$, $k = 1, \ldots, k$. So, if the family $\mathcal{A}$ is not primitive, then there is a partition of the set of basis vectors, common to all matrices from $\mathcal{A}$, such that each matrix $A \in \mathcal{A}$ defines a permutation of this partition. If $\mathcal{A}$ is primitive, we formally set $r = 1$ and the partition is trivial $\Omega_1 = \Omega$. For one matrix, this fact is a part of Perron-Frobenius theorem. Moreover, in this case the permutation is cyclic. For families of matrices, these permutations can be arbitrary.

Despite the simple formulation, the proof of Theorem 3 is surprisingly long and technical. Now in the literature there are at least five different proofs of this theorem based on different ideas. The authors proof from [16] is based on geometry of convex polytopes. In that work the problem of finding a purely geometrical and possibly simpler proof was left. The problem seems to be reasonable in view of the combinatorial nature of the theorem. The first successful responds to this challenge was made by Alpin and Alpina in [1] and by Blondel, Jungers, and Olshevsky in [4]. Then Alpin and Alpina [2] suggested another construction. All those works presented (different!) combinatorial proofs, although still rather long. In 2015 Voynov and Protasov [17] noted that Theorem 3 can be actually derived by the same idea of contraction families of affine operators (Theorem 1), which gives the shortest known proof of this result. That proof may be called analytic, since Theorem 1 is established by analytic methods, with functional equations.

Before giving the proof we make a couple of observations. Nothing changes in Theorem 3 if we replace the family $\mathcal{A}$ by the family of all column-stochastic matrices with the same supports as matrices from $\mathcal{A}$. We assume that this replacement is already done and we keep the same notation for the new family. Thus, all matrices from $\mathcal{A}$ generate affine operators on the affine hyperplane V

and respect the simplex $\Delta \subset V$. Under the assumptions of Theorem 3, primitivity of the family $\mathcal{A}$ is equivalent to contractivity of this family on V. This fact is rather simple, its proof can be found in [14].

Proof of Theorem 3. Sufficiency is obvious. To prove the necessity we apply Theorem 1 to the family $\mathcal{A}|_V$ of affine operators on V and to the convex body $G = \Delta$. If $\mathcal{A}$ non-contracting, then the operators from $\mathcal{A}|_V$ share a common invariant affine subspace $L \subset V$, $0 \leq \dim L \leq n - 2$, intersecting Δ. Due to the irreducibility of $\mathcal{A}$, the subspace L intersects the interior of Δ, i.e., contains a positive point $\boldsymbol{a} \in \Delta$. Consider the following relation on the set Ω: $i \sim j$, if the vector $\boldsymbol{e}_i - \boldsymbol{e}_j$ belongs to $\tilde{L}$ (the linear part of L). This is an equivalence relation splitting Ω to classes $\Omega_1, \dots, \Omega_r$. Since $\dim L \leq n - 2$, we have $r \geq 2$. Let us show that for every $i \in \Omega$ and $A \in \mathcal{A}$, the supports of all vectors of the set $M_i(A) = \{A\boldsymbol{e}_i, \ A \in \mathcal{A}\}$ lie in one set Ω_k. It suffices to prove that the difference of each two elements of the set $M_i(A)$ lies in $\tilde{L}$. Take arbitrary $A, B \in \mathcal{A}$. Let a matrix $C \in \mathcal{A}$ have the same ith column as the matrix A and all other columns as the matrix B. For a positive vector $\boldsymbol{a} = \sum_i a_i \boldsymbol{e}_i \in L$, we have

$$a_i(A\boldsymbol{e}_i - B\boldsymbol{e}_i) = a_i(C\boldsymbol{e}_i - B\boldsymbol{e}_i) = C(a_i\boldsymbol{e}_i) - B(a_i\boldsymbol{e}_i) = C\boldsymbol{a} - B\boldsymbol{a} \in \tilde{L}.$$

Since $C\boldsymbol{a} \in L$ and $B\boldsymbol{a} \in L$, we have $C\boldsymbol{a} - B\boldsymbol{a} \in \tilde{L}$, and hence $A\boldsymbol{e}_i - B\boldsymbol{e}_i \in \tilde{L}$. Thus, the supports of ith columns of all matrices from $\mathcal{A}$ belong to one set Ω_k. Then the supports of all columns with indices from the equivalence class of the index i (say, Ω_j) lie in the same class Ω_k. Consequently, the matrix A defines a map $\sigma(j) = k$. Since A does not have zero rows, it follows that σ is a permutation. $\qquad \square$

3   Contractive Families and Functional Equations

The proof of Theorem 1 is realized by applying the theory of fractal curves and equations of self-similarity. This idea originated in [17], here we slightly simplify that proof. Let us have a finite family of affine operators $\mathcal{B} = \{B_1, \dots, B_m\}$. The *self-similarity equation* is the equation on function $\boldsymbol{v} : [0, 1] \to \mathbb{R}^d$:

$$\boldsymbol{v}(t) = B_k \boldsymbol{v}(mt - k + 1), \quad t \in \left[\frac{k-1}{m}, \frac{k}{m}\right] \qquad k = 1, \dots, m. \qquad (1)$$

This equation plays an exceptional role in the theory of subdivision schemes, compactly supported wavelets, etc. see [5, 6] and references therein. A advantage of those equations is an existence and uniqueness theorem for solutions. We will restrict ourselves to L_2-solutions. The solvability and smoothness of the solution are expressed in terms of the so-called L_2-*spectral radius* or in short, *2-radius* of linear or affine operators: $\rho_2(\mathcal{B}) = \lim_{k \to \infty} [m^{-k} \sum_{\Pi \in \mathcal{B}^k} \|\Pi\|^{1/k}]^{1/m^k}$. Here we denote by $\mathcal{B}^k$ the set of products of length k of operators from the family $\mathcal{B}$ (repetitions permitted). If there is a convex body G such that $BG \in G$ for all $B \in \mathcal{B}$, then, of course, all norms of products $\|\Pi\|$ are uniformly bounded, and hence $\rho_2 \leq 1$. It turns out that the family $\mathcal{B}$ is contrative precisely when $\rho_2 < 1$. This fact is rather simple, its proof can be found in [17]. Now we formulate the main result on equations of self-similarity.

Theorem 4. *[13] Suppose a finite family of affine operators $\mathcal{B}$ does not possess a common invariant affine subspace; then Eq. (1) possesses an L_2-solution if and only if $\rho_2(\mathcal{B}) < 1$. In this case the solution is unique and continuously depends on the family $\mathcal{B}$.*

This theorem implies Theorem 1 on contractive families.

Proof of Theorem 1. We show only the existence of a common invariant subspace. For the proof that this subspace intersects G, see [17]. It suffices to prove the theorem for finite families $\mathcal{B}$. Indeed, if $\mathcal{B}$ is non-contractive, then so is every its finite subset, and consequently every finite subset of $\mathcal{B}$ possesses an invariant affine subspace intersecting G. In this case the whole family $\mathcal{B}$ possesses such an invariant subspace. Thus, we assume $\mathcal{B} = \{B_1, \ldots, B_m\}$ and that this family is not contractive. Take an arbitrary point $a \in \operatorname{int} G$ and consider the operator B_1^ε defined as $B_1^\varepsilon x = (1 - \varepsilon) B_1 x + \varepsilon x$. For every $\varepsilon \in (0, 1)$, the operator B_1^ε maps G to $\operatorname{int} G$, hence $\rho(B_1^\varepsilon) < 1$. Therefore, the family $\mathcal{B}^\varepsilon = \{B_1^\varepsilon, B_2, \ldots, B_m\}$ is contractive, and consequently $\rho_2(\mathcal{B}^\varepsilon) < 1$. Hence, the self-similarity Eq. (1) with the family $\mathcal{B}^\varepsilon$ possesses a unique L_2-solution v_ε. We have $v_\varepsilon(t) \in G$ for almost all $t \in [0, 1]$. Taking an arbitrary sequence $\varepsilon_k \to 0$ we obtain a bounded sequence v_{ε_k}, which has a weak-* limit v. This limit satisfies Eq. (1) with the initial family of operators $\mathcal{B}$. Therefore, if the family $\mathcal{B}$ does not have an invariant affine subspace, then $\rho_2(\mathcal{B}) < 1$. Hence, $\mathcal{B}$ is contractive which is a contradiction. $\square$

4 m-primitivity and m-syncronising Automata

In this section we compare the main results on primitive and on m-primitive families and rise a question whether the analytic method can be extended to the study of m-primitivity. Then we define m-synchronizing automata, prove that the existence of a synchronising m-tuple can be decided in polynomial time and leave several open problems. For example, how to find the synchronising m-tuple and what is an upper bound for its length?

The concept of m-primitivity was introduced in 1990 by Fornasini [7] and then studied by Fornasini and Valcher [8,9]. Now there is an extensive literature on this subject, see [10,12,15] and references therein. The main application of m-primitivity is the multivariate Markov chains [7,8], although there are some natural generalization to the graph theory, dynamical systems and large networks. In the notation, m is the number of matrices, that is why we say not "k-primitive" as in some works, but "m-primitive". It would also be natural to use the term "Hurwitz primitive".

Let us have a finite family of nonnegative $n \times n$-matrices $\mathcal{A} = \{A_1, \ldots, A_m\}$. For a given m-tuple $\alpha = (\alpha_1, \ldots, \alpha_m)$ of nonnegative integers, $\sum_{i=1}^m \alpha_i = k \geq 1$, called also *colour vector* we denote by $\mathcal{A}^\alpha$ the sum of all products of m matrices from $\mathcal{A}$, in which every product contains exactly α_i factors equal to A_i. The number k will be referred to as the length of α and denoted by $|\alpha|$. For example, if $\mathcal{A} = \{A_1, A_2, A_3\}$, then $\mathcal{A}^{(1,3,0)} = A_1 A_2^3 + A_2 A_1 A_2^2 + A_2^2 A_1 A_2 + A_2^3 A_1$. Such sums are called in the literature Hurwitz products, although they are not actually products but sums of products. A family of non-negative matrices is m-*primitive* if there exists a strictly positive Hurwitz product of those matrices. This property is weaker than primitivity: primitivity implies m-primitivity, but not vice versa.

This notion has an obvious combinatorial interpretation. Suppose there are n villages, some of them are connected by one-way roads colored in m colors (two villages may be connected by several roads). The m-primitivity means that there exists a colored vector $\alpha = (\alpha_1, \ldots, \alpha_m)$ such that every two villages are connected by a path of length $|\alpha|$ that for each $i = 1, \ldots, m$ contains precisely α_i roads of the ith color. The criterion of m-primitivity was proved in [15] in 2013:

Theorem 5. *[15] Let $\mathcal{A}$ be an irreducible family of non-negative matrices. Suppose all matrices of $\mathcal{A}$ have no zero columns; then $\mathcal{A}$ is not primitive if and only if there exists a partition of the set $\Omega = \{1, \ldots, n\}$ to $r \geq 2$ nonempty subsets $\{\Omega_k\}_{k=1}^r$, on which all the matrices from $\mathcal{A}$ act as permutations and all those permutations commute.*

If $\mathcal{A}$ is primitive, we formally set $r = 1$ and the partition is trivial $\Omega_1 = \Omega$. This criterion almost literally repeats the criterion of primitivity in Theorem 3. The main difference is the following: if for non-primitivity the matrices A_i can define arbitrary permutations of the sets $\{\Omega_j\}_{j=1}^r$, for non m-primitivity those permutations have to commute. This criterion quite naturally shows the common properties and the difference between those two concepts. Another difference is that the criterion of Theorem 5 does not require the absence of zero rows and columns, as it was in Theorem 3, but just zero columns. This condition is much less restrictive, since it is always satisfied for column stochastic matrices and for matrices of automata.

As a corollary of Theorem 5 one can obtain that the m-primitivity is polynomially decidable. In [15] an algorithm was presented to find the partition $\{\Omega_j\}_{j=1}^r$. If $r = 1$, then the family is m-primitive. The complexity of the algorithm is $O(mn^3 + m^2 n^2)$.

Now let us formulate three open problems.

Problem 1. *Can Theorem 5 be derived by a geometrical of function-analytic argument, similar to Theorem 1?*

In spite of similarity of Theorems 3 and 5, their proofs are totally different. The only known proof of Theorem 5 is combinatorial and has nothing in common with all five known proofs of Theorem 3.

Problem 2. *If the family is primitive, how to find its positive Hurwitz product within polynomial time?*

The algorithm from [15] based on Theorem 5 is not constructive. It decides whether the family is m-primitive by finding the partition $\{\Omega_j\}_{j=1}^r$, but does not find any positive Hurwitz product. The greedy algorithm for finding positive product of a primitive family seems not to work here.

Problem 3. *What is a sharp upper bound for the exponent of m-primitivity?*

The exponent of m-primitivity is the minimal number $k = k(n, m)$ such that every m-primitive family of matrices of size n has a positive Hurwitz product with a colour vector of length at most k.

Conjecture 1. *Under the assumption that all matrices of the family $\mathcal{A}$ do not have zero columns, the exponent of m-primitivity is polynomial in n and m.*

Examples from the works [10, 12] show that the exponents of m-primitivity can be exponential in m. In example from [12], we have $k(n, m) \geq Cn^{m+1}$. However, in all those examples the matrices have zero columns.

Similarly to m-primitivity, we can define m-synchronising automata. Let us have deterministic finite automaton with n states and with m actions. The automaton is called m-synchronizing if there exists an m-tuple $\alpha = (\alpha_1, \ldots, \alpha_m)$ such that for every state there exists a reset word of length $|\alpha|$ with α_i commands of the ith action, $i = 1, \ldots, m$. This means that for every starting state there is a word (may be different for different states) with the m-tuple α that leaves the automaton on a prescribed particular state, one for all starting states. Thus the reset words may be different for different states, but with the same *reset m-tuple*.

In terms of graphs, this means that there is a path from each vertex to the particular vertex that has exactly α_i edges of the ith colour, $i = 1, \ldots, m$. The synchronising m-tuple has the following meaning. Assume realisation of each action is not free, it requires some resources. It is possible to leave a stock set of resources, α_i units for the ith action, so that it is always possible to reset the system using this stock? Let us stress that we can control the sequences of actions to reset the system from a given state, but we are not able to control the stock, it is the same for all states.

The existence of reset m-tuples can be decided within polynomial time, at least for irreducible automata, i.e., whose sets of matrices are irreducible.

Theorem 6. *There is a polynomial time algorithm to decide if a given automaton is m-synchronising. The algorithm spends less than $mn^2(\log_2 n + \frac{m+4}{2})$ arithmetic operations.*

Proof. Let $\mathcal{A} = \{A_1, \ldots, A_m\}$ be a family of matrices of the automaton. An m-tuple is synchronising precisely when the corresponding Hurwitz product of matrices $A_1, \ldots, A_m$ possesses a positive row. Since all matrices A_i are column stochastic, we can apply Theorem 5. Using algorithm from [15], we find the partition $\{\Omega_j\}_{j=1}^r$. If $r = 1$, then $\mathcal{A}$ is m-primitive and hence the corresponding Hurwitz product is strictly positive and so has a positive row. If $r \geq 2$, then there is no Hurwitz product with a positive row. Indeed, since all permutations of the set $\{\Omega_j\}_{j=1}^r$ defined by matrices from $\mathcal{A}$ commute, it follows that all products $\Pi = A_{d_1} \cdots A_{d_k}$ corresponding to one m-tuple $\boldsymbol{\alpha}$ define the same permutation. Therefore, they have the same block structure corresponding to the partition $\{\Omega_j\}_{j=1}^r$ and hence their sum does not have a positive row. Thus, the automaton is m-synchronizing if and only if $r = 1$. By [15, Theorem 2], the algorithm spends less than $mn^2(2p + \log_2 n + \frac{m}{2})$ operations, where p is the maximal number of positive components in columns of the matrices from $\mathcal{A}$. In our case, $p = 1$, which completes the proof. $\qquad\square$

Thus, for irreducible automata, the existence of reset m-tuple can be decided by a polynomial algorithm. But this algorithm does not find the reset m-tuple.

Problem 4. *If an automaton in m-primitive, how to find its reset m-tuple within polynomial time?*

Problem 5. *What is a sharp upper bound for the minimal length of the reset m-tuple?*

For synchronising automata this bound is known to be $O(n^3)$ and there is a long standing Černý conjecture that the minimal upper bound is actually $(n-1)^2$. What are the bounds for m-synchronizing automata?

References

1. Alpin, Yu.A., Alpina, V.S.: Combinatorial properties of irreducible semigroups of nonnegative matrices. J. Math. Sci. (N. Y.) **191**(1), 4–9 (2013)
2. Alpin, Yu.A., Alpina, V.S.: A new proof of the Protasov-Voynov theorem on semigroups of nonnegative matrices. Math. Notes **105**, 805–811 (2019)
3. Berlinkov, M.V., Szykula, M.: Algebraic synchronization criterion and computing reset words. Inf. Sci. **369**, 718–730 (2016)
4. Blondel, V.D., Jungers, R.M., Olshevsky, A.: On primitivity of sets of matrices. Automatica **61**, 80–88 (2015)
5. Cabrelli, C.A., Heil, C., Molter, U.M.: Self-similarity and Multiwavelets in Higher Dimensions, vol. 170, no. 807. Memoirs of the American Mathematical Society (2004)
6. Cavaretta, A.S., Dahmen, W., Micchelli, C.A.: Stationary Subdivision, vol. 93, no. 453. Memoirs of the American Mathematical Society (1991)
7. Fornasini, E.: 2D Markov chains. Linear Algebra Appl. **140**, 101–127 (1990)
8. Fornasini, E., Valcher, M.: Directed graphs, 2D state models and characteristic polynomials of irreducible matrix pairs. Linear Alg. Appl. **263**, 275–310 (1997)
9. Fornasini, E., Valcher, M.: A polynomial matrix approach to the structural properties of 2D positive systems. Linear Algebra Appl. **413**(2–3), 458–473 (2006)
10. Gao, Y., Shao, Y.: Generalized exponents of primitive two-colored digraphs. Linear Algebra Appl. **430**(5–6), 1550–1565 (2009)
11. Gerencsér, B., Gusev, V.V., Jungers, R.M.: Primitive sets of nonnegative matrices and synchronizing automata. SIAM J. Matr. Anal. **39**(1), 83–98 (2018)
12. Olesky, D.D., Shader, B., van den Driessche, P.: Exponents of tuples of nonnegative matrices. Linear Algebra Appl. **356**(1–3), 123–134 (2002)

13. Protasov, V.Yu.: Extremal L_p-norms of linear operators and self-similar functions. Linear Alg. Appl. **428**(10), 2339–2356 (2008)
14. Protasov, V.Yu.: Semigroups of non-negative matrices. Russian Math. Surv. **65**(6), 1186–1188 (2010)
15. Protasov, V.Yu.: Classification of k-primitive sets of matrices. SIAM J. Matrix Anal. **34**(3), 1174–1188 (2013)
16. Protasov, V.Yu., Voynov, A.S.: Sets of nonnegative matrices without positive products. Linear Alg. Appl. **437**(3), 749–765 (2012)
17. Voynov, A.S., Protasov, V.Yu.: Compact noncontraction semigroups of affine operators. Sb. Math. **206**(7), 921–940 (2015)

Reaching Out Towards Fully Verified Autonomous Systems

Sriram Sankaranarayanan[1](✉)(iD), Souradeep Dutta[1](iD), and Sergio Mover[2](iD)

[1] University of Colorado, Boulder, USA
{srirams,souradeep.dutta}@colorado.edu
[2] Ecole Polytechnique, Institut Polytechnique de Paris, Palaiseau, France
smover@lix.polytechnique.fr

Abstract. Autonomous systems such as "self-driving" vehicles and closed-loop medical devices increasingly rely on learning-enabled components such as neural networks to perform safety critical perception and control tasks. As a result, the problem of verifying that these systems operate correctly is of the utmost importance. We will briefly examine the role of neural networks in the design and implementation of autonomous systems, and how various verification approaches can contribute towards engineering verified autonomous systems. In doing so, we examine promising initial solutions that have been proposed over the past three years and the big challenges that remain to be tackled.

Keywords: Formal verification · Autonomous systems · Constraint solvers

1 Introduction

This paper presents a brief overview of recent progress towards the verification of autonomous systems. A system is defined as *autonomous* if it can operate in a reliable manner without requiring "frequent" human intervention. As such, the definition encompasses a wide variety of autonomous systems that are characterized by varying levels of human involvement, including teleoperated surgical robotic systems that *translate* the surgeon's actions from a remote terminal into precise movements of the surgical instruments placed inside the body of the patient [31]; closed loop medical devices such as pacemakers and artificial insulin delivery systems; autonomous "self-driving" cars, and unmanned aerial vehicles (UAVs). The examples mentioned above clearly demonstrate that autonomous systems are *safety critical*: even as we expect these systems to operate with limited human intervention, we also expect them to perform in a provably safe manner despite uncertainties about the environment and the numerous limitations on the system's ability to sense, compute and actuate.

Over the past decade, machine learning approaches have become default "go-to" approaches for building autonomous systems. These approaches use a variety of mathematical and computational models that are trained during design time

© Springer Nature Switzerland AG 2019
E. Filiot et al. (Eds.): RP 2019, LNCS 11674, pp. 22–32, 2019.
https://doi.org/10.1007/978-3-030-30806-3_3

using input-output examples in a supervised manner, or continuously learn and adapt from "past mistakes" using ideas such as reinforcement learning. In both cases, the use of *neural network* models has become quite popular due to the ability of neural networks to approximate complex nonlinear functions and the availability of powerful optimization tools that can infer these models from the given data. Neural network models have been widely applied in a variety of tasks. For instance, feedforward neural network models are widely used to build *perception stacks* for autonomous vehicles that can be used to process large amounts of sensor data from cameras, Lidars and other sensors to recognize other vehicles, pedestrians, road signs and traffic lights [17,22]. Current "end-to-end" driving pipelines seek to go from raw sensor data directly to steering and throttle commands that can help drive the vehicle, skipping the need for a human designed controller [5]. In applications such as robotic surgery, neural networks can be potentially applied to enable decision support by monitoring pre-operative, intra-operative and post-operative data to minimize the overall risk at each stage of patient care [19]. Neural networks have also been used to predict future blood glucose levels to help make real-time treatment decisions for patients with type-1 diabetes [14].

The key challenge in all of these applications lies in building systems with neural network components that are also guaranteed to satisfy key safety and liveness properties, even in the presence of significant uncertainties in the environment. This challenge is significant, since autonomous systems are often too large and complex to reason about manually. Furthermore, besides the system, the operating environment can also be complex and uncertain. Finally, it is challenging to arrive at well-defined specifications for such systems. For instance, it is highly challenging to specify a deep neural-network based object detection component for a self driving car. Such a specification must describe a stream of images from a road scene and the output of the object detector for these scenes: a task that has not proven easy, to date.

Despite these challenges, the broad area of verified autonomous systems has rapidly gained prominence in the formal verification community. We will briefly examine existing approaches, their advantages and drawbacks. Despite these promising steps, a lot more work needs to be carried out to move this area forward.

2   Preliminaries: Neural Networks

In this section, we will briefly explain background on feedforward neural networks, their role in learning-enabled autonomous systems. Our presentation will be brief and at a high level. We refer the reader to standard textbooks for further details [18].

2.1   Neural Networks

Neural networks belong to a class of *connectionist* models that are loosely inspired by the way neurons are connected to each other in human and ani-

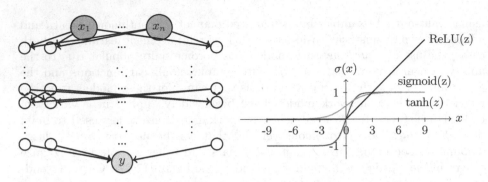

Fig. 1. (Left) A schematic diagram of a feedforward neural network with n inputs $x_1, \ldots, x_n$ and a single output y. Intermediate nodes are shown as unfilled circles. (Right) Commonly used activation functions.

mal brains. There are many types of neural networks, some including units that can store information. We classify neural networks broadly into two types: (a) *feedforward* networks: that do not have internal states; and (b) *recurrent* networks: that include units that can store information internally to the network. The difference between feedforward and recurrent networks is (roughly speaking) analogous to that between combinatorial boolean circuits and sequential circuits in digital logic. Most of our discussions in this paper will be centered around feedforward neural networks.

A feedforward network can be seen as a directed acyclic graph that represents the output as a function of the input. The nodes of this graph can be input nodes, output nodes or intermediate nodes. Each edge of the network is a directed edge from some node i to another node j with an associated real-valued weight $w_{i,j}$. The inputs to the network are fed to the input nodes, which do not have incoming edges. Likewise, the outputs are available at output nodes, which do not have any outgoing edges. Figure 1 shows a schematic diagram of a feedforward neural network.

Each intermediate node j of a feedforward network is associated with an activation function σ_j computed as follows:

1. Let $(i_1, j), \ldots, (i_k, j)$ be the incoming edges at node j, with associated weights $w_1, \ldots, w_k$ respectively.
2. Let $y_1, \ldots, y_k$ be the values computed at nodes $i_1, \ldots, i_k$, respectively.
3. The output at node j is given by $\sigma_j(\sum_{i=1}^{k} w_i y_i + b_j)$, wherein b_j is a constant called the *bias* at node j.

The activation functions associated with nodes are typically nonlinear functions. Popularly used functions are depicted in Fig. 1.

1. **ReLU:** The ReLU unit is defined by the activation function $\sigma(z)$: $\max(z, 0)$.
2. **Sigmoid:** The sigmoid unit is defined by the activation function $\sigma(z)$: $\frac{1}{1+e^{-z}}$.
3. **Tanh:** The activation function for this unit is $\sigma(z)$: $\tanh(z)$.

Note also that besides intermediate nodes with such activation functions, neural networks (especially networks used in image classification) employ specialized nodes such as *max-pooling* and *softmax* nodes that are not discussed here. They are explained in detail elsewhere [18]. A neural network computes a function of its inputs as follows: (a) the value of the input nodes are set according to the inputs to the network; (b) each intermediate node is enabled as soon as values are available at the target nodes for its incoming edges; and (c) once enabled, a node computes its output by applying its activation function. The computation terminates as soon as all output nodes are evaluated. Note that since the network is a DAG, a topological ordering of the nodes can be used to identify an evaluation order of the nodes in the network.

Neural networks have many desirable properties as universal function approximators: they can uniformly approximate any given continuous function f over a compact domain C to any desired accuracy [11]. Neural networks are used primarily for two important tasks: (a) *classify* an input into one of many discrete categories: for instance, categorize an image of a road sign as being a stop sign, a speed limit sign or a pedestrian crossing sign; and (b) *represent* a function from inputs to outputs learned from data through regression. Neural networks are applied in other ways besides just classification. For instance, networks can be used to identify a bounding box around objects of interest in a given image. Since the networks are too complex to design by hand, they are constructed by machine learning techniques that learn the weights and biases of the network given the *topology* of the network that includes the nodes, edges, the activation functions at each node; the input/output data in terms of training examples and a loss function that penalizes discrepancies between the output predicted by the neural network and the actual output in the training data.

There are many algorithms for "learning" the network weights and biases from given training data [18]. The most popular algorithms use variants of a strategy called *stochastic gradient descent* that updates the weights by calculating the gradient over a randomly chosen batch from the training data in order to achieve a local minimum for the loss function. Often, activation functions such as the ReLU function discussed above are *smoothed* in order to make it differentiable. There are many popular tools that automate the training process, notably TensorFlow and PyTorch [1,25]. These tools allow the user to create a neural network topology with unknown weights and biases, specify a loss function and perform the stochastic gradient descent. The networks are then evaluated on a "held-back" test data set that is not part of the data over which it was trained to evaluate its ability to generalize. The recent advent of GPUs that can perform rapid vector and matrix calculations along with the availability of large amounts of data has led to *deep neural networks* with hundreds of thousands of nodes.

3   Verification of Neural Networks

Even though deep neural networks, are essentially acyclic computation graphs formed by composing simple activation functions, the overall behavior of the network can be exceedingly complex and highly non-linear. In this section, we present a brief overview of the existing verification tools and techniques for neural networks and systems that incorporate neural networks in them.

In general, neural networks are used as components inside a closed loop autonomous system. As a result, verification problems have involved component-wise specification involving just the neural network or an end-to-end approach that studies the network in composition with other parts of the system. We distinguish different but closely related verification problems over neural networks: (a) BNNs have been shown to be quite effective for regression and classification tasks. The unit weights also yield computational savings and are amenable to implementation as digital circuits. One of the first attempts at verifying BNN's was proposed by Narodytska et al. [23]. Another recent approach proposed by Shih et al. [3] learns an Ordered Binary Decision Diagram (OBDD) locally to abstract parts of the neural networks. Cheng et al. [9] reduce the problem of BNN verification to hardware verification problem, and have reported speed ups in performance.

3.1   Abstract Interpretation for Neural Networks

Abstract interpretation originally formalized by Cousot and Cousot was developed to systematically propagate sets of reachable states of a program through individual program statements in order to establish properties of a program as a whole [10]. Such techniques rely on abstract domains to represent the reachable set of states [24]. This idea can be applied to neural networks which represent loop free computations involving the application of nonlinear activation functions.

Vechev et al. use zonotopes as an abstract domain to perform image computation across a neural network [16]. In particular, zonotopes are used to over-approximation the non-convex set of possible outputs for each layer of the network. This allows for a layer-by-layer analysis to compute sound over-approximations for the output of the neural network.

Xiang et al. that computes the output ranges as a union of convex polytopes [36]. This approach does not use SMT or MILP solvers unlike other approaches and thus can lead to highly accurate estimates of the output range. However, judging from preliminary evaluation reported, the cost of manipulating polyhedra is quite expensive, and thus, the approach is currently restricted to smaller networks when compared to SMT/MILP-based approaches.

Range computations using symbolic intervals were attempted in Reluval [33], which essentially relied on affine arithmetic techniques to reduce the over-approximation errors, and handle the case splitting imposed by ReLU units. Likewise, Cheng et al. [8] propose a heuristic approach to compute tight ranges for individual neurons.

3.2 Training with Robustness

Verification approaches have been incorporated to improve the process of learning networks from data [32,34]. For instance Jana et al. use the output set estimates computed by verification tools in order to incorporate robustness in the training phase wherein the network is rendered somewhat immune to small perturbations of the input. This has been proposed as a means to defend against any adversarial perturbations of the input. However, the computational cost can be orders magnitude more expensive than standard approaches to adversarially robust training that do not involve expensive verification tools in the loop.

3.3 Closed Loop Verification

Until this point we have been interested in verifying properties of a single neural network *in isolation*. However, as mentioned previously, autonomous systems employ neural networks as components in a closed loop that controls a physical process. Such physical processes can often be described by ordinary differential equations (ODEs). The simplest such situation involves a neural network that applies a feedback control to a physical process modeled as an ODE. This setup has been studied recently in order to perform reachability analysis of the resulting closed loop behaviors [13,21,30,35].

Dutta et al. [13], propose a technique to compute Taylor models (polynomial + error) as approximations of the behavior of the neural network in a compact domain. This was then used in conjunctions with standard reachability tools like Flow* [6] to compute reachable set of states of the closed loop involving an ODE and a neural network. A followup approach [20] approximates the neural network controller with Bernstein polynomials to deal with activation functions that are more general than ReLU. Ivanov et al. [21] propose a technique whereby activation functions such as sigmoid and tanh are modeled using differential equations evolving over time to encode a network as an ODE itself. This allows the transformation of a single layer of the neural network into a hybrid system. Which could then be used in standard reachability analysis tools for such systems. Another recent work by Xiang et al. considers the combination of neural networks in feedback with piecewise linear dynamical systems [37] using the techniques presented in [36].

Barrier certificates serve as an important approach to establish safety properties of dynamical systems [26]. Tuncali et al. [30] present an approach to synthesize barrier certificates using an SMT solver to prove properties of ODEs with neural networks as feedback.

However, neural networks are also employed in autonomous systems to classify a large volume of sensor data from cameras and LIDAR sensors. It is an enormous challenge to specify the behavior of these sensors with respect to changes in the environment and the vehicle. Shoukry et al. present a recent step towards verifying robotic systems that employ LIDAR sensors by means of simplifying the LIDAR system to consider a finite set of angles along with the system finds ranges [28]. The approach also "hard codes" a fixed environment with obstacles having fixed positions and geometries. The authors use a SMT based approach to construct a finite state abstraction of the closed loop system using fixed set of predicates to partition the state-space. This abstraction is then used to check reachability properties.

3.4 Falsification and Testing

We have focused our attention entirely on the use of formal verification approaches to prove properties of autonomous systems with neural network components. The problems of "best-effort" falsification to find counterexamples and that of systematic testing have also received a lot of attention. We mention a few representative approaches that relate closely to the verification approaches mentioned above without claiming to be a comprehensive survey on falsification/testing approaches for autonomous systems. An important line of work (e.g., [12,29,38]) focuses on the falsification problem for systems containing neural network components, as autonomous vehicles. The falsification problem consist of finding an execution of the system that violates a requirement, and the falsification algorithms for cyber-physical systems (e.g., S-TaLiRo [4]) implement efficient heuristics to search for a system's input that can falsify a requirement.

One challenge addressed in [38] is to find adversarial examples, a perturbation of the input that falsifies a temporal logic formula, for a closed loop control system formed by a neural network controller and a dynamical system. The proposed solution tries to find an adversarial example minimizing the robustness function of the Signal Temporal Logic (STL) formula via gradient descent.

Recent approaches also address the falsification of autonomous vehicles where neural networks are used in the perception stack. Dreossi et al. [12] propose an approach that falsifies STL formulas compositionally, first falsifying an abstraction of the neural network component and the cyber-physical system, and then confirming the counterexample in the neural network component. An alternative approach proposed by Fainekos et al. [29] focuses on perturbing driving scenarios for autonomous vehicles that can result in reaching undesired state (e.g., a crash). The scenarios are expressed in STL, and the approach generates input test cases from different combinations of discrete parameters of the system.

4   Challenges

We conclude our discussion by briefly mentioning some of the important challenges that remain to be tackled in this rapidly emerging area.

Specification: Despite initial approaches to verifying properties of neural networks in isolation, or as part of larger closed loops, the problem of formally specifying the behavior of these systems remains largely open for *perception* systems that classify sensor data including images and LIDAR data. The key challenge here lies in specifying what a valid image is in a logical formalism that is compatible with existing verification tools. This in turn requires a specification of the environment, and the imaging/sensing processes. To make matters more complicated, small changes to the orientation/pose of the vehicle can drastically alter the image generated. Current approaches sidestep functional specifications in favor of requiring the classifier to be "robust" to perturbations around some selected training examples. Alternatively, one may simplify the sensor's capabilities to make modeling easier. Another popular alternative uses generative models that specify inputs at a high level. Fremont et al. propose an approach that uses generative models for creating road scenes corresponding to simple programmatic specifications for the purposes of testing [15]. Extending such formalisms to verification problems remains an important challenge.

Scalability: Scalability of verification approaches remains yet another challenge. Simply put, the current state-of-the art networks are 100x or 1000x larger than the most efficient verification tools available. This gap needs to be considerably narrowed before verification approaches can be used on realistic systems. This challenge may requires to improve the existing verification techniques, for example improving the underlying constraint solvers by specializing them to handle neural networks. Alternative approaches such as using abstractions that are sufficient precise to show the correctness of the neural network can also be useful. The challenge lies in the definition of these abstractions and how they can be obtained for large networks without resorting to expensive verification tools in the first place.

Recurrent Networks: Another important challenge lies in tackling recurrent networks that involve units such as *long short term memory* (LSTM) with internal state. These networks are widely used in applications such as data-driven modeling and natural language processing. Verification of such networks is highly challenging for existing tools and techniques.

Runtime Verification: Runtime verification provides an important alternative to everything mentioned here that focuses on static/pre-deployment verification. The use of real-time monitors to predict and act against imminent property violations form the basis for runtime assurance using L1-Simplex architectures that switch between a lower performance but formally validated control when an impending failure is predicted [27]. However, the key issue lies in how impending failures are to be predicted. An alternative approach to verification to guarantee safety is shielding [2,40] that uses a supervisor (or so-called shield) to monitor the execution of the autonomous system and intervene to enforce temporal logic properties if a violation is imminent. Chen et al. present a different approach based on monitoring *viability* rather than safety in order to sidestep the need to reason about the controller [7]. Instead, their approach can perform lightweight reasoning just over the behavior of the plant model. A recent application of their approach involves monitoring geofences for unmanned aerial vehicle [39].

5   Conclusion

In conclusion, we have attempted to classify the rapidly emerging area of verifying autonomous systems involving neural networks. Our presentation has focused on some of the current successes and future challenges in this area.

Acknowledgments. This work was supported in part by the Air Force Research Laboratory (AFRL) and by the US NSF under Award # 1646556.

References

1. Abadi, M., Agarwal, A., Barham, P., et al.: TensorFlow: large-scale machine learning on heterogeneous systems (2015). https://www.tensorflow.org/
2. Alshiekh, M., Bloem, R., Ehlers, R., Könighofer, B., Niekum, S., Topcu, U.: Safe reinforcement learning via shielding (2018). https://aaai.org/ocs/index.php/AAAI/AAAI18/paper/view/17211
3. Shih, A., Darwiche, A., Choi, A.: Verifying binarized neural networks by local automaton learning (2019). http://reasoning.cs.ucla.edu/fetch.php?id=193&type=pdf
4. Annpureddy, Y., Liu, C., Fainekos, G., Sankaranarayanan, S.: S-TaLiRo: a tool for temporal logic falsification for hybrid systems. In: Abdulla, P.A., Leino, K.R.M. (eds.) TACAS 2011. LNCS, vol. 6605, pp. 254–257. Springer, Heidelberg (2011). https://doi.org/10.1007/978-3-642-19835-9_21
5. Bojarski, M., et al.: End to end learning for self-driving cars. CoRR abs/1604.07316 (2016). http://arxiv.org/abs/1604.07316
6. Chen, X., Ábrahám, E., Sankaranarayanan, S.: Flow*: an analyzer for non-linear hybrid systems. In: Sharygina, N., Veith, H. (eds.) CAV 2013. LNCS, vol. 8044, pp. 258–263. Springer, Heidelberg (2013). https://doi.org/10.1007/978-3-642-39799-8_18
7. Chen, X., Sankaranarayanan, S.: Model-predictive real-time monitoring of linear systems. In: IEEE Real-Time Systems Symposium (RTSS), pp. 297–306. IEEE Press (2017)
8. Cheng, C., Nührenberg, G., Ruess, H.: Maximum resilience of artificial neural networks. CoRR abs/1705.01040 (2017). http://arxiv.org/abs/1705.01040
9. Cheng, C., Nührenberg, G., Ruess, H.: Verification of binarized neural networks. CoRR abs/1710.03107 (2017). http://arxiv.org/abs/1710.03107
10. Cousot, P., Cousot, R.: Abstract interpretation: a unified lattice model for static analysis of programs by construction or approximation of fixpoints. In: ACM Principles of Programming Languages, pp. 238–252 (1977)
11. Cybenko, G.: Approximation by superpositions of a sigmoidal function. Math. Sig. Syst. **2**, 303–314 (1989)
12. Dreossi, T., Donzé, A., Seshia, S.A.: Compositional falsification of cyber-physical systems with machine learning components. In: Barrett, C., Davies, M., Kahsai, T. (eds.) NFM 2017. LNCS, vol. 10227, pp. 357–372. Springer, Cham (2017). https://doi.org/10.1007/978-3-319-57288-8_26
13. Dutta, S., Chen, X., Sankaranarayanan, S.: Reachability analysis for neural feedback systems using regressive polynomial rule inference. In: Proceedings of the Hybrid Systems: Computation and Control (HSCC), HSCC 2019, pp. 157–168. ACM, New York (2019)

14. Dutta, S., Kushner, T., Sankaranarayanan, S.: Robust data-driven control of artificial pancreas systems using neural networks. In: Češka, M., Šafránek, D. (eds.) CMSB 2018. LNCS, vol. 11095, pp. 183–202. Springer, Cham (2018). https://doi.org/10.1007/978-3-319-99429-1_11
15. Fremont, D.J., Dreossi, T., Ghosh, S., Yue, X., Sangiovanni-Vincentelli, A.L., Seshia, S.A.: Scenic: a language for scenario specification and scene generation. In: Proceedings of the ACM Programming Language Design and Implementation (PLDI), pp. 63–78 (2019)
16. Gehr, T., Mirman, M., Drachsler-Cohen, D., Tsankov, P., Chaudhuri, S., Vechev, M.: Ai2: safety and robustness certification of neural networks with abstract interpretation. In: 2018 IEEE Symposium on Security and Privacy (SP), pp. 3–18, May 2018
17. Geiger, A., Lenz, P., Urtasun, R.: Are we ready for autonomous driving? The Kitti vision benchmark suite. In: 2012 IEEE Conference on Computer Vision and Pattern Recognition, pp. 3354–3361, June 2012
18. Goodfellow, I., Bengio, Y., Courville, A.: Deep Learning. MIT Press (2016). http://www.deeplearningbook.org
19. Hashimoto, D.A., Rosman, G., Rus, D., Meireles, O.: Artificial intelligence in surgery: promises and perils. Ann. Surg. **268**, 70–76 (2018)
20. Huang, C., Fan, J., Li, W., Chen, X., Zhu, Q.: Reachnn: reachability analysis of neural-network controlled systems. CoRR abs/1906.10654 (2019). http://arxiv.org/abs/1906.10654
21. Ivanov, R., Weimer, J., Alur, R., Pappas, G.J., Lee, I.: Verisig: verifying safety properties of hybrid systems with neural network controllers. In: Proceedings of the Hybrid Systems: Computation and Control (HSCC), HSCC 2019, pp. 169–178. ACM, New York (2019)
22. LeCun, Y., Kavukcuoglu, K., Farabet, C.: Convolutional networks and applications in vision. In: Proceedings of 2010 IEEE International Symposium on Circuits and Systems, pp. 253–256, May 2010. https://doi.org/10.1109/ISCAS.2010.5537907
23. Narodytska, N., Kasiviswanathan, S.P., Ryzhyk, L., Sagiv, M., Walsh, T.: Verifying properties of binarized deep neural networks. CoRR abs/1709.06662 (2017). http://arxiv.org/abs/1709.06662
24. Nielson, F., Nielson, H.R., Hankin, C.: Principles of Program Analysis. Springer, Heidelberg (1999). https://doi.org/10.1007/978-3-662-03811-6
25. Paszke, A., et al.: Automatic differentiation in PyTorch. In: NIPS Workshop on Automatic Differentiation (2017). https://openreview.net/forum?id=BJJsrmfCZ
26. Prajna, S., Jadbabaie, A.: Safety verification using barrier certificates. In: Proceedings of the HSCC 2004, vol. 2993, pp. 477–492 (2004)
27. Sha, L.: Using simplicity to control complexity. IEEE Softw. **18**(4), 20–28 (2001)
28. Sun, X., Khedr, H., Shoukry, Y.: Formal verification of neural network controlled autonomous systems. In: Proceedings of the Hybrid Systems: Computation and Control (HSCC), HSCC 2019, pp. 147–156. ACM, New York (2019)
29. Tuncali, C.E., Fainekos, G., Ito, H., Kapinski, J.: Simulation-based adversarial test generation for autonomous vehicles with machine learning components. In: 2018 IEEE Intelligent Vehicles Symposium, pp. 1555–1562 (2018)
30. Tuncali, C.E., Kapinski, J., Ito, H., Deshmukh, J.V.: Reasoning about safety of learning-enabled components in autonomous cyber-physical systems. In: Proceedings of the Design Automation Conference, DAC 2018, pp. 30:1–30:6 (2018)
31. U.S Food and Drug Administration: Computer-assisted surgical systems (2019). https://www.fda.gov/medical-devices/surgery-devices/computer-assisted-surgical-systems. Accessed July 2019

32. Wang, S., Chen, Y., Abdou, A., Jana, S.: Mixtrain: scalable training of formally robust neural networks. CoRR abs/1811.02625 (2018). http://arxiv.org/abs/1811.02625

33. Wang, S., Pei, K., Whitehouse, J., Yang, J., Jana, S.: Formal security analysis of neural networks using symbolic intervals. CoRR abs/1804.10829 (2018). http://arxiv.org/abs/1804.10829

34. Wong, E., Kolter, J.Z.: Provable defenses against adversarial examples via the convex outer adversarial polytope. In: Proceedings of the International Conference on Machine Learning, ICML, pp. 5283–5292 (2018). http://proceedings.mlr.press/v80/wong18a.html

35. Xiang, W., Tran, H., Johnson, T.T.: Reachable set computation and safety verification for neural networks with relu activations. CoRR abs/1712.08163 (2017). http://arxiv.org/abs/1712.08163

36. Xiang, W., Tran, H.D., Johnson, T.T.: Reachable set computation and safety verification for neural networks with relu activations (2107). https://arxiv.org/pdf/1712.08163.pdf. Posted on arxiv December 2017

37. Xiang, W., Tran, H.D., Rosenfeld, J.A., Johnson, T.T.: Reachable set estimation and verification for a class of piecewise linear systems with neural network controllers (2018). To Appear in the American Control Conference (ACC), invited session on Formal Methods in Controller Synthesis

38. Yaghoubi, S., Fainekos, G.: Gray-box adversarial testing for control systems with machine learning components. In: Proceedings of Hybrid Systems: Computation and Control, pp. 179–184 (2019)

39. Yoon, H., Chou, Y., Chen, X., Frew, E., Sankaranarayanan, S.: Predictive runtime monitoring for linear stochastic systems and applications to geofence enforcement for UAVs (2019). In: Proceedings of the Runtime Verification 2019, October 2019 (to appear)

40. Zhu, H., Xiong, Z., Magill, S., Jagannathan, S.: An inductive synthesis framework for verifiable reinforcement learning. In: ACM Programming Language Design and Implementation (PLDI), pp. 686–701 (2019)

On the m-eternal Domination Number
of Cactus Graphs

Václav Blažej, Jan Matyáš Křišť'an[✉], and Tomáš Valla

Faculty of Information Technology, Czech Technical University in Prague,
Prague, Czech Republic
matyas.kristan@gmail.com

Abstract. Given a graph G, guards are placed on vertices of G. Then
vertices are subject to an infinite sequence of attacks so that each attack
must be defended by a guard moving from a neighboring vertex. The
m-eternal domination number is the minimum number of guards such
that the graph can be defended indefinitely. In this paper we study the
m-eternal domination number of cactus graphs, that is, connected graphs
where each edge lies in at most one cycle, and we consider three variants of the m-eternal domination number: first variant allows multiple
guards to occupy a single vertex, second variant does not allow it, and
in the third variant additional "eviction" attacks must be defended. We
provide a new upper bound for the m-eternal domination number of cactus graphs, and for a subclass of cactus graphs called Christmas cactus
graphs, where each vertex lies in at most two biconnected components,
we prove that these three numbers are equal. Moreover, we present a
linear-time algorithm for computing them.

1 Introduction

Let us have a graph G whose vertices are occupied by guards. The graph is
subject to an infinite sequence of vertex attacks. The guards may move to any
neighboring vertex after each attack. After moving, a vertex attack is defended
if the vertex is occupied by a guard. The task is to come up with a strategy such
that the graph can be defended indefinitely.

Defending a graph from attacks using guards for an infinite number of steps
was introduced by Burger et al. [2]. In this paper we study the concept of the
m-eternal domination, which was introduced by Goddard et al. [4] (eternal domination was originally called eternal security).

The m-eternal guarding number $\Gamma_m^\infty(G)$ is the minimum number of guards
which tackle all attacks in G indefinitely. Here the (slightly confusing) notion of
the letter "m" emphasizes that multiple guards may move during each round.
The m-eternal domination number $\gamma_m^\infty(G)$ is the minimum number of guards
which tackle all attacks indefinitely, with the restriction that no two guards may

V. Blažej and T. Valla acknowledge the support of the OP VVV MEYS funded project
CZ.02.1.01/0.0/0.0/16_019/0000765 "Research Center for Informatics".

E. Filiot et al. (Eds.): RP 2019, LNCS 11674, pp. 33–47, 2019.
https://doi.org/10.1007/978-3-030-30806-3_4

occupy a single vertex simultaneously. We also introduce the m-eternal domination number with eviction $\gamma_{me}^{\infty}(G)$, which is similar to γ_{m}^{∞} with the additional requirement, that during each round one can decide to either attack a vertex or choose an "evicted" vertex or edge, which must be cleared of guards in the next round. There is also a variant of the problem studied by Goddard et al. [4] where only one guard may move during each round, which is not considered in our paper. We will define all concepts formally at the end of this section.

Goddard et al. [4] established γ_{m}^{∞} exactly for paths, cycles, complete graphs and complete bipartite graphs, showing that $\gamma_{m}^{\infty}(P_n) = \lceil n/2 \rceil$, $\gamma_{m}^{\infty}(C_n) = \lceil n/3 \rceil$, $\gamma_{m}^{\infty}(K_n) = 1$ and $\gamma_{m}^{\infty}(K_{m,n}) = 2$. The authors also provide several bounds for general graphs, most notably $\gamma(G) \leq \gamma_{m}^{\infty}(G) \leq \alpha(G)$, where $\alpha(G)$ denotes the size of the maximum independent set in G and $\gamma(G)$ is the size of the smallest dominating set in G. Since that several results focused on finding bounds of γ_{m}^{∞} in different conditions or graph classes.

Henning, Klostermeyer and MacGillivray [7] explored the relationship between γ_{m}^{∞} and the minimum degree $\delta(G)$ of a graph G: If G is a connected graph with minimum degree $\delta(G) \geq 2$ and has $n \neq 4$ vertices, then $\gamma_{m}^{\infty}(G) \leq \lfloor (n-1)/2 \rfloor$, and this bound is tight.

Finbow, Messinger and van Bommel [3] proved the following result for $3 \times n$ grids. For $n \geq 2$

$$\gamma_{m}^{\infty}(P_3 \square P_n) \leq \lceil 6n/7 \rceil + \begin{cases} 1 & \text{if } n \equiv 7, 8, 14 \text{ or } 15 \bmod 21, \\ 0 & \text{otherwise.} \end{cases}$$

Here $G \square H$ denotes the Cartesian product of graphs G and H.

Van Bommel and van Bommel [12] showed for $5 \times n$ grids that

$$\left\lfloor \frac{6n+9}{5} \right\rfloor \leq \gamma_{m}^{\infty}(P_5 \square P_n) \leq \left\lfloor \frac{4n+4}{3} \right\rfloor.$$

For a good survey on other related results and topics see Klostermeyer and Mynhardt [9].

Very little is known regarding the algorithmic aspects of m-eternal domination. The decision problem (asking if $\gamma_{m}^{\infty}(G) \leq k$) is obviously NP-hard and belongs to EXPTIME, however, it is not known whether it lies in the class PSPACE (see [9]). On the positive side, there is a linear algorithm for computing γ_{m}^{∞} for trees by Klostermeyer and MacGillivray [8]. Braga, de Souza and Lee [1] showed that $\gamma_{m}^{\infty}(G) = \alpha(G)$ in all proper-interval graphs. Very recently Gupta et al. [5] showed that the maximum independent set in an interval graph on n vertices can be solved in time $\mathcal{O}(n \log n)$, or $\mathcal{O}(n)$ in the case when endpoints of the intervals are already sorted. We can thus compute $\gamma_{m}^{\infty}(G)$ efficiently on proper-interval graphs.

In this paper we contribute to the positive side and provide an extension of the result by Klostermeyer and MacGillivray [8]. *Cactus* is a graph that is connected and its every edge lies on at most one cycle. An equivalent definition is that it is connected and any two cycles have at most one vertex in common.

Christmas cactus graph is a cactus in which each vertex is in at most two 2-connected components. Christmas cactus graphs were introduced by Leighton and Moitra [10] in the context of greedy embeddings, where Christmas cactus graphs play an important role in the proof that every polyhedral graph has a greedy embedding in the Euclidean plane.

Our main result is summarized in the following theorem.

Theorem 1. *Let G be a Christmas cactus graph. Then $\Gamma_m^\infty(G) = \gamma_m^\infty(G) = \gamma_{me}^\infty(G)$ and there exists a linear-time algorithm which computes these values.*

Using Theorem 1 we are able to devise a new bound on the m-eternal domination number of cactus graphs, which is stated in Theorem 3 in Sect. 3. In Sect. 4 we provide the linear-time algorithm for computing γ_m^∞ of Christmas cactus graphs.

Let us now introduce all concepts formally. For an undirected graph G let a *configuration* be a multiset $C = \{c_1, c_2, \ldots, c_k : c_i \in V(G)\}$. We will refer to the elements of configurations as *guards*. *Movement* of a guard $u \in C$ means changing u to some element $v \in N_G[u]$ of its closed neighborhood and we denote it by $u \to v$. Two configurations C_1 and C_2 of G are mutually *traversable* in G if it is possible to move each guard of C_1 to obtain C_2. A *strategy* in G is a graph $S_G = (\mathbb{C}, \mathbb{F})$ where $\mathbb{C}$ is a set of configurations in G of same size and $\mathbb{F} = \{\{C_1, C_2\} \in \mathbb{C}^2 \mid C_1 \text{ and } C_2 \text{ are mutually traversable in } G\}$. The *order* of a strategy is the number of guards in each of its configurations.

We now define the variants of the problem which we study in our paper. For the purpose of the proof of our main result we devise a variant of the problem, where a vertex or an edge can be "evicted" during a round, that means, no guard may be left on the respective vertex or edge. We call the strategy S_G to be *defending against vertex attacks* if for any $C \in \mathbb{C}$ the configuration C and its neighbors in S_G cover all vertices of G, i.e., when a vertex $v \in V(G)$ is "attacked" one can always respond by changing to a configuration which has a guard at the vertex v. Note that every configuration in a strategy which defends against vertex attacks induces a dominating set in G. We call a strategy S_G to be *evicting vertices* if for any $C \in \mathbb{C}$ and any $u \in V(G)$ the configuration C has a neighbor C' in S_G such that $u \notin C'$, i.e., when a vertex v is "to be evicted" one can respond by changing to a configuration where no guard is present at v. We call a strategy S_G to be *evicting cycle edges* if for any $C \in \mathbb{C}$ and any edge $\{u, v\} \in E(G)$ lying in some cycle in G the configuration C has a neighbor C' in S_G such that $u, v \notin C'$. That means, when an edge is "to be evicted" one can respond by changing to a configuration where no guards are incident to the edge.

Let the *m-eternal guard strategy* in G be a strategy defending against vertex attacks in G. Let the m-eternal guard configuration number $\Gamma_m^\infty(G)$ be the minimum order among all m-eternal guard strategies in G. Let the *m-eternal dominating strategy* in G be a strategy in G such that none of its configurations has duplicates and is also defending against vertex attacks. The *m-eternal dominating set* in G is a configuration, which is contained in some m-eternal dominating strategy in G. Let the m-eternal dominating number $\gamma_m^\infty(G)$ be the minimum order of m-eternal dominating strategy in G. Let the *m-eternal dominating strategy with eviction* in G be a strategy such that none of its configurations has duplicates, is defending vertex attacks, is evicting vertices, and is evicting cycle edges in G. Let the m-eternal dominating with eviction number $\gamma_{me}^\infty(G)$ be the minimum order of m-eternal dominating strategy evicting vertices and edges in G.

A cycle in G is a *leaf cycle* if exactly one of its vertices has degree greater than 2. By P_n we denote a path with n edges and $n + 1$ vertices. By $G[U]$ we denote the subgraph of G induced by the set of vertices $U \subseteq V(G)$.

2   The m-eternal Domination of Christmas Cactus Graphs

In this section we prove that $\Gamma_m^\infty(G) = \gamma_m^\infty(G) = \gamma_{me}^\infty(G)$ for Christmas cactus graphs by showing the optimal strategy. The main idea is to repeatedly use reductions of the Christmas cactus graph G to produce smaller Christmas cactus graph I. We prove that the optimal strategy for G uses a constant number of guards more than the optimal strategy for I.

This will be one part of the proof of Theorem 1. Before we describe the reductions, we present several technical tools that are used in the proofs of validity of the reductions and that give a hint into the machinery of the proof. We defer the technical proofs of reductions correctness to the full version of this paper which is available online.[1]

Observation 1. Every strategy used in the m-eternal domination with eviction can be applied in an m-eternal domination strategy, and every m-eternal domination strategy can be applied as an m-eternal guard configuration strategy. Every configuration in each of these strategies must induce a dominating set, therefore, they are all lower bound by the domination number γ. We see that the following inequality holds for every graph G.

$$\gamma(G) \leq \Gamma_m^\infty(G) \leq \gamma_m^\infty(G) \leq \gamma_{me}^\infty(G)$$

Note that we can prove bounds on all of these strategies by showing that for G and its reduction I it holds that $\gamma_{me}^\infty(G) \leq \gamma_{me}^\infty(I) + k$ and $\Gamma_m^\infty(I) \leq \Gamma_m^\infty(G) - k$ for some integer constant k. If we have an exact result for I the reduction gives us an exact bound on G as well. This is summed up in the following lemma.

Lemma 1. *Let us assume that for graphs G, I, and an integer constant k*

$$\gamma_{me}^\infty(G) \leq \gamma_{me}^\infty(I) + k, \tag{1}$$
$$\Gamma_m^\infty(G) \geq \Gamma_m^\infty(I) + k, \tag{2}$$
$$\gamma_{me}^\infty(I) = \Gamma_m^\infty(I). \tag{3}$$

Then $\gamma_{me}^\infty(G) = \Gamma_m^\infty(G)$.

---

[1] https://arxiv.org/abs/1907.07910

Proof. Given the assumptions, we get $\gamma_{me}^\infty(G) \leq \Gamma_m^\infty(G)$ in the following manner.

$$\gamma_{me}^\infty(G) \leq^{(1)} \gamma_{me}^\infty(I) + k =^{(3)} \Gamma_m^\infty(I) + k \leq^{(2)} \Gamma_m^\infty(G)$$

Recall Observation 1 where we saw that $\Gamma_m^\infty(G) \leq \gamma_{me}^\infty(G)$ holds, giving us the desired equality.   □

Let us have a graph G with a strategy. By *simulating* a vertex attack, a vertex eviction, or an edge eviction on G, we mean performing the attack on G and retrieving the strategy's response configuration. Simulating attacks is useful mainly in merging several strategies over subgraphs into a strategy for the whole graph.

In the following theorem we introduce a general upper bound applicable to the m-eternal dominating strategy with eviction.

Lemma 2. *Let G be a graph with an articulation v such that $G \setminus v$ has two connected components H and I' such that there are exactly two vertices u and w in $V(I')$ which are neighbors of v. Let $I = \big(V(I'), E(I') \cup \{\{u, w\}\}\big)$. If $\{u, w\}$ lies on a cycle in I then*

$$\gamma_{me}^\infty(G) \leq \gamma_{me}^\infty(H) + \gamma_{me}^\infty(I).$$

Proof. We will show that having two separate strategies for H and I we can merge them into one strategy for G without using any additional guards (Fig. 1).

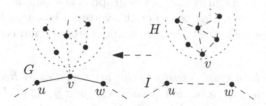

Fig. 1. Decomposition of G into H and I

We will create a strategy which keeps the invariant that in all its configurations either v is occupied by a guard or the pair of vertices u and w are evicted. This will ensure that whenever v is not occupied due to the strategy of H needing a guard from v to defend other vertices of H, the strategy of I will be in a configuration where no guard can traverse the $\{u, w\}$ edge.

Let the initial configuration be a combination of a configuration of H which defends v and a configuration of I which evicts $\{u, w\}$. The final strategy will consist of configurations which are unions of configurations of H and I which we choose in the following manner.

The vertices of G were partitioned among H and I so a vertex attack can be distinguished by the target component. Whenever a vertex z of G is attacked, choose a configuration of respective component which defends z. If H was not attacked then simulate an attack on v. If I was not attacked then simulate an edge eviction on $\{u, w\}$. By the configuration of the non-attacked component we ensure the invariant is true. Whenever the $\{u, w\}$ edge might be traversed by a guard in the I's strategy, we use the fact that v is occupied and instead of performing $u \rightarrow w$ we move the guards $u \rightarrow v$ and $v \rightarrow w$ which has the same effect considering guard configuration of I.

The eviction of vertices and edges present in H and I is solved in the same way as vertex attacks. The only attack which remains to be solved is an edge eviction of $\{u, v\}$ or $\{w, v\}$. Both of these are defended by simulating an eviction of v in H and $\{u, w\}$ in I. The two strategies will ensure there are no guards on either u, v, nor w and the invariant is still true. $\square$

We may now proceed with the reductions.

Lemma 3. *Let C_k be a cycle on k vertices. Then $\gamma(C) = \Gamma_{\mathrm{m}}^{\infty}(C) = \gamma_{\mathrm{m}}^{\infty}(C) = \gamma_{\mathrm{me}}^{\infty}(C) = \lceil \frac{k}{3} \rceil$.*

Reduction 1. Let G be a Christmas cactus graph and u be a leaf vertex which is connected to a vertex v of degree 2. Remove u and v from G.

Lemma 4. *Let G be a graph satisfying the prerequisites of Reduction 1. Let I be G after application of Reduction 1. Then I is a Christmas cactus graph and $\gamma_{\mathrm{me}}^{\infty}(G) = \Gamma_{\mathrm{m}}^{\infty}(G) = \Gamma_{\mathrm{m}}^{\infty}(I) + 1 = \gamma_{\mathrm{me}}^{\infty}(I) + 1$.*

Reduction 2. Let G be a Christmas cactus graph and u be a leaf vertex which is connected to a vertex v of degree 3. Let the vertex v has neighbors u, x, y such that x and y are not connected. Remove u and v from G and connect x, y by an edge.

Lemma 5. *Let G be a graph satisfying the prerequisites of Reduction 2. Let I be G after application of Reduction 2 on u, v, x and y. Then I is a Christmas cactus graph and $\gamma_{\mathrm{me}}^{\infty}(G) = \Gamma_{\mathrm{m}}^{\infty}(G) = \Gamma_{\mathrm{m}}^{\infty}(I) + 1 = \gamma_{\mathrm{me}}^{\infty}(I) + 1$.*

Reduction 3. Let G be a Christmas cactus graph and C be a leaf cycle on n vertices where $n \in \{3k, 3k + 2 \mid k \geq 1\}$. Let v be the only articulation on this cycle. Remove $C \setminus v$ and create a new vertex u and the edge $\{v, u\}$ in G.

Lemma 6. *Let G be a graph satisfying the prerequisites of Reduction 3. Let I be G after application of Reduction 3 with C. Then I is a Christmas cactus graph and $\gamma_{\mathrm{me}}^{\infty}(G) = \Gamma_{\mathrm{m}}^{\infty}(G) = \Gamma_{\mathrm{m}}^{\infty}(I) + k - 1 = \gamma_{\mathrm{me}}^{\infty}(I) + k - 1$.*

Reduction 4. Let G be a Christmas cactus graph and C be a leaf cycle on $3k + 1$ vertices. Let v be the only articulation on this cycle. Remove $C \setminus v$ from G.

Lemma 7. *Let G be a graph satisfying the prerequisites of Reduction 4. Let I be G after application of Reduction 4 substituting C by K_1. Then I is a Christmas cactus graph and $\gamma_{\mathrm{me}}^{\infty}(G) = \Gamma_{\mathrm{m}}^{\infty}(G) = \Gamma_{\mathrm{m}}^{\infty}(I) + k = \gamma_{\mathrm{me}}^{\infty}(I) + k$.*

Reduction 5. Let G be a Christmas cactus graph and C be a cycle on three vertices $\{v, x, y\}$, let x', y' be leafs in G, such that x' connects to x, y' to y, and v is connected to the rest of the graph (no other edges are incident to C). We call the $C \cup \{x', y'\}$ subgraph a *bull graph*. Remove $\{x, y, x', y'\}$ from G.

Lemma 8. *Let G be a Christmas cactus graph. Let K be a bull graph connected to the rest of G via a vertex of degree 2. Let I be G after application of Reduction 5 on K. Then I is a Christmas cactus graph and $\gamma_{\text{me}}^{\infty}(G) = \Gamma_{\text{m}}^{\infty}(G) = \Gamma_{\text{m}}^{\infty}(I) + 2 = \gamma_{\text{me}}^{\infty}(I) + 2$.*

Let the 3-*pan graph* be a K_3 with one leaf attached.

Reduction 6. Let G be a Christmas cactus graph and C be a cycle on three vertices $\{v, x, y\}$, let x' be a leaf in G, such that x' connects to x and v is connected to the rest of the graph (no other edges are incident to C). The $C \cup \{x'\}$ is a 3-pan graph. Remove $\{x, x'\}$ from G.

Lemma 9. *Let G be a Christmas cactus graph. Let K be a 3-pan graph connected to rest of the graph via a vertex of degree 2. Let I be G after application of Reduction 6 on K. Then I is a Christmas cactus graph and $\gamma_{\text{me}}^{\infty}(G) = \Gamma_{\text{m}}^{\infty}(G) = \Gamma_{\text{m}}^{\infty}(I) + 1 = \gamma_{\text{me}}^{\infty}(I) + 1$.*

Using the reductions we are ready to prove the part of Theorem 1 stating that $\Gamma_{\text{m}}^{\infty}(G) = \gamma_{\text{m}}^{\infty}(G) = \gamma_{\text{me}}^{\infty}(G)$ for all Christmas cactus graphs.

A *block* or a *2-connected component* of graph G is a maximal 2-connected subgraph of G.

Lemma 10. *In a non-elementary christmas cactus graph with no leaf cycles, no leaf vertices connected to a vertex of degree 2, and no leaf vertices connected to a block of size bigger than 3, there is at least one leaf bull or one leaf 3-pan graph.*

Proof. Let us call the blocks of size 3 triangles. Removing edges of all the triangle subgraphs would split the christmas cactus into connected components of blocks. Let us choose a triangle and traverse the graph in the following way. If the current triangle is a bull or a 3-pan we end the traversal and have a positive result. Otherwise, choose the component we have not visited yet and find a different triangle graph incident to it. Such triangle must exist otherwise it would be a leaf component. Mark this component as visited and repeat the process. See Fig. 2. □

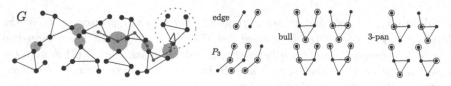

Theorem 2. *Let G be a Christmas cactus graph. Then $\Gamma_{\text{m}}^{\infty}(G) = \gamma_{\text{m}}^{\infty}(G) = \gamma_{\text{me}}^{\infty}(G)$.*

Proof. A Christmas cactus graph G always contains either a leaf or a leaf cycle. This will be shown by contradiction. If all vertices have degree at least two and each cycle has at least two neighboring blocks then the chain of blocks would either never end or it must close itself, creating another big cycle, contradicting that the graph is a cactus.

We will use reductions until we obtain an elementary graph for which the optimal strategy is known. The graph is called *elementary* if it is a cycle, single edge, a path on three vertices, a bull, or a 3-pan. The proper reduction is chosen repeatedly in the following manner, which is also depicted in Fig. 2.

- If G is elementary we return the optimal strategy.
- If there is a leaf cycle on $k \geq 3$ vertices:
  - If $k \not\equiv 1 \mod 3$ use Reduction 3,
  - otherwise $k \equiv 1 \mod 3$ and then use Reduction 4.
- Otherwise there is a leaf vertex u in G and its neighbor v is an articulation.
- If the vertex v is connected to the rest of the graph by only one edge then use Reduction 1,
- Vertex v is connected to two vertices x and y which are different from u.
- If there is no edge between x and y then they must be connected by a path in G, otherwise, v would be in more than two blocks. Use Reduction 2.
- If there is an edge between x and y then it cannot be on any other cycle than $\{v, x, y\}$. Note that vertices v, x, y form a triangle which is be connected to at most 3 other blocks.
- Now, the previous evaluation can be done on every leaf vertex u. If no of the previous cases is applicable, it means by Lemma 10 that there is a leaf bull (use Reduction 5) or a leaf 3-pan graph (use Reduction 6).

Using the reductions we eventually end up in a situation where the Christmas cactus graph is an elementary graph. The optimal strategy for cycle was shown in Lemma 3, all the configurations of optimal strategies of all the remaining graphs are depicted in Fig. 2.

In each of these elementary graphs, allowing eviction attacks does not increase the necessary number of guards. Also, allowing more guards at one vertex does not add any advantage and does not decrease the necessary number of guards. Therefore, for all of these cases it holds that $\gamma_{\mathrm{me}}^{\infty} = \gamma_{\mathrm{m}}^{\infty} = \Gamma_{\mathrm{m}}^{\infty}$. □

3   Upper Bound on the m-eternal Domination Number of Cactus Graphs

Definition 1. *Let us have a cactus graph G. Let us color vertices of G in the following way. Let a vertex be colored red if it is contained in more than two 2-connected components of G, otherwise it is colored black. Let $R(G)$ denote the number of red vertices, and $Rg(G)$ denote the number of red connected components (e.g. $R(G) = 7$ and $Rg(G) = 3$ in Fig. 3).*

Let G' be a graph created from G by contracting each red connected components into a red vertex.

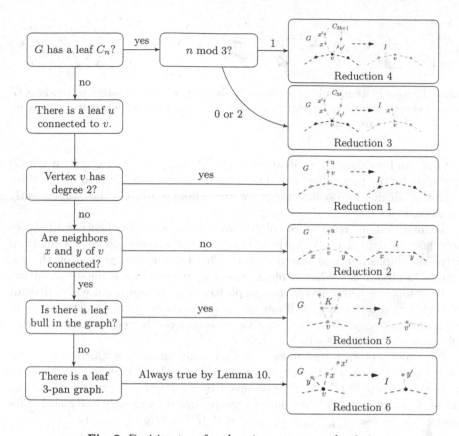

Fig. 2. Decision tree for choosing a proper reduction

Let $B_{G'}$ be a set of maximal connected components of black vertices in G'. Let $R_{G'} = \{b \cup N(b) \mid b \in B_{G'}\}$. Note that $N(b)$ contains only red vertices. Let the Christmas cactus decomposition $Op(G)$ be a disjoint union of graphs induced by $G[r]$ for all $r \in R_{G'}$. See Fig. 3.

Theorem 3. *The m-eternal domination number of a cactus graph G is bounded by*

$$\gamma_m^\infty(G) \le \sum_{H \in Op(G)} \left(\gamma_m^\infty(H) - R(H) \right) + R(G) + Rg(G),$$

where $Op(G)$ are the components of the Christmas cactus decomposition, $R(G)$ is the number of red vertices in G and $Rg(G)$ is the number of connected components of red vertices.

Proof. Let G' be a graph where all connected components of red vertices are contracted creating one red group vertex for each component as shown in Fig. 3. Let $Op(G)$ be the Christmas cactus decomposition of G.

Fig. 3. Process of transforming a cactus G by contracting red edges to produce G' and subsequently duplicating red vertices for each black connected component to get $Op(G)$. (Color figure online)

First, find an optimal strategy for each Christmas cactus graph in $Op(G)$ separately by the process presented in Sect. 2. We will show how to merge these disjoint strategies into one strategy for the whole graph G' and subsequently generalize it for G.

Assume that all red vertices of G' are always occupied, hence we have to show how to defend black vertices. Let us reverse the process of Christmas cactus decomposition and merge disjoint Christmas cactus graphs by the red vertices to obtain G'. When a black vertex is attacked we simulate an attack on the respective Christmas cactus graph to get a configuration which defends the vertex as shown in Fig. 4. If any of the red vertices of the Christmas cactus graph are not occupied then we simulate an attack on the red vertex in all other Christmas cactus graphs which contain it.

The process ensures that in each configuration all but one Christmas cactus graph incident to each red vertex has a guard on it. A red vertex v incident to k black components in G' is always occupied by exactly $k - 1$ guards. So we can remove $k - 2$ guards and it remains always occupied by exactly one guard. This strategy for G' uses $\sum_{H \in Op(G)} \left(\gamma_m^\infty(H) - R(H) \right) + 2Rg(G)$ guards.

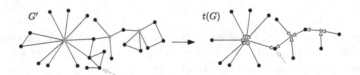

Fig. 4. Simulating attacks to get the right amount of guards on each red vertex. (Color figure online)

Now we get G from G' by expanding the red vertices back into the original red connected components. Add guards such that all red vertices are occupied. The strategy will be altered slightly. When we defend G' by moving a guard from red vertex v then another guard from a different Christmas cactus component is forced to move to v by a simulated attack. However in G the left vertex u and the attacked vertex u' might not coincide so we pick a path from u to u' in the red component and move all the guards along the path.

The change in number of guards can be imagined as removing guards on red vertices of G' and adding guards on all red vertices of G. We devised a strategy for G which uses $\sum_{H \in Op(G)} \left(\gamma_m^\infty(H) - R(H) \right) + R(G) + Rg(G)$ guards. $\square$

4   Linear-Time Algorithm

We present a description of a linear-time algorithm, which computes γ_m^∞ in Christmas cactus graphs in linear time. The algorithm applies previously presented reductions on the block-cut tree of the input graph.

Definition 2 (Harary [6]). *Let the* block-cut *tree of a graph G be a graph $BC(G) = (A \cup \mathcal{B}, E')$, where A is the set of articulations in G and $\mathcal{B}$ is the set of biconnected components in G. A vertex $a \in A$ is connected by an edge to a biconnected component $B \in \mathcal{B}$ if and only if $a \in B$ in G.*

The high level description of the algorithm is as follows.

1. Construct the Christmas cactus decomposition of G and iterate the following for each component $H_1, H_2, \ldots, H_c$.
   (a) Construct the block-cut tree BC of the Christmas cactus H_i.
   (b) Repeatedly apply the reductions on the leaf components of BC.
   (c) If a reduction can by applied, appropriately modify BC in constant time, so that it represents H_i with the chosen reduction applied. At the same time, increase the resulting γ_m^∞ appropriately.
   (d) If BC is empty return the resulting γ_m^∞ and end the process.
2. Use Theorem 3 to get the upper bound.

This result is summed up in the following theorem. We also present the detailed pseudocode of the linear-time algorithm for finding the m-eternal domination number for Christmas cactus graphs.

Theorem 4. *Let G be a cactus on n vertices and m edges. Then there exists an algorithm which computes an upper bound on $\gamma_m^\infty(G)$ in time $\mathcal{O}(n + m)$. Moreover, this algorithm computes the γ_m^∞ of Christmas cactus graphs exactly.*

Proof. First step of the algorithm is to create the Christmas cactus decomposition of the input graph G. For each of these Christmas cactus graphs we run Algorithm 1 and then output the answer devised by Theorem 3.

The construction of a Christmas cactus decomposition of G is done by constructing the block-cut tree of G and coloring red all the vertices which are present in at least 3 blocks, and coloring black all other vertices. We use the DFS algorithm to find all the connected components of red vertices and contract each of these components into a single red vertex. Next, we use the DFS algorithm to retrieve all connected components of black vertices along with their incident red vertices. Note that every edge of G is contained at most once in the Christmas cactus decomposition, hence the total number of vertices and edges in the decomposition is bounded by $\mathcal{O}(|E(G)|)$.

We run the Algorithm 1 for each component separately. Now we will show its correctness. The algorithm performs reductions in the while loop at line 10. In each iteration it processes one leaf block vertex in BC.

First, we argue that the algorithm correctly counts the number of guards on all elementary graphs. In case BC consists of a single block, it is detected on line

Algorithm 1. γ_m^∞ of a Christmas cactus graph, Part 1

1: **procedure** M-EDN-CHRISTMAS-CACTUS(G)
2:     $BC = (V', E', size, deg) \leftarrow$ the block-cut tree of G
3:     $stack \leftarrow \emptyset$ ▷ $stack$ keeps track of all leaf blocks in BC
4:     **for** $v \in V'$ **do**
5:         **if** $deg(v) \leq 1$ **then**   ▷ All leaf blocks, or the only block
6:             add v on top of the $stack$
7:         **end if**
8:     **end for**
9:     $g \leftarrow 0$ ▷ The resulting $\gamma_m^\infty(G)$
10:    **while** $stack \neq \emptyset$ **do**
11:        $v \leftarrow$ retrieve and remove an element from top of the $stack$
12:        **if** $deg(v) = 0$ **then** ▷ Block is an elementary cycle, an edge, or one vertex
13:            $(g, stack, size) \leftarrow$ REMOVE-LEAF-BLOCK($v, stack$)
14:        **else**
15:            $a \leftarrow$ the articulation incident to v ▷ $deg(a) = 2$ in Ch. cactus graphs
16:            $u \leftarrow$ the second block neighbor of a other than v
17:            $(g, stack, size) \leftarrow$ BLOCK($u, v, g, stack, size$)
18:        **end if**
19:    **end while**
20:    **return** g
21: **end procedure**

---

12. The block is removed and the number of guards increases by $\lceil size(v)/3 \rceil$ on line 58, which is consistent with the result for elementary cycle and edge. In the other case the BC consists of several blocks. If BC represents a path on three vertices, one guard will added by line 27 and one by line 58 and both blocks are removed. If BC represents a 3-pan it will count 2 guards by first reducing one block by line 31 or 27, and then reducing a single block of size at most 2 on line 58. If BC represents a bull the algorithm will find a leaf block of size 2 reducing the bull to a path on 3 vertices on line 31, counting correctly 3 guards.

Now we show that if the algorithm performs an operation on a leaf block it resolves in the correct number of guards at the end. Let v be a leaf block vertex processed in the loop. Note that each block of a Christmas cactus graph is either a cycle or an edge.

Consider the case where v is a leaf cycle. If the cycle has $size(v) \equiv 1 \mod 3$, then the block is removed entirely adding $(size(v) - 1)/3$ guards on line 40, exactly as in Reduction 4. Otherwise the cycle is contracted to a block of size 2 and $\lceil size(v)/3 \rceil - 1$ guards are added on line 44, as in Reduction 3.

Consider the case where the leaf block v has $size(v) = 2$, representing a leaf vertex. Let u be the block that shares an articulation with v. If block u has $size(u) = 2$ then we remove both of these blocks and add a guard on line 27, as in Reduction 1. If block u has $size(u) \geq 3$ then v is a leaf vertex connected to a cycle. On line 31 v is removed and $size(u)$ is decreased by one. This is consistent with Reductions 2, 5, and 6. Note that reducing a leaf incident to a

Algorithm 1. γ_m^∞ of a Christmas cactus graph, Part 2

22: **procedure** BLOCK($u, v, g, stack, size$)
23: **if** $size(v) \geq 3$ **then**   ▷ Leaf cycle of size at least 3
24:    ($g, stack, size$) ← LEAF-CYCLE($v, g, stack, size$)
25: **else if** $size(v) = 2$ **then**
26:    **if** $size(u) = 2$ **then**   ▷ Reduction 2
27:      $g \leftarrow g + 1$
28:      ($g, stack$) ← REMOVE-LEAF-BLOCK($v, g, stack, size$)
29:      ($g, stack$) ← REMOVE-LEAF-BLOCK($u, g, stack, size$)
30:    **else if** $size(u) \geq 3$ **then**   ▷ Partial Reductions 5 and 6
31:      $g \leftarrow g + 1$
32:      ($g, stack$) ← REMOVE-LEAF-BLOCK($v, g, stack, size$)
33:      $size(u) \leftarrow size(u) - 1$
34:    **end if**
35: **end if**
36: **return** ($g, stack, size$)
37: **end procedure**
38: **procedure** LEAF-CYCLE($v, g, stack, size$)
39: **if** $size(v) \equiv 0 \mod 3$ or $size(v) \equiv 2 \mod 3$ **then** ▷ Reduction 3
40:    $g \leftarrow g + \lceil size(v)/3 \rceil - 1$
41:    $size(v) \leftarrow 2$
42:    add v on top of the $stack$ ▷ The block still remains a leaf
43: **else if** $size(v) \equiv 1 \mod 3$ **then**   ▷ Reduction 4
44:    $g \leftarrow g + (size(v) - 1)/3$
45:    ($g, stack$) ← REMOVE-LEAF-BLOCK($v, g, stack, size$)
46: **end if**
47: **return** ($g, stack, size$)
48: **end procedure**
49: **procedure** REMOVE-LEAF-BLOCK($v, g, stack, size$)
50: **if** $deg(v) \geq 1$ **then**
51:    $a \leftarrow$ the articulation incident to v ▷ $deg(a) = 2$ in Ch. cactus graphs
52:    $u \leftarrow$ the second block neighbor of a other than v
53:    remove a from the neighbor list in u and erase v and a from BC
54:    **if** $deg(u) \leq 1$ **then**
55:      add u on top of the $stack$ ▷ Vertex can become a leaf only here
56:    **end if**
57: **else if** $deg(v) = 0$ **then**
58:    $g \leftarrow g + \lceil size(v)/3 \rceil$
59:    erase v from BC
60: **end if**
61: **return** ($g, stack$)
62: **end procedure**

---

component on three vertices first, yields the same result as reducing the graph first and waiting for the leaf be reduced in either Reduction 2, 5, or 6.

The algorithm performs reductions which were proved to be correct. This concludes the proof of correctness.

Let n' be the number of vertices of the Christmas cactus graph, and m' be the number of its edges. Now we show that Algorithm 1 runs in time $\mathcal{O}(n' + m')$. Using Tarjan's algorithm [11], we can find the blocks of a graph in linear time. By the straight-forward augmentation of the algorithm we obtain the block-cut tree BC where every block v contains additional information $size(v)$ with the number of vertices it contains. Note that the number of vertices and edges of BC is bounded by $2n'$. Therefore $|V(BC)| = \mathcal{O}(n')$ and $|E(BC)| = \mathcal{O}(n')$.

Now consider the while loop at line 10. We claim that every vertex in BC is processed at most twice in the loop and every iteration takes constant time. Let v be the currently processed vertex. During one iteration of the main cycle either a block of size at most 2 is deleted or a block of size at least 3 is either shrunk to size 2 or deleted. Therefore, the algorithm for Christmas cactus graphs runs in time $\mathcal{O}(n' + m')$.

For the cactus graph we need to create the Christmas cactus decomposition which uses Tarjan's algorithm [11] for creating the block-cut tree, and the DFS, which both runs in $\mathcal{O}(n + m)$. As stated, the running time of the algorithm is linear in the size of the Christmas cactus. Therefore, the total running time is bound by sum of their sizes $\mathcal{O}(n + m)$. □

This result together with Theorem 2 implies Theorem 1.

5   Future Work

The computational complexity of the decision variant of the m-eternal domination problem is still mostly unknown as mentioned in the introduction.

The natural extension of the algorithm from cactus graphs is to the more general case of graphs with treewidth 2. It is also an interesting question if we can design an algorithm, whose running time is parameterized by the treewidth of the input graph.

Acknowledgments. We would like to thank Martin Balko and an anonymous referee for their valuable comments and insights.

References

1. Braga, A., de Souza, C.C., Lee, O.: The eternal dominating set problem for proper interval graphs. Inf. Process. Lett. **115**(6), 582–587 (2015)
2. Burger, A.P., Cockayne, E.J., Gründlingh, W.R., Mynhardt, C.M., van Vuuren, J.H., Winterbach, W.: Infinite order domination in graphs. J. Comb. Math. Comb. Comput. **50**, 179–194 (2004)
3. Finbow, S., Messinger, M.-E., van Bommel, M.F.: Eternal domination on 3 × n grid graphs. Australas. J. Comb. **61**, 156–174 (2015)
4. Goddard, W., Hedetniemi, S.M., Hedetniemi, S.T.: Eternal security in graphs. J. Comb. Math. Comb. Comput. **52**, 169–180 (2005)
5. Gupta, U.I., Lee, D.-T., Leung, J.Y.-T.: Efficient algorithms for interval graphs and circular arc graphs. Networks **12**(4), 459–467 (1982)

6. Harary, F.: Graph Theory. Addison-Wesley, Boston (1969)
7. Henning, M.A., Klostermeyer, W.F., MacGillivray, G.: Bounds for the m-eternal domination number of a graph. Contrib. Discrete Math. **12**(2), 91–103 (2017). ISSN 1715-0868
8. Klostermeyer, W.F., MacGillivray, G.: Eternal dominating sets in graphs. J. Comb. Math. Comb. Comput. **68**, 97–111 (2009)
9. Klostermeyer, W.F., Mynhardt, C.M.: Protecting a graph with mobile guards. Appl. Anal. Discrete Math. **10**, 1–29 (2014)
10. Leighton, T., Moitra, A.: Some results on greedy embeddings in metric spaces. Discrete Comput. Geom. **44**(3), 686–705 (2010)
11. Tarjan, R.: Depth-first search and linear graph algorithms. SIAM J. Comput. **1**(2), 146–160 (1972)
12. van Bommel, C.M., van Bommel, M.F.: Eternal domination numbers of 5 × n grid graphs. J. Comb. Math. Comb. Comput. **97**, 83–102 (2016)

On Relevant Equilibria in Reachability Games

Thomas Brihaye[1], Véronique Bruyère[1], Aline Goeminne[1,2(✉)],
and Nathan Thomasset[1,3]

[1] Université de Mons (UMONS), Mons, Belgium
{thomas.brihaye,veronique.bruyere,aline.goeminne}@umons.ac.be
[2] Université libre de Bruxelles (ULB), Brussels, Belgium
[3] ENS Paris-Saclay, Université Paris-Saclay, Cachan, France
nathan.thomasset@ens-paris-saclay.fr

Abstract. We study multiplayer reachability games played on a finite
directed graph equipped with target sets, one for each player. In those
reachability games, it is known that there always exists a Nash equilibrium (NE) and a subgame perfect equilibrium (SPE). But sometimes
several equilibria may coexist such that in one equilibrium no player
reaches his target set whereas in another one several players reach it. It
is thus very natural to identify "relevant" equilibria. In this paper, we
consider different notions of relevant equilibria including Pareto optimal
equilibria and equilibria with high social welfare. We provide complexity
results for various related decision problems.

Keywords: Multiplayer non-zero-sum games played on graphs ·
Reachability objectives · Relevant equilibria · Social welfare ·
Pareto optimality

1 Introduction

Two-player zero-sum games played on graphs are commonly used to model *reactive systems* where a system interacts with its environment [16]. In such setting
the system wants to achieve a goal - to respect a certain property - and the environment acts in an antagonistic way. The underlying game is defined as follows:
the two players are the system and the environment, the vertices of the graph
are all the possible configurations in which the system can be and an infinite
path in this graph depicts a possible sequence of interactions between the system
and its environment. In such a game, each player chooses a *strategy*: it is the
way he plays given some information about the game and past actions of the
other player. Following a strategy for each player results in a *play* in the game.
Finding how the system can ensure that a given property is satisfied amounts

Research partially supported by the PDR project "Subgame perfection in graph games"
(F.R.S.-FNRS) and by COST Action 16228 "GAMENET" (European Cooperation in
Science and Technology).

E. Filiot et al. (Eds.): RP 2019, LNCS 11674, pp. 48–62, 2019.
https://doi.org/10.1007/978-3-030-30806-3_5

to finding, if it exists, a *winning strategy* for the system in this game. For some situations, this kind of model is too restrictive and a setting with more than two agents such that each of them has his own not necessarily antagonistic objective is more realistic. These games are called *multiplayer non zero-sum games*. In this setting, the solution concept of winning strategy is not suitable anymore and different notions of *equilibria* can be studied.

In this paper, we focus on *Nash equilibrium* (NE) [14]: given a strategy for each player, no player has an incentive to deviate unilaterally from his strategy. We also consider the notion of *subgame perfect equilibrium* (SPE) well suited for games played on graphs [15]. We study these two notions of equilibria on *reachability games*. In reachability games, we equip each player with a subset of vertices of the graph game that he wants to reach. We are interested in both the *qualitative* and *quantitative* settings. In the qualitative setting, each player only aims at reaching his target set, unlike the quantitative setting where each player wants to reach his target set as soon as possible.

It is well known that both NEs and SPEs exist in both qualitative and quantitative reachability games. But, equilibria such that no player reaches his target set and equilibria such that some players reach it may coexist. This observation has already been made in [17,18]. In such a situation, one could prefer the second situation to the first one. In this paper, we study different versions of *relevant equilibria*.

Contributions. For quantitative reachability games, we focus on the following three kinds of relevant equilibria: *constrained equilibria*, *equilibria optimizing social welfare* and *Pareto optimal equilibria*. For constrained equilibria, we aim at minimizing the cost of each player *i.e.*, the number of steps it takes to reach his target set (Problem 1). For equilibria optimizing social welfare, a player does not only want to minimize his own cost, he is also committed to maximizing the *social welfare* (Problem 2). For Pareto optimal equilibria, we want to decide if there exists an equilibrium such that the tuple of the costs obtained by players following this equilibrium is *Pareto optimal* in the set of all the possible costs that players can obtain in the game (Problem 3). We consider the decision variant of Problems 1 and 2; and the qualitative adaptations of the three problems.

Our main contributions are the following. *(i)* We study the complexity of the three decision problems. Our results gathered with previous works are summarized in Table 1. *(ii)* In case of a positive answer to any of the three decision problems, we prove that finite-memory strategies are sufficient. Our results and others from previous works are given in Table 1. *(iii)* We identify a subclass of reachability games in which there always exists an SPE where each player reaches his target set. *(iv)* Given a play, we provide a *characterization* which guarantees that this play is the outcome of an NE. This characterization is based on the values in the associated two-player zero-sum games called *coalitional games*.

Table 1. Complexity classes and memory results

Complexity	Qual. Reach.		Quant. Reach.		Memory	Qual. Reach.		Quant. Reach.	
	NE	SPE	NE	SPE		NE	SPE	NE	SPE
Prob. 1	NP-c [10]	PSPACE-c[4]	NP-c	PSPACE-c[5]	Prob. 1	Poly.[10]	Expo.[4]	Poly.	Expo.
Prob. 2	NP-c	PSPACE-c	NP-c	PSPACE-c	Prob. 2	Poly.	Expo.	Poly.	Expo.
Prob. 3	NP-h/Σ_2^P	PSPACE-c	NP-h/Σ_2^P	PSPACE-c	Prob. 3	Poly.	Expo.	Poly.	Expo.

Related Work. There are many results on NEs and SPEs played on graphs, we refer the reader to [9] for a survey and an extended bibliography. Here we focus on the results directly related to our contributions.

Regarding Problem 1, for NEs, it is shown to be NP-complete only in the qualitative setting [10]; for SPEs it is shown to be PSPACE-complete in both the qualitative and quantitative settings in [4–6]. Notice that in [18], variants of Problem 1 for games with Streett, parity or co-Büchi winning conditions are shown NP-complete and decidable in polynomial time for Büchi.

Regarding Problem 2, in the setting of games played on matrices, deciding the existence of an NE such that the expected social welfare is at most k is NP-hard [11]. Moreover, in [1] it is shown that deciding the existence of an NE which maximizes the social welfare is undecidable in concurrent games in which a cost profile is associated only with terminal nodes.

Regarding Problem 3, in the setting of zero-sum two-player multidimensional mean-payoff games, the *Pareto-curve* (the set of maximal thresholds that a player can force) is studied in [2] by giving some properties on the geometry of this set. The authors provide a Σ_2^P algorithm to decide if this set intersects a convex set defined by linear inequations.

Regarding the memory, in [7] it is shown that there always exists an NE with memory at most $|V| + |\Pi|$ in quantitative reachability games, without any constraint on the cost of the NE. It is shown in [17] that, in multiplayer games with ω-regular objectives, there exists an SPE with a given payoff if and only if there exists an SPE with the same payoff but with finite memory. Moreover, in [4] it is claimed that it is sufficient to consider strategies with an exponential memory to solve Problem 1 for SPE in qualitative reachability games.

Finally, we can find several kinds of outcome characterizations for Nash equilibria and variants, *e.g.*, in multiplayer games equipped with prefix-linear cost functions and such that the vertices in coalitional games have a value (summarized in [9]), in multiplayer games with prefix-independent Borel objectives [18], in multiplayer games with classical ω-regular objectives (as reachability) by checking if there exists a play which satisfies an LTL formula [10], in concurrent games [12], etc. Such characterizations are less widespread for subgame perfect equilibria, but one can recover one for quantitative reachability games thanks to a value-iteration procedure [5].

Structure of the Paper. We decide to only detail results for quantitative reachability games while results for qualitative reachability games are only summarized in Table 1. The proofs for the qualitative reachability setting are not given because they are in the same spirit as for the quantitative setting. In Sect. 2, we introduce the needed background and define the different studied problems. In Sect. 3, we identify families of reachability games for which there always exists a relevant equilibrium, for different notions of relevant equilibria. In Sect. 4, we provide the main ideas necessary to obtain our complexity results (see Table 1). The detailed proofs for the quantitative reachability setting, together with additional results on qualitative reachability games, are provided in the full version of the paper available at https://arxiv.org/abs/1907.05481.

2   Preliminaries and Studied Problems

Arena, Game and Strategies. An *arena* is a tuple $\mathcal{A} = (\Pi, V, E, (V_i)_{i \in \Pi})$ such that: *(i)* Π is a finite set of players; *(ii)* V is a finite set of vertices; *(iii)* $E \subseteq V \times V$ is a set of edges such that for all $v \in V$ there exists $v' \in V$ such that $(v, v') \in E$ and *(iv)* $(V_i)_{i \in \Pi}$ is a partition of V between the players.

A *play* in $\mathcal{A}$ is an infinite sequence of vertices $\rho = \rho_0 \rho_1 \ldots$ such that for all $k \in \mathbb{N}$, $(\rho_k, \rho_{k+1}) \in E$. A *history* is a finite sequence $h = h_0 h_1 \ldots h_k$ with $k \in \mathbb{N}$ defined similarly. The *length* $|h|$ of h is the number k of its edges. We denote the set of plays by Plays and the set of histories by Hist. Moreover, the set Hist_i is the set of histories such that their last vertex v is a vertex of player i, i.e. $v \in V_i$.

Given a play $\rho \in$ Plays and $k \in \mathbb{N}$, the *prefix* $\rho_0 \rho_1 \ldots \rho_k$ of ρ is denoted by $\rho_{\leq k}$ and its *suffix* $\rho_k \rho_{k+1} \ldots$ by $\rho_{\geq k}$. A play ρ is called a *lasso* if there exists $h\ell \in$ Hist such that $\rho = h\ell^\omega$. The *length* of this lasso is the length of $h\ell$. Notice that ℓ is not necessarily a simple cycle.

A *game* $\mathcal{G} = (\mathcal{A}, (\text{Cost}_i)_{i \in \Pi})$ is an arena equipped with a cost function profile $(\text{Cost}_i)_{i \in \Pi}$ such that for all $i \in \Pi$, $\text{Cost}_i : \text{Plays} \to \mathbb{N} \cup \{+\infty\}$ is a *cost function* which assigns a cost to each play ρ for player i. We also say that the play ρ has *cost profile* $(\text{Cost}_i(\rho))_{i \in \Pi}$. Given two cost profiles $c, c' \in (\mathbb{N} \cup \{+\infty\})^{|\Pi|}$, we say that $c \leq c'$ if and only if for all $i \in \Pi$, $c_i \leq c'_i$.

An initial vertex $v_0 \in V$ is often fixed, and we call $(\mathcal{G}, v_0)$ an *initialized game*. A play (resp. a history) of $(\mathcal{G}, v_0)$ is then a play (resp. a history) of $\mathcal{G}$ starting in v_0. The set of such plays (resp. histories) is denoted by $\text{Plays}(v_0)$ (resp. $\text{Hist}(v_0)$). The notation $\text{Hist}_i(v_0)$ is used when these histories end in a vertex $v \in V_i$.

Given a game $\mathcal{G}$, a *strategy* for player i is a function $\sigma_i : \text{Hist}_i \to V$. It assigns to each history hv, with $v \in V_i$, a vertex v' such that $(v, v') \in E$. In an initialized game $(\mathcal{G}, v_0)$, σ_i needs only to be defined for histories starting in v_0. We denote by Σ_i the set of strategies for Player i. A play $\rho = \rho_0 \rho_1 \ldots$ is *consistent* with σ_i if for all $\rho_k \in V_i$, $\sigma_i(\rho_0 \ldots \rho_k) = \rho_{k+1}$. A strategy σ_i is *positional* if it only depends on the last vertex of the history, *i.e.*, $\sigma_i(hv) = \sigma_i(v)$ for all $hv \in \text{Hist}_i$. It is *finite-memory* if it can be encoded by a finite-state machine.

A *strategy profile* is a tuple $\sigma = (\sigma_i)_{i \in \Pi}$ of strategies, one for each player. Given an initialized game $(\mathcal{G}, v_0)$ and a strategy profile σ, there exists a unique play from v_0 consistent with each strategy σ_i. We call this play the *outcome* of σ and denote it by $\langle \sigma \rangle_{v_0}$. We say that σ has cost profile $(\text{Cost}_i(\langle \sigma \rangle_{v_0}))_{i \in \Pi}$.

Quantitative Reachability Games. In this article, we are interested in *reachability games*: each player has a target set of vertices that he wants to reach.

Definition 1. *A* quantitative reachability game $\mathcal{G} = (\mathcal{A}, (\mathrm{Cost}_i)_{i \in \Pi}, (F_i)_{i \in \Pi})$ *is a game enhanced with a target set $F_i \subseteq V$ for each player $i \in \Pi$ and for all $i \in \Pi$ the cost function Cost_i is defined as follows: for all $\rho = \rho_0 \rho_1 \ldots \in$ Plays:* $\mathrm{Cost}_i(\rho) = k$ *if $k \in \mathbb{N}$ is the least index such that $\rho_k \in F_i$ and $\mathrm{Cost}_i(\rho) = +\infty$ if such index does not exist.*

In *quantitative reachability games*, players have to pay a cost equal to the number of edges until visiting their own target set or $+\infty$ if it is not visited. Thus each player aims at *minimizing* his cost.

Solution Concepts. In the multiplayer game setting, the solution concepts usually studied are *equilibria*. We recall the concepts of Nash equilibrium and subgame perfect equilibrium.

Let $\sigma = (\sigma_i)_{i \in \Pi}$ be a strategy profile in an initialized game $(\mathcal{G}, v_0)$. When we highlight the role of player i, we denote σ by (σ_i, σ_{-i}) where σ_{-i} is the profile $(\sigma_j)_{j \in \Pi \setminus \{i\}}$. A strategy $\sigma_i' \neq \sigma_i$ is a *deviating* strategy of Player i, and it is a *profitable deviation* for him if $\mathrm{Cost}_i(\langle \sigma \rangle v_0) > \mathrm{Cost}_i(\langle \sigma_i', \sigma_{-i} \rangle v_0)$.

The notion of Nash equilibrium is classical: a strategy profile σ in an initialized game $(\mathcal{G}, v_0)$ is a *Nash equilibrium* (NE) if no player has an incentive to deviate unilaterally from his strategy, i.e. no player has a profitable deviation.

Definition 2 (Nash equilibrium). *Let $(\mathcal{G}, v_0)$ be an initialized quantitative reachability game. The strategy profile σ is an NE if for each $i \in \Pi$ and each deviating strategy σ_i' of Player i, we have* $\mathrm{Cost}_i(\langle \sigma \rangle v_0) \leq \mathrm{Cost}_i(\langle \sigma_i', \sigma_{-i} \rangle v_0)$.

When considering games played on graphs, a useful refinement of NE is the concept of *subgame perfect equilibrium* (SPE). An SPE is a strategy profile that is an NE in each subgame. Formally, given a game $\mathcal{G} = (\mathcal{A}, (\mathrm{Cost}_i)_{i \in \Pi})$, an initial vertex v_0, and a history $hv \in \mathrm{Hist}(v_0)$, the initialized game $(\mathcal{G}_{\upharpoonright h}, v)$ such that $\mathcal{G}_{\upharpoonright h} = (\mathcal{A}, (\mathrm{Cost}_{i \upharpoonright h})_{i \in \Pi})$ where $\mathrm{Cost}_{i \upharpoonright h}(\rho) = \mathrm{Cost}_i(h\rho)$ for all $i \in \Pi$ and $\rho \in V^\omega$ is called a *subgame* of $(\mathcal{G}, v_0)$. Notice that $(\mathcal{G}, v_0)$ is a subgame of itself. Moreover if σ_i is a strategy for player i in $(\mathcal{G}, v_0)$, then $\sigma_{i \upharpoonright h}$ denotes the strategy in $(\mathcal{G}_{\upharpoonright h}, v)$ such that for all histories $h' \in \mathrm{Hist}(v)$, $\sigma_{i \upharpoonright h}(h') = \sigma_i(hh')$. Similarly, from a strategy profile σ in $(\mathcal{G}, v_0)$, we derive the strategy profile $\sigma_{\upharpoonright h}$ in $(\mathcal{G}_{\upharpoonright h}, v)$.

Definition 3 (Subgame perfect equilibrium). *Let $(\mathcal{G}, v_0)$ be an initialized game. A strategy profile σ is an SPE in $(\mathcal{G}, v_0)$ if for all $hv \in \mathrm{Hist}(v_0)$, $\sigma_{\upharpoonright h}$ is an NE in $(\mathcal{G}_{\upharpoonright h}, v)$.*

Clearly, any SPE is an NE and it is stated in Theorem 2.1 in [3] that there always exists an SPE (and thus an NE) in quantitative reachability games.

Studied Problems. We conclude this section with the problems studied in this article. Let us first recall the concepts of social welfare and Pareto optimality. Let $(\mathcal{G}, v_0)$ be an initialized quantitative reachability game with $\mathcal{G} = (\mathcal{A}, (\mathrm{Cost}_i)_{i \in \Pi}, (F_i)_{i \in \Pi})$. Given $\rho = \rho_0 \rho_1 \ldots \in \mathrm{Plays}(v_0)$, we denote by $\mathrm{Visit}(\rho)$ the set of players who visit their target set along ρ, i.e., $\mathrm{Visit}(\rho) = \{i \in \Pi \mid$ there exists $n \in \mathbb{N}$ st. $\rho_n \in F_i\}$.[1] The *social welfare* of ρ, denoted by $\mathrm{SW}(\rho)$, is the pair $(|\mathrm{Visit}(\rho)|, \sum_{i \in \mathrm{Visit}(\rho)} \mathrm{Cost}_i(\rho))$. Note that it takes into account both the number of players who visit their target set and their accumulated cost to reach those sets. Finally, let $P = \{(\mathrm{Cost}_i(\rho))_{i \in \Pi} \mid \rho \in \mathrm{Plays}(v_0)\} \subseteq (\mathbb{N} \cup \{+\infty\})^{|\Pi|}$. A cost profile $p \in P$ is *Pareto optimal in* $\mathrm{Plays}(v_0)$ if it is minimal in P with respect to the componentwise ordering $\leq$ on P.[2]

Let us now state the studied decision problems. The first two problems are classical: they ask whether there exists a solution (NE or SPE) σ satisfying certain requirements that impose bounds on either $(\mathrm{Cost}_i(\langle\sigma\rangle_{v_0}))_{i \in \Pi}$ or on $\mathrm{SW}(\langle\sigma\rangle_{v_0})$.

Problem 1 (Threshold decision problem). Given an initialized quantitative reachability game $(\mathcal{G}, v_0)$, given a threshold $y \in (\mathbb{N} \cup \{+\infty\})^{|\Pi|}$, decide whether there exists a solution σ such that $(\mathrm{Cost}_i(\langle\sigma\rangle_{v_0}))_{i \in \Pi} \leq y$.

The most natural requirements are to impose upper bounds on the costs that the players have to pay and no lower bounds. One might also be interested in imposing an interval $[x_i, y_i]$ in which the cost paid by Player i must lie.

In [5], Problem 1 with upper and lower bounds is already solved for SPEs.

Theorem 1 ([5]). *For SPEs, Problem 1 with upper (and lower) bounds is PSPACE-complete.*

In the second problem, constraints are imposed on the social welfare, with the aim to maximize it. We use the lexicographic ordering on $\mathbb{N}^2$ such that $(k, c) \succeq (k', c')$ if and only if (i) $k \geq k'$ or (ii) $k = k'$ and $c \leq c'$.

Problem 2 (Social welfare decision problem). Given an initialized quantitative reachability game $(\mathcal{G}, v_0)$, given two thresholds $k \in \{0, \ldots, |\Pi|\}$ and $c \in \mathbb{N}$, decide whether there exists a solution σ such that $\mathrm{SW}(\langle\sigma\rangle_{v_0}) \succeq (k, c)$.

Notice that with the lexicographic ordering, we want to first maximize the number of players who visit their target set, and then to minimize the accumulated cost to reach those sets. Let us now state the last studied problem.

Problem 3 (Pareto optimal decision problem). Given an initialized quantitative reachability game $(\mathcal{G}, v_0)$ decide whether there exists a solution σ in $(\mathcal{G}, v_0)$ such that $(\mathrm{Cost}_i(\langle\sigma\rangle_{v_0}))_{i \in \Pi}$ is Pareto optimal in $\mathrm{Plays}(v_0)$.

---

[1] We can easily adapt this definition to histories.

[2] For convenience, we prefer to say that p is Pareto optimal in $\mathrm{Plays}(v_0)$ rather than in P.

Remark 1. Problems 1 and 2 impose constraints with *non-strict* inequalities. We could also impose strict inequalities or even a mix of strict and non-strict inequalities. The results of this article can be easily adapted to those variants.

We conclude this section with an illustrative example.

Example 1. Consider the quantitative reachability game $(\mathcal{G}, v_0)$ of Fig. 1. We have two players such that the vertices of Player 1 (resp. Player 2) are rounded (resp. rectangular) vertices. For the moment, the reader should not consider the value indicated on the right of the vertices' labeling. Moreover $F_1 = \{v_3, v_4\}$ and $F_2 = \{v_1, v_4\}$. In this figure, an edge (v, v') labeled by x should be understood as a path from v to v' with length x. Observe that F_1 and F_2 are both reachable from the initial vertex v_0. Moreover the two Pareto optimal cost profiles are $(3, 3)$ and $(2, 6)$: take a play with prefix $v_0 v_2 v_4$ in the first case, and a play with prefix $v_0 v_2 v_3 v_0 v_1$ in the second case.

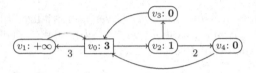

Fig. 1. A two-player quantitative reachability game such that $F_1 = \{v_3, v_4\}$ and $F_2 = \{v_1, v_4\}$

For this example, we claim that there is no NE (and thus no SPE) such that its cost profile is Pareto optimal (see Problem 3). Assume the contrary and suppose that there exists an NE σ such that its outcome ρ has cost profile $(3, 3)$, meaning that ρ begins with $v_0 v_2 v_4$. Then Player 1 has a profitable deviation such that after history $v_0 v_2$ he goes to v_3 instead of v_4 in a way to pay a cost of 2 instead of 3, which is a contradiction. Similarly assume that there exists an NE σ such that its outcome ρ has cost profile $(2, 6)$, meaning that ρ begins with $v_0 v_2 v_3 v_0 v_1$. Then Player 2 has a profitable deviation such that after history v_0 he goes to v_1 instead of v_2, again a contradiction. So there is no NE σ in $(\mathcal{G}, v_0)$ such that $(\mathrm{Cost}_i(\langle \sigma \rangle_{v_0}))_{i \in \Pi}$ is Pareto optimal in $\mathrm{Plays}(v_0)$.

The previous discussion shows that there is no NE σ such that $(0, 0) = x \leq (\mathrm{Cost}_i(\langle \sigma \rangle_{v_0}))_{i \in \Pi} \leq y = (3, 3)$ (see Problem 1). This is no longer true with $y = (6, 3)$. Indeed, one can construct an NE τ whose outcome has prefix $v_0 v_1 v_0 v_2 v_3$ and cost profile $(6, 3)$. This also shows that there exists an NE σ (the same τ as before) that satisfies $\mathrm{SW}(\langle \sigma \rangle_{v_0}) \succeq (k, c) = (2, 9)$ (with τ both players visit their target set and their accumulated cost to reach it equals 9).   □

3   Existence Problems

In this section, we show that for particular families of reachability games and requirements, there is no need to solve the related decision problems because they always have a positive answer in this case.

We begin with the family constituted by all reachability games with a *strongly connected* arena. The next theorem then states that there always exists a solution that visits all non-empty target sets.

Theorem 2. *Let $(\mathcal{G}, v_0)$ be an initialized quantitative reachability game such that its arena $\mathcal{A}$ is strongly connected. There exists an SPE σ (and thus an NE) such that its outcome $\langle\sigma\rangle_{v_0}$ visits all target sets F_i, $i \in \Pi$, that are non-empty.*

Let us comment this result. For this family of games, the answer to Problem 1 is always positive for particular thresholds. In case of quantitative reachability, take strict constraints $< +\infty$ if $F_i \neq \emptyset$ and non-strict constraints $\leq +\infty$ otherwise. We will see later that the strict constraints $< +\infty$ can be replaced by the non-strict constraints $\leq |V| \cdot |\Pi|$ (see Theorem 7). We will also see that, in this setting, the answer to Problem 2 is also always positive for thresholds $k = |\{i \mid F_i \neq \emptyset\}|$ and $c = |\Pi|^2 \cdot |V|$ (see Theorem 7).

In the statement of Theorem 2, as the arena is strongly connected, F_i is non-empty if and only if F_i is reachable from v_0. Also notice that the hypothesis that the arena is strongly connected is necessary. Indeed, it is easy to build an example with two players (Player 1 and Player 2) such that from v_0 it is not possible to reach both F_1 and F_2.

We now turn to the second result of this section. The next theorem states that even with only two players there exists an initialized quantitative reachability game that has no NE with a cost profile which is Pareto optimal. To prove this result, we only have to come back to the quantitative reachability game of Fig. 1. We explained in Example 1 that there is no NE in this game such that its cost profile is Pareto optimal.

Theorem 3. *There exists an initialized quantitative reachability game such that $|\Pi| = 2$ and that has no NE with a cost profile Pareto optimal in $\mathrm{Plays}(v_0)$.*

Notice that in the qualitative setting, in two-player games, there always exists an NE (resp. SPE) such that the gain profile[3] is Pareto optimal in $\mathrm{Plays}(v_0)$ however this existence result cannot be extended to three players.

4 Solving Decision Problems

In this section, we provide the complexity results for the different problems without any assumption on the arena of the game. Even if we provide complexity lower bounds, the main part of our contribution is to give the upper bounds. Roughly speaking the decision algorithms work as follows: they guess a path and check that it is the outcome of an equilibrium satisfying the relevant property (such as Pareto optimality). In order to verify that a path is an equilibrium outcome, we rely on the outcome characterization of equilibria, presented in

---

[3] In the qualitative setting, each player obtains a gain that he wants to maximize: either 1 (if he visits his target set) or 0 (otherwise), all definitions are adapted accordingly.

Sect. 4.2. These characterizations rely themselves on the notion of λ-consistent play, introduced in Sect. 4.1. As the guessed path should be finitely representable, we show that we can only consider λ-consistent lassoes, in Sect. 4.3. Finally, we expose the philosophy of the algorithms providing the upper bounds on the complexity of the three problems in Sect. 4.4.

4.1 λ-Consistent Play

We here define the *labeling function*, $\lambda : V \rightarrow \mathbb{N} \cup \{+\infty\}$ used to obtain the outcome characterization of equilibria. Given a vertex $v \in V$ along a play ρ, intuitively, the value $\lambda(v)$ represents the maximal number of steps within which the player who owns this vertex should reach his target set along ρ starting from v. A play which satisfies the constraints given by λ is called a λ-consistent play.

Definition 4 (λ-consistent play). *Let $(\mathcal{G}, v_0)$ be a quantitative reachability game and $\lambda : V \rightarrow \mathbb{N} \cup \{+\infty\}$ be a labeling function. Let $\rho \in$ Plays be a play, we say that $\rho = \rho_0\rho_1 \ldots$ is λ-consistent if for all $i \in \Pi$ and all $k \in \mathbb{N}$ such that $i \notin \text{Visit}(\rho_0 \ldots \rho_k)$ and $\rho_k \in V_i$: $\text{Cost}_i(\rho_{\geq k}) \leq \lambda(\rho_k)$.*

The link between λ-consistency and equilibrium is made in Sect. 4.2.

Example 2. Let us come back to Example 1 and assume that the values indicated on the right of the vertices' labeling represent the valuation of a labeling function λ. Let us first consider the play $\rho = (v_0v_2v_4)^\omega$ with cost profile $(3,3)$. We have that $\text{Cost}_2(\rho) = 3 \leq \lambda(v_0) = 3$ but $\text{Cost}_1(\rho_{\geq 1}) = \text{Cost}_1(v_2v_4(v_0v_2v_4)^\omega) = 2 > \lambda(v_2) = 1$. This means that $(v_0v_2v_4)^\omega$ is not λ-consistent. Secondly, one can easily see that the play $v_0v_1(v_0v_2v_3)^\omega$ is λ-consistent.

4.2 Characterizations

Outcome Characterization of Nash Equilibria. To define the labeling function λ which allows us to obtain this characterization, we need to study the rational behavior of one player playing against the *coalition* of the other players. In order to do so, with a quantitative reachability game $\mathcal{G} = (\mathcal{A}, (\text{Cost}_i)_{i \in \Pi}, (F_i)_{i \in \Pi})$, we can associate $|\Pi|$ *two-player zero-sum quantitative games* [7]. For each $i \in \Pi$, we depict by $\mathcal{G}_i$ the *(quantitative) coalitional game* associated with Player i. In such a game Player i (which becomes Player *Min*) wants to reach the target set $F = F_i$ within a minimum number of steps, and the coalition of all players except Player i (which forms one player called Player *Max*, aka $-i$) aims to avoid it or, if it is not possible, maximize the number of steps until reaching F.

Given a coalitional game $\mathcal{G}_i$ and a vertex $v \in V$, the *value* of $\mathcal{G}_i$ from v, depicted by $\text{Val}_i(v)$, allows us to know what is the lowest (resp. greatest) cost (resp. gain) that Player *Min* (resp. Player *Max*) can ensure to obtain from v. Moreover, as quantitative coalitional games are determined these values always exist and can be computed in polynomial time [7,8,13].

An *optimal strategy* for Player Min (resp. Player Max) in a coalitional game $\mathcal{G}_i$ is a strategy which ensures that, from all vertices $v \in V$, Player Min (resp. Player Max) will pay (resp. obtain) at most $\mathrm{Val}_i(v)$ by following this strategy whatever the strategy of the other player. For each $i \in \Pi$, we know that there always exist optimal strategies for both players in $\mathcal{G}_i$. Moreover, we can always find optimal strategies which are positional [7].

In our characterization, we show that the outcomes of NEs are exactly the plays which are Val-consistent, with the labeling function Val defined in this way: for all $v \in V$, $\mathrm{Val}(v) = \mathrm{Val}_i(v)$ if $v \in V_i$.

Theorem 4 (Characterization of NEs). *Let* $(\mathcal{G}, v_0)$ *be a quantitative reachability game and let* $\rho \in \mathrm{Plays}(v_0)$ *be a play, the next assertions are equivalent:*

1. *there exists an NE* σ *such that* $\langle\sigma\rangle_{v_0} = \rho$;
2. *the play* ρ *is* Val-consistent.

Additionally, if $\rho = h\ell^\omega$ *is a lasso, we can replace the first item by: there exists an NE* σ *with memory in* $\mathcal{O}(|h\ell| + |\Pi|)$ *and such that* $\langle\sigma\rangle_{v_0} = \rho$.

The main idea is that if the second assertion is false, then there exists a player i who has an incentive to deviate along ρ. Indeed, if there exists $k \in \mathbb{N}$ such that $\mathrm{Cost}_i(\rho_{\geq k}) > \mathrm{Val}_i(\rho_k)$ $(\rho_k \in V_i)$ it means that Player i can ensure a better cost for him even if the other players play in coalition and in an antagonistic way. Thus, Player i has a profitable deviation. For the second implication, the Nash equilibrium σ is defined as follows: all players follow the outcome ρ but if one player, assume it is Player i, deviates from ρ the other players form a coalition $-i$ and punish the deviator by playing the optimal strategy of player $-i$ in the coalitional game $\mathcal{G}_i$. Thus, if $\rho = h\ell^\omega$, a player has to remember: *(i)* $h\ell$ to know both what he has to play and if someone has deviated and *(ii)* who is the deviator.

Example 3. Let us go back to Example 2, in this example the used labeling function λ is in fact the labeling function Val. We proved in Example 2 that the play $(v_0 v_2 v_4)^\omega$ is not Val-consistent and so not the outcome of an NE by Theorem 4. On the contrary, we have seen that the play $v_0 v_1 (v_0 v_2 v_3)^\omega$ is Val-consistent and it means that it is the outcome of an NE (again by Theorem 4). Notice that we have already proved these two facts in Example 1.

Outcome Characterization of Subgame Perfect Equilibria. In the previous section, we proved that the set of plays which are Val-consistent is equal to the set of outcomes of NEs. We now want to have the same kind of characterization for SPEs. We may not use the notion of Val-consistent plays because there exist plays which are Val-consistent but which are not the outcome of an SPE. But, we can recover the characterization of SPEs thanks to a different labeling function defined in [5] that we depict by λ^*. Notice that, λ^* is not defined on the vertices of the game $\mathcal{G}$ but on the vertices of the *extended game* $\mathcal{X}$ associated with $\mathcal{G}$. Vertices in such a game are the vertices in $\mathcal{G}$ equipped with a subset of players who have already visited their target set. This game is also a reachability game thus all concepts and definitions introduced in Sect. 2 hold. Moreover, there is a one-to-one correspondence between SPEs in $\mathcal{G}$ and its extended game. This is the reason why we solve the different decision problems on the extended games $(\mathcal{X}, x_0)$, where $x_0 = (v_0, \text{Visit}(v_0))$, instead of $(\mathcal{G}, v_0)$. More details are given in [5]. However, it is very important to notice that some of our results depend on $|V|$ (resp. $|\Pi|$) that are the number of vertices (resp. players) in $\mathcal{G}$ and not in $\mathcal{X}$.

Theorem 5 ([5] Characterization of SPEs). *Let $(\mathcal{G}, v_0)$ be a quantitative reachability game and $(\mathcal{X}, x_0)$ be its extended game and let $\rho = \rho_0\rho_1 \cdots \in$ Plays(x_0) be a play in the extended game, the next assertions are equivalent:*

1. *there exists a subgame perfect equilibrium σ such that $\langle\sigma\rangle_{x_0} = \rho$;*
2. *the play ρ is λ^*-consistent.*

4.3 Sufficiency of Lassoes

In this section, we provide technical results which given a λ-consistent play produce an associated λ-consistent lasso. In the sequel, we show that working with these lassoes is sufficient for the algorithms.

The associated lassoes are built by eliminating some *unnecessary cycles* and then identifying a prefix $h\ell$ such that ℓ can be repeated infinitely often. An unnecessary cycle is a cycle inside of which no new player visits his target set. More formally, let $\rho = \rho_0\rho_1 \ldots \rho_k \ldots \rho_{k+\ell} \ldots$ be a play in $\mathcal{G}$, if $\rho_k = \rho_{k+\ell}$ and $\text{Visit}(\rho_0 \ldots \rho_k) = \text{Visit}(\rho_0 \ldots \rho_{k+\ell})$ then the cycle $\rho_k \ldots \rho_{k+\ell}$ is called an unnecessary cycle.

We call: (P1) the procedure which eliminates an unnecessary cycle, *i.e.*, let $\rho = \rho_0\rho_1 \ldots \rho_k \ldots \rho_{k+\ell} \ldots$ such that $\rho_k \ldots \rho_{k+\ell}$ is an unnecessary cycle, ρ becomes $\rho' = \rho_0 \ldots \rho_k\rho_{k+\ell+1} \ldots$ and (P2) the procedure which turns ρ into a lasso $\rho' = h\ell^\omega$ by copying ρ long enough for all players to visit their target set and then to form a cycle after the last player has visited his target set. If no player visits his target set along ρ, then (P2) only copies ρ long enough to form a cycle. Notice that, given $\rho \in$ Plays, applying (P1) or (P2) may involve a decreasing of the costs but for (P1) and (P2) $\text{Visit}(\rho) = \text{Visit}(\rho')$. Additionally, (P2) $\text{Visit}(h) = \text{Visit}(\rho')$. Additionally, applying (P1) until it is no longer possible and then (P2), leads to a lasso with length at most $(|\Pi| + 1) \cdot |V|$ and cost less than or equal to $|\Pi| \cdot |V|$ for players who have visited their target set.

Additionally, applying (P1) or (P2) on λ-consistent play preserves this property. This is stated in Lemma 1 which is in particular true for extended games.

Lemma 1. *Let $(\mathcal{G}, v_0)$ be a quantitative reachability game and $\rho \in$ Plays be a λ-consistent play for a given labeling function λ. If ρ' is the play obtained by applying (P1) or (P2) on ρ, then ρ' is λ-consistent.*

These properties on (P1) and (P2) allow us to claim that it is sufficient to deal with lassoes with polynomial length to solve Problems 1, 2 and 3. Moreover, it yields some bounds on the needed memory and the costs for each problem.

The next corollary is used to solve Problems 1 and 2.

Corollary 1. *Let σ be an NE (resp. SPE) in a quantitative reachability game $(\mathcal{G}, v_0)$ (resp. $(\mathcal{X}, x_0)$ its extended game). Let $w_0 = v_0$ (resp. $w_0 = x_0$). Then there exists τ an NE (resp. SPE) in $(\mathcal{G}, v_0)$ (resp. $(\mathcal{X}, x_0)$) such that:*

- *$\langle\tau\rangle_{w_0}$ is a lasso $h\ell^\omega$ such that $|h\ell| \leq (|\Pi|+1) \cdot |V|$;*
- *for each $i \in \text{Visit}(\langle\tau\rangle_{w_0})$, $\text{Cost}_i(\langle\tau\rangle_{w_0}) \leq |\Pi| \cdot |V|$;*
- *τ has memory in $\mathcal{O}((|\Pi|+1) \cdot |V|)$ (resp. $\mathcal{O}(2^{|\Pi|} \cdot |\Pi| \cdot |V|^{(|\Pi|+2)\cdot(|V|+3)+1}))$.*

Moreover, given $y \in (\mathbb{N} \cup \{+\infty\})^{|\Pi|}$, $k \in \{0, \ldots, |\Pi|\}$ and $c \in \mathbb{N}$:

- *If $(\text{Cost}_i(\langle\sigma\rangle_{w_0}))_{i\in\Pi} \leq y$, then $(\text{Cost}_i(\langle\tau\rangle_{w_0}))_{i\in\Pi} \leq y$;*
- *If $SW(\langle\sigma\rangle_{w_0}) \succeq (k, c)$, then $SW(\langle\tau\rangle_{w_0}) \succeq (k, c)$.*

The following corollary is used to solve Problem 3.

Corollary 2. *Let σ be an NE (resp. SPE) in a quantitative reachability game $(\mathcal{G}, v_0)$ (resp. $(\mathcal{X}, x_0)$ its extended game). Let $w_0 = v_0$ (resp. $w_0 = x_0$). If we have that $(\text{Cost}_i(\langle\sigma\rangle_{w_0}))_{i\in\Pi}$ is Pareto optimal in $\text{Plays}(w_0)$, then:*

- *for all $i \in \text{Visit}(\langle\sigma\rangle_{w_0})$, $\text{Cost}_i(\langle\sigma\rangle_{w_0}) \leq |V| \cdot |\Pi|$;*
- *there exists τ an NE (resp. SPE) such that $\langle\tau\rangle_{w_0} = h\ell^\omega$, $|h\ell| \leq (|\Pi|+1) \cdot |V|$ and $(\text{Cost}_i(\langle\sigma\rangle_{w_0}))_{i\in\Pi} = (\text{Cost}_i(\langle\tau\rangle_{w_0}))_{i\in\Pi}$;*
- *τ has memory in $\mathcal{O}((|\Pi|+1) \cdot |V|)$ (resp. $\mathcal{O}(2^{|\Pi|} \cdot |\Pi| \cdot |V|^{(|\Pi|+2)\cdot(|V|+3)+1})$).*

4.4   Algorithms

In this section, we provide the main ideas behind our algorithms.

We first focus on algorithms to solve Problems 1, 2 and 3 for NEs. Each algorithm works as follows:

1. it guesses a lasso of polynomial length;
2. it verifies that the cost profile of this lasso satisfies the conditions[4] given by the problem;
3. it verifies that the lasso is the outcome of an NE.

Let us comment the different steps of these algorithms.

- Step 1: For Problems 1 and 2 (resp. Problem 3), it is sufficient to consider plays which are lassoes with polynomial length thanks to Corollary 1 (resp. Corollary 2).
- Step 3: This property is verified thanks to Theorem 4. This is done in polynomial time as the lasso has a polynomial length and the values of the coalitional games are computed in polynomial time.
- Step 2: For Problems 1 and 2, this verification can be obviously done in polynomial time. For Problem 3, we need to have an oracle allowing us to know if the cost profile of the lasso is Pareto optimal. As a consequence, we study Problem 4 which lies in co-NP.

---

[4] Satisfying the conditions is either satisfying the constraints (Problems 1 and 2) or having a cost profile which is Pareto optimal (Problem 3).

Problem 4. Given a reachability game $(\mathcal{G}, v_0)$ (resp. its extended game $(\mathcal{X}, x_0)$) and a lasso $\rho \in \text{Plays}(v_0)$ (resp. $\rho \in \text{Plays}(x_0)$), decide whether $(\text{Cost}_i(\rho))_{i \in \Pi}$ is Pareto optimal in $\text{Plays}(v_0)$ (resp. $\text{Plays}(x_0)$).

Proposition 1. *Problem 4 lies in co-NP.*

Now, we explain our algorithms to solve the three decision problems for SPEs. As Problem 1 is already solved in PSPACE (see Theorem 1), we here focus only on Problems 2 and 3. Each algorithm works as follows:

1. it guesses a lasso of polynomial length;
2. it verifies that the cost profile p of this lasso satisfies the conditions given by the problem;
3. it checks, whether there exists an SPE with cost profile equal to p.

The explanations for the first and the second steps are the same as for the algorithms for NEs. Finally, we know that the third step can be done in PSPACE (Theorem 1).

4.5   Results

Thanks to the previous discussions in Sect. 4.4, we obtain the following results. Notice that we do not provide the proof for the NP-hardness (resp. PSPACE-hardness) as it is very similar to the one given in [10] (resp. [5]).

Theorem 6. *Let $(\mathcal{G}, v_0)$ be a quantitative reachability game.*

- *For NEs: Problems 1 and 2 are NP-complete while Problem 3 is NP-hard and belongs to Σ_2^P.*
- *For SPEs: Problems 1, 2 and 3 are PSPACE-complete.*

Theorem 7. *Let $(\mathcal{G}, v_0)$ be a quantitative reachability game.*

- *For NEs: for each decision problem, if the answer is positive, then there exists a strategy profile σ with memory in $\mathcal{O}((|\Pi| + 1) \cdot |V|)$ which satisfies the conditions.*
- *For SPEs: for each decision problem, if the answer is positive, then there exists a strategy profile σ with memory in $\mathcal{O}(2^{|\Pi|} \cdot |\Pi| \cdot |V|^{(|\Pi|+2) \cdot (|V|+3)+1})$ which satisfies the conditions.*

  Moreover, for both NEs and SPEs:

- *for Problems 1 and 3, σ is such that: if $i \in \text{Visit}(\langle \sigma \rangle_{v_0})$, then $\text{Cost}_i(\langle \sigma \rangle_{v_0}) \leq |\Pi| \cdot |V|$;*
- *for Problem 2, σ is such that: $\sum_{i \in \text{Visit}(\langle \sigma \rangle_{v_0})} \text{Cost}_i(\langle \sigma \rangle_{v_0}) \leq |\Pi|^2 \cdot |V|$.*

References

1. Bouyer, P., Markey, N., Stan, D.: Mixed Nash equilibria in concurrent terminal-reward games. In: 34th International Conference on Foundation of Software Technology and Theoretical Computer Science, FSTTCS 2014, 15–17 December 2014, New Delhi, India, pp. 351–363 (2014)
2. Brenguier, R., Raskin, J.-F.: Pareto curves of multidimensional mean-payoff games. In: Kroening, D., Păsăreanu, C.S. (eds.) CAV 2015. LNCS, vol. 9207, pp. 251–267. Springer, Cham (2015). https://doi.org/10.1007/978-3-319-21668-3_15
3. Brihaye, T., Bruyère, V., De Pril, J., Gimbert, H.: On subgame perfection in quantitative reachability games. Logical Methods Comput. Sci. 9(1) (2012)
4. Brihaye, T., Bruyère, V., Goeminne, A., Raskin, J.: Constrained existence problem for weak subgame perfect equilibria with ω-regular Boolean objectives. In: Proceedings Ninth International Symposium on Games, Automata, Logics, and Formal Verification, GandALF 2018, 26–28th September 2018, Saarbrücken, Germany, pp. 16–29 (2018)
5. Brihaye, T., Bruyère, V., Goeminne, A., Raskin, J., van den Bogaard, M.: The complexity of subgame perfect equilibria in quantitative reachability games. CoRR abs/1905.00784 (2019). http://arxiv.org/abs/1905.00784
6. Brihaye, T., Bruyère, V., Goeminne, A., Raskin, J., van den Bogaard, M.: The complexity of subgame perfect equilibria in quantitative reachability games. CONCUR 2019 (2019)
7. Brihaye, T., De Pril, J., Schewe, S.: Multiplayer cost games with simple Nash equilibria. In: Artemov, S., Nerode, A. (eds.) LFCS 2013. LNCS, vol. 7734, pp. 59–73. Springer, Heidelberg (2013). https://doi.org/10.1007/978-3-642-35722-0_5
8. Brihaye, T., Geeraerts, G., Haddad, A., Monmege, B.: Pseudopolynomial iterative algorithm to solve total-payoff games and min-cost reachability games. Acta Inf. 54(1), 85–125 (2017)
9. Bruyère, V.: Computer aided synthesis: a game-theoretic approach. In: Charlier, É., Leroy, J., Rigo, M. (eds.) DLT 2017. LNCS, vol. 10396, pp. 3–35. Springer, Cham (2017). https://doi.org/10.1007/978-3-319-62809-7_1
10. Condurache, R., Filiot, E., Gentilini, R., Raskin, J.F.: The complexity of rational synthesis. In: Chatzigiannakis, I., Mitzenmacher, M., Rabani, Y., Sangiorgi, D. (eds.) 43rd International Colloquium on Automata, Languages, and Programming (ICALP 2016). Leibniz International Proceedings in Informatics (LIPIcs), vol. 55, pp. 121:1–121:15. Schloss Dagstuhl-Leibniz-Zentrum fuer Informatik, Dagstuhl (2016)
11. Conitzer, V., Sandholm, T.: Complexity results about Nash equilibria. CoRR cs.GT/0205074 (2002). http://arxiv.org/abs/cs.GT/0205074
12. Haddad, A.: Characterising Nash equilibria outcomes in fully informed concurrent games. http://web1.ulb.ac.be/di/verif/haddad/H16.pdf
13. Khachiyan, L., et al.: On short paths interdiction problems: total and node-wise limited interdiction. Theory Comput. Syst. 43(2), 204–233 (2008)
14. Nash, J.F.: Equilibrium points in n-person games. In: PNAS, vol. 36, pp. 48–49. National Academy of Sciences (1950)
15. Osborne, M.: An Introduction to Game Theory. Oxford University Press, Oxford (2004)
16. Pnueli, A., Rosner, R.: On the synthesis of a reactive module. In: POPL, pp. 179–190. ACM Press (1989)

17. Ummels, M.: Rational behaviour and strategy construction in infinite multiplayer games. In: Arun-Kumar, S., Garg, N. (eds.) FSTTCS 2006. LNCS, vol. 4337, pp. 212–223. Springer, Heidelberg (2006). https://doi.org/10.1007/11944836_21

18. Ummels, M.: The complexity of Nash equilibria in infinite multiplayer games. In: Amadio, R. (ed.) FoSSaCS 2008. LNCS, vol. 4962, pp. 20–34. Springer, Heidelberg (2008). https://doi.org/10.1007/978-3-540-78499-9_3

Partial Solvers for Generalized Parity Games

Véronique Bruyère[1]([✉]), Guillermo A. Pérez[2], Jean-François Raskin[3], and Clément Tamines[1]

[1] University of Mons (UMONS), Mons, Belgium
veronique.bruyere@umons.ac.be
[2] University of Antwerp (UAntwerp), Antwerp, Belgium
[3] Université libre de Bruxelles (ULB), Brussels, Belgium

Abstract. Parity games have been broadly studied in recent years for their applications to controller synthesis and verification. In practice, partial solvers for parity games that execute in polynomial time, while incomplete, can solve most games in publicly available benchmark suites. In this paper, we combine those partial solvers with the classical algorithm for parity games due to Zielonka. We also extend partial solvers to generalized parity games that are games with conjunction of parity objectives. We have implemented those algorithms and evaluated them on a large set of benchmarks proposed in the last LTL synthesis competition.

Keywords: Parity games · Generalized parity games · Partial solvers

1 Introduction

Since the early nineties, parity games have been attracting a large attention in the formal methods and theoretical computer science communities for two main reasons. First, parity games are used as intermediary steps in the solution of several relevant problems like, among others, the reactive synthesis problem from LTL specifications [18] or the emptiness problem for tree automata [9]. Second, their exact complexity is a long standing open problem: while we know that they are in $NP \cap coNP$ [9] (and even in $UP \cap coUP$ [16]), we do not yet have a polynomial time algorithm to solve them. Indeed, the best known algorithm so far has a worst-case complexity which is quasi-polynomial [3].

The classical algorithm for reactive synthesis from LTL specifications is as follows: from an LTL formula ϕ whose propositional variables are partitioned

Work partially supported by the PDR project *Subgame perfection in graph games* (F.R.S.-FNRS), the ARC project *Non-Zero Sum Game Graphs: Applications to Reactive Synthesis and Beyond* (Fédération Wallonie-Bruxelles), the EOS project *Verifying Learning Artificial Intelligence Systems* (F.R.S.-FNRS & FWO), the COST Action 16228 *GAMENET* (European Cooperation in Science and Technology). The full version of this article is available at https://arxiv.org/abs/1907.06913.

© Springer Nature Switzerland AG 2019
E. Filiot et al. (Eds.): RP 2019, LNCS 11674, pp. 63–78, 2019.
https://doi.org/10.1007/978-3-030-30806-3_6

into inputs (controllable by the environment) and outputs (controlled by the system), construct a deterministic parity automaton (DPA) A_ϕ that recognizes the set of traces that are models of ϕ. This DPA can then be seen as a two player graph game where the two players choose in turn the values of the input variables (Player 1) and of the output variables (Player 0). The winning condition in this game is the parity acceptance condition of the DPA. The two main difficulties with this approach are that the DPA may be doubly exponential in the size of ϕ and that its parity condition may require exponentially many priorities. So the underlying parity game may be hard to solve with the existing (not polynomial) algorithms. These difficulties have triggered two series of results.

First, incomplete algorithms that partially solve parity games in polynomial time have been investigated [1,13,14]. Although they are incomplete, experimental results show that they behave well on benchmarks generated with a random model and on examples that are forcing the worst-case behavior of the classical recursive algorithm for solving parity games due to Zielonka [19]. The latter algorithm has a worst-case complexity which is exponential in the number of priorities of the parity condition. Second, compositional approaches to generate the automata from LTL specifications have been advocated, when the LTL formula ϕ is a conjunction of smaller formulas, i.e., $\phi = \phi_1 \wedge \cdots \wedge \phi_n$. In this case, the procedure constructs a DPA A_i for each subformula ϕ_i. The underlying game is a product of the automata A_i; the winning condition, the conjunction (for Player 0) of the parity conditions of each automaton. Those games are thus generalized parity games, that are known to be co-NP-complete [6].

In this paper, we contribute to these lines of research in several ways. First, we show how to extend the partial solvers for parity games to generalized parity games. In the generalized case, we show how antichain-based data structures can be used to retain efficiency. Second, we show how to combine partial solvers for parity games and generalized parity games with the classical recursive algorithms [6,19]. In this combination, the recursive algorithm is only executed on the portion of the game graph that was not solved by the partial solver, and this is repeated at each recursive call. Third, we provide for the first time extensive experiments that compare all those algorithms on benchmarks that are generated from LTL specifications used in the LTL synthesis competition [15]. For parity games, our experiments show behaviors that differ largely from the behaviors observed on experiments done on random graphs only. Indeed Zielonka's algorithm is faster than partial solvers on average which was not observed on random graphs in [14]. Equally interestingly, we show that there are instances of our benchmarks of generalized parity games that cannot be solved by the classical recursive algorithm or by any of the partial solvers alone, but that can be solved by algorithms that combine them. We also show that when combined with partial solvers, the performances of the classical recursive algorithms are improved on a large portion of our benchmarks for both the parity and generalized parity cases.

2 Preliminaries

Game Structures. A *game structure* is a tuple $G = (V_0, V_1, E)$ where *(i)* (V, E) is a finite directed graph, with $V = V_0 \cup V_1$ the set of vertices and $E \subseteq V \times V$ the set of edges such that[1] for each $v \in V$, there exists some $(v, v') \in E$, *(ii)* (V_0, V_1) forms a partition of V where V_i is the set of vertices controlled by player i. Given $U \subseteq V$, if $G_{\upharpoonright U} = (V_0 \cap U, V_1 \cap U, E \cap (U \times U))$ has no deadlock, then $G_{\upharpoonright U}$ is called the *subgame structure* induced by U.

A *play* in G is an infinite sequence of vertices $\pi = v_0 v_1 \ldots \in V^\omega$ such that $(v_j, v_{j+1}) \in E$ for all $j \in \mathbb{N}$. *Histories* in G are finite sequences $h = v_0 \ldots v_j \in V^+$ defined similarly. We denote by $\mathsf{Plays}(G)$ the set of plays in G and by $\mathsf{Plays}(v_0)$ the set of plays starting in a given *initial vertex* v_0. Given a play $\pi = v_0 v_1 \ldots$, the set $\mathsf{Occ}(\pi)$ denotes the set of vertices that occur in π, and the set $\mathsf{Inf}(\pi)$ denotes the set of vertices that occur infinitely often in π.

A *strategy* σ_i for player i is a function $\sigma_i \colon V^* V_i \to V$ assigning to each history $hv \in V^* V_i$ a vertex $v' = \sigma_i(hv)$ such that $(v, v') \in E$. Given a strategy σ_i of player i, a play $\pi = v_0 v_1 \ldots$ is *consistent* with σ_i if $v_{j+1} = \sigma_i(v_0 \ldots v_j)$ for all $j \in \mathbb{N}$ such that $v_j \in V_i$. Consistency is naturally extended to histories.

Objectives. An *objective for player i* is a set of plays $\Omega \subseteq \mathsf{Plays}(G)$. A *game* (G, Ω) is composed of a game structure G and an objective Ω for *player* 0. A play π is *winning* for player 0 if $\pi \in \Omega$, and losing otherwise. The games that we study are *zero-sum*: player 1 has the opposite objective $\overline{\Omega} = V^\omega \setminus \Omega$, meaning that a play π is winning for player 0 if and only if it is losing for player 1. Given a game (G, Ω) and an initial vertex v_0, a strategy σ_0 for player 0 is *winning from v_0* if all the plays $\pi \in \mathsf{Plays}(v_0)$ consistent with σ_0 belong to Ω. We say that v_0 is *winning* for player 0 and that player 0 is winning from v_0. We denote by $\mathsf{Win}(G, 0, \Omega)$ the set of such winning vertices v_0. Similarly we denote by $\mathsf{Win}(G, 1, \overline{\Omega})$ the set of vertices from which player 1 can ensure his objective $\overline{\Omega}$.

A game (G, Ω) is *determined* if each vertex of G belongs to either $\mathsf{Win}(G, 0, \Omega)$ or $\mathsf{Win}(G, 1, \overline{\Omega})$. Martin's theorem [17] states that all games with Borel objectives are determined. The problem of *solving a game* (G, Ω) means to decide, given an initial vertex v_0, whether player 0 is winning from v_0 for Ω (or dually whether player 1 is winning from v_0 for $\overline{\Omega}$ when the game is determined). The sets $\mathsf{Win}(G, 0, \Omega)$ and $\mathsf{Win}(G, 1, \overline{\Omega})$ are also called the *solutions* of the game.

Parity and Generalized Parity Objectives. Let G be a game structure and $d \in \mathbb{N}$ be an integer. Let $\alpha \colon V \to [d]$, with $[d] = \{0, 1, \ldots, d\}$, be a *priority* function that associates a priority with each vertex. The *parity* objective $\Omega = \mathsf{EvenParity}(\alpha)$ asks that the maximum priority seen infinitely often along a play is even, i.e., $\mathsf{EvenParity}(\alpha) = \{\pi \in \mathsf{Plays}(G) \mid \max_{v \in \mathsf{Inf}(\pi)} \alpha(v) \text{ is even}\}$. Games $(G, \mathsf{EvenParity}(\alpha))$ are called *parity games*. In those games, player 1 has the opposite objective $\overline{\Omega}$ equal to $\{\pi \in \mathsf{Plays}(G) \mid \max_{v \in \mathsf{Inf}(\pi)} \alpha(v) \text{ is odd}\}$. We denote $\overline{\Omega}$ by $\mathsf{OddParity}(\alpha)$. In the sequel, an *$i$-priority* means an even priority

---

[1] This condition guarantees that there is no deadlock.

if $i = 0$ and an odd priority if $i = 1$. For convenience, we also use notation iParity(α) such that 0Parity(α) = EvenParity(α) and 1Parity(α) = OddParity(α).

The *generalized parity* objective Ω = ConjEvenParity($\alpha_1, \ldots, \alpha_k$) is the conjunction of $k \geq 1$ parity objectives, that is, $\Omega = \bigcap_{\ell=1}^{k}$ EvenParity(α_ℓ), where each $\alpha_\ell : V \to [d_\ell]$ is a priority function. The opposite objective $\overline{\Omega}$ for player 1 is equal to DisjOddParity = $\bigcup_{\ell=1}^{k}$ OddParity(α_ℓ). Games $(G, \text{ConjEvenParity}(\alpha_1, \ldots, \alpha_k))$ are called *generalized parity games*.

Parity games and generalized parity games are determined because their objectives are ω-regular and thus Borel. Solving parity games is in UP $\cap$ co-UP [16] and solving generalized parity games is co-NP-complete [6].

Partial Solvers. We study *partial solvers* for parity games and generalized parity games. A partial solver returns two partial sets of winning vertices $Z_0 \subseteq$ Win($G, 0, \Omega$) and $Z_1 \subseteq$ Win($G, 1, \overline{\Omega}$) such that $G_{\restriction U}$ is a subgame structure with $U = V \setminus (Z_0 \cup Z_1)$. In the next sections, we present the polynomial time partial solvers proposed in [13,14] for parity games and show how to extend them to generalized parity games.

Other ω-Regular Objectives. We recall some other useful ω-regular objectives. Let G be a game structure and $U, U_1, \ldots, U_k \subseteq V$ be subsets: the *reachability objective* Reach(U) = $\{\pi \in \text{Plays}(G) \mid \text{Occ}(\pi) \cap U \neq \emptyset\}$ asks to visit U at least once; the *safety objective* Safe(U) = $\{\pi \in \text{Plays}(G) \mid \text{Occ}(\pi) \cap U = \emptyset\}$ asks to avoid visiting U; the *Büchi objective* Büchi(U) = $\{\pi \in \text{Plays}(G) \mid \text{Inf}(\pi) \cap U \neq \emptyset\}$ asks to visit infinitely often a vertex of U; the *co-Büchi objective* CoBüchi(U) = $\{\pi \in \text{Plays}(G) \mid \text{Inf}(\pi) \cap U = \emptyset\}$ asks to avoid visiting infinitely often U; the *generalized Büchi objective* GenBüchi($U_1, \ldots, U_k$) is equal to the intersection $\bigcap_{\ell=1}^{k}$ Büchi(U_ℓ). The next theorem summarizes the time complexities for solving those games as implemented in our prototype tool.[2]

Theorem 1. *For solving games (G, Ω), we have the following time complexities.*

- *Reachability, safety objectives: $O(|E|)$ [12].*
- *Büchi, co-Büchi, Büchi $\cap$ safety objectives: $O(|V| \cdot |E|)$ [12].*
- *Generalized Büchi, generalized Büchi $\cap$ safety objectives: $O(k \cdot |V| \cdot |E|)$.*[3]

Attractors. Let G be a game structure. The *controllable predecessors* for player i of a set $U \subseteq V$, denoted by Cpre$_i(G, U)$, is the set of vertices from which player i can ensure to visit U in *one step*. Formally, Cpre$_i(G, U)$ is equal to

$$\{v \in V_i \mid \exists (v, v') \in E, v' \in U\} \cup \{v \in V_{1-i} \mid \forall (v, v') \in E, v' \in U\}. \qquad (1)$$

The *attractor* Attr$_i(G, U)$ *for player i* is the set of vertices from which he can ensure to visit U in *any* number of steps (including zero steps). It is constructed

---

[2] A better algorithm in $O(|V|^2)$ for Büchi objectives is proposed in [5], and in $O(k \cdot |V|^2)$ for generalized Büchi objectives in [4].

[3] This result is obtained with a classical reduction to games with Büchi objectives [2].

as follows: $\mathsf{Attr}_i(G, U) = \bigcup_{j \geq 0} X_j$ such that $X_0 = U$, and for all $j \in \mathbb{N}$, $X_{j+1} = X_j \cup \mathsf{Cpre}_i(G, X_j)$. It is thus the winning set $\mathsf{Win}(G, i, \mathsf{Reach}(U))$. The *positive attractor* $\mathsf{PAttr}_i(G, U)$ is the set of vertices from which player i can ensure to visit U in any *positive* number of steps, that is, $\mathsf{PAttr}_i(G, U) = \bigcup_{j \geq 0} X_j$ with:

$$X_0 = \mathsf{Cpre}_i(G, U), \quad X_{j+1} = X_j \cup \mathsf{Cpre}_i(G, X_j \cup U) \text{ for all } j \in \mathbb{N}. \qquad (2)$$

Given $U \subseteq V$, we say that U is an *i-trap* if for all $v \in U \cap V_i$ and all $(v, v') \in E$, we have $v' \in U$ (player i cannot leave U), and for all $v \in U \cap V_{1-i}$, there exists $(v, v') \in E$ such that $v' \in U$ (player $1-i$ can ensure to stay in U). Therefore $G_{\restriction U}$ is a subgame structure. When $V \setminus U$ is an *i-trap*, we also use the notation $G \setminus U$ (instead of $G_{\restriction V \setminus U}$) for the subgame structure induced by $V \setminus U$. It is well-known that the set $V \setminus \mathsf{Attr}_i(G, U)$ is an *i-trap* [12].

3 Zielonka's Algorithm with Partial Solvers

The classical algorithm used to solve parity games is the algorithm proposed by Zielonka in [19]. Despite its relatively bad theoretical $O(|V|^d)$ time complexity, it is known to outperform other algorithms in practice [7,11]. This algorithm solves parity games $(G, \mathsf{EvenParity}(\alpha))$ by working in a divide-and-conquer manner, combining solutions of subgames to obtain the solution of the whole game. It returns two sets $\{W_0, W_1\}$ such that $W_i = \mathsf{Win}(G, i, i\mathsf{Parity}(\alpha))$ is the winning set for player i. See Algorithm 1 in which no call to a partial solver is performed (therefore line 4 is to be replaced by $\{Z_0, Z_1\} = \{\emptyset, \emptyset\}$).

Let us explain how Zielonka's algorithm can be combined with a partial solver for parity games (see Algorithm 1). When V is not empty, we first execute the partial solver. If it solves the game completely, we are done. Otherwise, let $\overline{G}$ be the subgame of G that was not solved. We then execute the Zielonka instructions on $\overline{G}$ and return the union of the partial solutions obtained by the partial solver with the solutions obtained for $\overline{G}$. Proposition 2 below guarantees the soundness of this approach under the hypothesis that if some player wants to escape from $\overline{G}$, then he necessarily goes to the partial solution of the other player.

Proposition 2. *Suppose that the partial solver used in Algorithm 1 computes partial solutions Z_0, Z_1 such that for all $(v, v') \in E$ and $i \in \{0, 1\}$, if $v \in \overline{V} \cap V_i$ and $v' \notin \overline{V}$, then $v' \in Z_{1-i}$. Then Algorithm Ziel&PSolver correctly computes the sets $\mathsf{Win}(G, i, i\mathsf{Parity}(\alpha))$, for $i \in \{0, 1\}$.*

An extension of Zielonka's algorithm to generalized parity games[4] is introduced in [6]. This algorithm, that we call GenZielonka, has $O(|E| \cdot |V|^{2D})\binom{D}{d_1, \ldots, d_k}$ time complexity where $D = \Sigma_{\ell=1}^k d_\ell$. While more complex, it has the same behavior with respect to the recursive call: an attractor X is computed as part of the solution of one player and a recursive call is executed on the subgame $G \setminus X$. Therefore, this algorithm can be combined with a partial solver for generalized parity games, as long as the latter satisfies the assumptions of Proposition 2.

---

[4] This algorithm is referred to as "the classical algorithm" in [6].

Algorithm 1. Ziel&PSolver(G, α)

---

1 **if** $V = \emptyset$ **then return** $\{W_0, W_1\} = \{\emptyset, \emptyset\}$
2 **else**
3      $\{Z_0, Z_1\} = \text{PSolver}(G, \alpha)$
4      $\overline{G} = G \setminus (Z_0 \cup Z_1); \quad \overline{V} = V \setminus (Z_0 \cup Z_1)$
5   **if** $\overline{V} = \emptyset$ **then return** $\{Z_0, Z_1\}$
6   **else**
7          $p = \max\{\alpha(v) \mid v \in \overline{V}\}; \quad i = p \bmod 2$
8          $U = \{v \in \overline{V} \mid \alpha(v) = p\}; \quad X = \text{Attr}_i(\overline{G}, U)$
9          $\{W_i', W_{1-i}'\} = \text{Ziel\&PSolver}(\overline{G} \setminus X, \alpha)$
10   **if** $W_{1-i}' = \emptyset$ **then**
11             $W_i = Z_i \cup W_i' \cup X; \quad W_{1-i} = Z_{1-i}$
12   **else**
13             $X = \text{Attr}_{1-i}(\overline{G}, W_{1-i}')$
14             $\{W_i'', W_{1-i}''\} = \text{Ziel\&PSolver}(\overline{G} \setminus X, \alpha)$
15             $W_i = Z_i \cup W_i''; \quad W_{1-i} = Z_{1-i} \cup W_{1-i}'' \cup X$
16   **return** $\{W_i, W_{1-i}\}$

In the next sections, we present three partial solvers for parity games and their extension to generalized parity games. They all satisfy the assumptions of Proposition 2 since the partial solutions that they compute are composed of one or several attractors.

4   Algorithms BüchiSolver and GenBüchiSolver

In this section, we present simple partial solvers for parity games and generalized parity games. More elaborate and powerful partial solvers are presented in the next two sections. These first partial solvers are based on Propositions 3 and 4 that are direct consequences of the definition of parity games and generalized parity games. The first proposition states that for parity games, if player i can ensure to visit infinitely often an i-priority without visiting infinitely often a greater $(1 - i)$-priority, then he is winning for $i\text{Parity}(\alpha)$.

Proposition 3. *Given a parity game* $(G, \text{EvenParity}(\alpha))$ *and an i-priority $p \in [d]$, let $U = \{v \in V \mid \alpha(v) = p\}$ and $U' = \{v \in V \mid \alpha(v)$ is a $(1 - i)$-priority and $\alpha(v) > p\}$. If $v_0 \in \text{Win}(G, i, \text{Büchi}(U) \cap \text{CoBüchi}(U'))$, then $v_0 \in \text{Win}(G, i, i\text{Parity}(\alpha))$.*

The second proposition states that for generalized parity games, *(i)* if player 0 can ensure to visit infinitely often a 0-priority p_ℓ without visiting infinitely often a 1-priority greater than p_ℓ on *all* dimensions ℓ, then he is winning in the generalized parity game, and *(ii)* if player 1 can ensure to visit infinitely often a 1-priority p_ℓ without visiting infinitely often a 0-priority greater than p_ℓ on *some* dimension ℓ, then he is also winning.

Proposition 4. *Let* $(G, \text{ConjEvenParity}(\alpha_1, \ldots, \alpha_k))$ *be a generalized parity game.*

- Let $p = (p_1, \ldots, p_k)$, with $p_\ell \in [d_\ell]$, be a vector of 0-priorities. For all ℓ, let $U_\ell = \{v \in V \mid \alpha_\ell(v) = p_\ell\}$. Let $U' = \{v \in V \mid \exists \ell, \alpha_\ell(v) \text{ is a 1-priority and } \alpha_\ell(v) > p_\ell\}$. Then for all $v_0 \in V$,

$$v_0 \in \mathsf{Win}(G, 0, \mathsf{GenB\ddot{u}chi}(U_1, \ldots, U_k) \cap \mathsf{CoB\ddot{u}chi}(U'))$$
$$\implies v_0 \in \mathsf{Win}(G, 0, \mathsf{ConjEvenParity}(\alpha_1, \ldots, \alpha_k)).$$

- Let $\ell \in \{1, \ldots, k\}$ and $p_\ell \in [d_\ell]$ be a 1-priority. Let $U = \{v \in V \mid \alpha_\ell(v) = p_\ell\}$ and $U' = \{v \in V \mid \alpha_\ell(v) \text{ is a 0-priority and } \alpha_\ell(v) > p_\ell\}$. Then for all $v_0 \in V$,

$$v_0 \in \mathsf{Win}(G, 1, \mathsf{B\ddot{u}chi}(U) \cap \mathsf{CoB\ddot{u}chi}(U'))$$
$$\implies v_0 \in \mathsf{Win}(G, 1, \mathsf{DisjOddParity}(\alpha_1, \ldots, \alpha_k)).$$

Partial Solver for Parity Games. A partial solver for parity games is easily derived from Proposition 3. The polynomial time algorithm BüchiSolver (see Algorithm 2) computes winning vertices for each player by applying this proposition as follows. Let us denote by W_p the computed winning set for priority p (line 5) and let us suppose for simplicity that the loop in line 1 treats the priorities from the highest one d to the lowest one 0. This algorithm computes W_d, $W_{d-1}, \ldots$, until finding $W_p \neq \emptyset$. At this stage, it was able to find some winning vertices. It then repeats the process on $G \setminus W_p$ to find other winning vertices. At the end of the execution, it returns the computed partial solutions $\{Z_0, Z_1\}$.

Algorithm 2. BüchiSolver(G, α)

---

1 **for each** $p \in [d]$ **do**
2      $i = p \bmod 2$
3      $U = \{v \in V \mid \alpha(v) = p\}$
4      $U' = \{v \in V \mid \alpha(v) \text{ is a } (1 - i)\text{-priority} > p\}$
5      $W = \mathsf{Win}(G, i, \mathsf{B\ddot{u}chi}(U) \cap \mathsf{CoB\ddot{u}chi}(U'))$
6   **if** $W \neq \emptyset$ **then**
7          $\{Z_i, Z_{1-i}\} = \mathsf{B\ddot{u}chiSolver}(G \setminus W, \alpha)$
8   **return** $\{Z_i \cup W, Z_{1-i}\}$
9 **return** $\{\emptyset, \emptyset\}$

Notice that we could replace line 5 by $W' = \mathsf{Win}(G, i, \mathsf{B\ddot{u}chi}(U) \cap \mathsf{Safe}(U'))$ and $W = \mathsf{Attr}_i(G, W')$ since $\mathsf{Attr}_i(G, W') \subseteq \mathsf{Win}(G, i, \mathsf{B\ddot{u}chi}(U) \cap \mathsf{CoB\ddot{u}chi}(U'))$ (as the parity objective is closed under attractor; and computing the attractor is necessary to get a subgame for the recursive call). This variant is investigated in [13, 14] under the name of Algorithm psolB.

Partial Solver for Generalized Parity Games. Similarly, a partial solver for generalized parity games can be derived from Proposition 4. This algorithm is called GenBüchiSolver (see Algorithm 3). Instead of considering all $p \in [d]$ as done in line 1 of Algorithm 2, we here have a loop on all elements, stored in a list L (see line 1), that are either a 1-*priority* p for some given *dimension* ℓ (case of player 1), or a *vector* $(p_1, \ldots, p_k)$ with 0-*priorities* (case of player 0). As in Algorithm BüchiSolver, objective CoBüchi(U') can be replaced by Safe(U') in lines 5 and 12 (with the addition of an attractor computation). This modification yields an algorithm in $O((\Pi_{\ell=1}^{k} \frac{d_\ell}{2}) \cdot k \cdot |V|^2 \cdot |E|)$ time by Theorem 1.

Algorithm 3. GenBüchiSolver$(G, \alpha_1, \ldots, \alpha_k, L)$

1 **for each** *element* $\in L$ **do**
2   **if** *element* $= (p, \ell)$ **then**
3      $U = \{v \in V \mid \alpha_\ell(v) = p\}$
4      $U' = \{v \in V \mid \alpha_\ell(v) \text{ is a 0-priority and } \alpha_\ell(v) > p\}$
5      $W = \text{Win}(G, 1, \text{Büchi}(U) \cap \text{CoBüchi}(U'))$
6   **if** $W \neq \emptyset$ **then**
7        $\{Z_0, Z_1\} = \text{GenBüchiSolver}(G \setminus W, \alpha_1, \ldots, \alpha_k, L)$
8   **return** $\{Z_0, Z_1 \cup W\}$
9   **else**   {We know that *element* $= (p_1, \ldots, p_k)$}
10   **for each** ℓ **do** $U_\ell = \{v \in V \mid \alpha_\ell(v) = p_\ell\}$
11     $U' = \{v \in V \mid \exists \ell, \alpha_\ell(v) \text{ is a 1-priority and } \alpha_\ell(v) > p_\ell\}$
12     $W = \text{Win}(G, 0, \text{GenBüchi}(U_1, \ldots, U_k) \cap \text{CoBüchi}(U'))$
13   **if** $W \neq \emptyset$ **then**
14       $\{Z_0, Z_1\} = \text{GenBüchiSolver}(G \setminus W, \alpha_1, \ldots, \alpha_k, L)$
15   **return** $\{Z_0 \cup W, Z_1\}$
16 **return** $\{\emptyset, \emptyset\}$

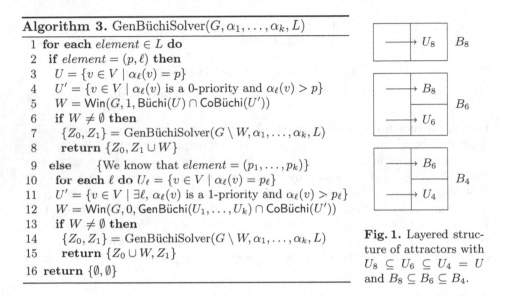

Fig. 1. Layered structure of attractors with $U_8 \subseteq U_6 \subseteq U_4 = U$ and $B_8 \subseteq B_6 \subseteq B_4$.

5 Algorithms GoodEpSolver and GenGoodEpSolver

In this section, we present a second polynomial time partial solver for parity games as proposed in [14]. We then explain how to extend it for partially solving generalized parity games. We finally explain how the latter solver can be transformed into a more efficient (in practice) antichain-based algorithm.

Partial Solver for Parity Games. We consider the *extended* game structure $G \times M$ with $M = [d]$ such that $m \in M$ records the *maximum visited priority*. More precisely, the set of vertices of $G \times M$ is equal to $V \times M$ (where $V_i \times M$ are the vertices controlled by player i), and the set E_M of its edges is composed of all pairs $((v, m), (v', m'))$ such that $(v, v') \in E$ and $m' = \max\{m, \alpha(v)\}$. Clearly, with this construction, in G, player i can ensure to visit v from v_0 such that the maximum visited priority (v excluded) is an i-priority if and only if in $G \times M$, he can ensure to visit (v, m) from $(v, \alpha(v_0))$ for some i-priority m.

Given a player i, we then compute the following fixpoint $F^{(i)} = \bigcap_{j \geq 0} F_j$. Initially $F_0 = V$ and for $j \geq 1$, F_j is computed from F_{j-1} as follows:

$$T_j = \{(v, m) \in V \times M \mid v \in F_{j-1} \text{ and } m \text{ is an } i\text{-priority}\} \tag{3}$$

$$A_j = \text{PAttr}_i(G \times M, T_j) \tag{4}$$

$$F_j = \{v \in V \mid (v, \alpha(v)) \in A_j\} \cap F_{j-1}. \tag{5}$$

Algorithm 4. GoodEpSolver(G, α)

1 **for each** $i \in \{0, 1\}$ **do**
2 $\quad W = \mathsf{GoodEp}_i(G, \alpha)$
3 $\quad$ **if** $W \neq \emptyset$ **then**
4 $\quad\quad X = \mathsf{Attr}_i(G, W)$
5 $\quad\quad \{Z_i, Z_{1-i}\} = \mathrm{GoodEpSolver}(G \backslash X, \alpha)$
6 $\quad\quad$ **return** $\{Z_i \cup X, Z_{1-i}\}$
7 **return** $\emptyset, \emptyset$

Algorithm 5. LaySolver(G, α)

1 **for each** $P_{\geq q}$ with $q \in [d]$ **do**
2 $\quad i = $ parity of the priorities in $P_{\geq q}$
3 $\quad W = \mathsf{LayEp}_i(G, \alpha, P_{\geq q})$
4 $\quad$ **if** $W \neq \emptyset$ **then**
5 $\quad\quad X = \mathsf{Attr}_i(G, W)$
6 $\quad\quad \{Z_i, Z_{1-i}\} = \mathrm{LaySolver}(G \backslash X, \alpha)$
7 $\quad\quad$ **return** $\{Z_i \cup X, Z_{1-i}\}$
8 **return** $\emptyset, \emptyset$

Intuitively, if $v_0 \in F_j$, then player i has a strategy to ensure to visit some vertex $v \in F_{j-1}$ such that the maximum visited priority along the consistent history hv from v_0 to v (priority of v excluded) is some i-priority. We say that h is a *good episode*. Notice that h is non empty since each A_j is a positive attractor.

We denote the fixpoint $F^{(i)}$ by $\mathsf{GoodEp}_i(G, \alpha)$. From a vertex v_0 in this fixpoint, player i can ensure a succession of good episodes in which the maximum visited priority is an i-priority. Thus he is winning from v_0 for $i\mathsf{Parity}(\alpha)$ as formalized in the next proposition (notice that if v_0 belongs to the attractor for player i of $\mathsf{GoodEp}_i(G, \alpha)$, then v_0 still belongs to $\mathsf{Win}(G, i, i\mathsf{Parity}(\alpha))$). From Proposition 5, we derive a polynomial time algorithm called GoodEpSolver (see Algorithm 4). This algorithm is called psolC in [14].

Proposition 5 ([14]). *Let* $(G, \mathsf{EvenParity}(\alpha))$ *be a parity game. For all* $v_0 \in V$, *if* $v_0 \in \mathsf{Attr}_i(G, F^{(i)})$, *then* $v_0 \in \mathsf{Win}(G, i, i\mathsf{Parity}(\alpha))$.

Partial Solver for Generalized Parity Games. The same approach can be applied to generalized parity games with some adaptations that depend on the player. For player 1, instead of applying Proposition 3, we now apply the Good-EpSolver approach by computing $W = \mathsf{GoodEp}_1(G, \alpha_\ell)$ for each dimension ℓ.

For player 0, we have to treat vectors of 0-priorities. We thus consider the extended game structure $G \times M_1 \times \ldots \times M_k$ such that for all ℓ, M_ℓ is equal to $[d_\ell]$, and where $m_\ell \in M_\ell$ records the maximum visited priority in *dimension* ℓ according to function α_ℓ. Hence, the edges of this game structure have the form $((v, m_1, \ldots, m_k), (v', m'_1, \ldots, m'_k))$ such that $(v, v') \in E$ and $m'_\ell = \max\{m_\ell, \alpha_\ell(v)\}$ for all ℓ. A good episode is now a history h such that for all ℓ, the maximum priority visited along h in dimension ℓ is a 0-priority. And the related set $\mathsf{GoodEp}_0(G, \alpha_1, \ldots, \alpha_\ell)$ equal to $\bigcap_{j \geq 0} F_j$ is computed with Eqs. (3–5) modified as expected. The resulting algorithm, called GenGoodEpSolver, has a time complexity in $O((\Pi_{\ell=1}^k d_\ell) \cdot |V|^2 \cdot |E|)$ by Theorem 1.

Partial Solver with Antichains. Algorithm GenGoodEpSolver has exponential time complexity due to the use of the game structure $G \times M_1 \times \ldots \times M_k$, and in particular to the computation of $\mathsf{GoodEp}_0(G, \alpha_1, \ldots, \alpha_k)$. We here show that the vertices of this extended game can be partially ordered in a way to obtain an antichain-based algorithm for $\mathsf{GoodEp}_0(G, \alpha_1, \ldots, \alpha_k)$. The antichains allow compact representation and efficient manipulation of partially ordered sets [8].

Let us first recall the basic notions about antichains. Consider a *partially ordered set* $(S, \preceq)$ where S is a finite set and $\preceq \subseteq S \times S$ is a partial order on S. Given $s, s' \in S$, we write $s \sqcap s'$ their *greatest lower bound* if it exists. A *lower semilattice* is a partially ordered set such that this greater lower bound always exists for all $s, s' \in S$. Given two subsets $R, R' \subseteq S$, we denote by $R \sqcap R'$ the set $\{s \sqcap s' \mid s \in R, s' \in R'\}$. An *antichain* $A \subseteq S$ is a set composed of pairwise incomparable elements with respect to $\preceq$. Given a subset $R \subseteq S$, we denote $\lceil R \rceil$ the set of its *maximal* elements (which is thus an antichain). We say that R is *closed* if whenever $s \in R$ and $s' \preceq s$, then $s' \in R$. If A is an antichain, we denote by $\downarrow A$ the closed set that it *represents*, that is, $A = \lceil \downarrow A \rceil$. Hence when R is closed, we have $R = \downarrow \lceil R \rceil$. The benefit of antichains is that they provide a *compact representation* of closed sets. Moreover some operations on those closed sets can be done at the level of their antichains:

Proposition 6 ([8]). *Let* $(S, \preceq)$ *be a lower semilattice, and* $R, R' \subseteq S$ *be two closed sets represented by their antichains* $A = \lceil R \rceil, A' = \lceil R' \rceil$. *Then*

– *for all* $s \in S$, $s \in R$ *if and only if there exists* $s' \in A$ *such that* $s \preceq s'$,
– $R \cup R'$, $R \cap R'$ *are closed, and* $\lceil R \cup R' \rceil = \lceil A \cup A' \rceil$, $\lceil R \cap R' \rceil = \lceil A \sqcap A' \rceil$.

For *simplicity*, we focus on the extended structure $G \times M$ associated to a parity game and begin to explain an antichain-based algorithm for the computation of the set $\mathsf{GoodEp}_0(G, \alpha)$ (this algorithm is inspired from [10]). We explain later what are the required adaptations to compute $\mathsf{GoodEp}_0(G, \alpha_1, \ldots, \alpha_k)$ in generalized parity games. We equip $V \times M$ with the following partial order:

Definition 7. *We define the strict partial order* $\prec$ *on* $V \times M$ *such that* $(v', m') \prec (v, m)$ *if and only if* $v = v'$ *and (i) either* m, m' *are even and* $m' > m$, *(ii) or* m, m' *are odd and* $m' < m$, *(iii) or* m *is odd and* m' *is even. We define* $(v', m') \preceq (v, m)$ *if either* $(v', m') = (v, m)$ *or* $(v', m') \prec (v, m)$.

For instance, if $[d] = [4]$, then $(v, 4) \prec (v, 2) \prec (v, 0) \prec (v, 1) \prec (v, 3)$. With this definition, two elements $(v, m), (v', m')$ are incomparable as soon as $v \neq v'$. It follows that in Proposition 6, if $R \subseteq \{v\} \times M$ and $R' \subseteq \{v'\} \times M$ with $v \neq v'$, then the union $A \cup A'$ of their antichains is already an antichain.

The computation of $\mathsf{GoodEp}_0(G, \alpha)$ is based on Eqs. (3–5). Equation (4) involves the computation of positive attractors over $G \times M$ thanks to Eqs. (1–2). Let us show that $\mathsf{GoodEp}_0(G, \alpha)$ is a closed set and that so are all the intermediate sets used to compute it. We already know from Proposition 6 that the family of closed sets is stable under union and intersection. So we now focus on the operation $\mathsf{Cpre}_0(G \times M, U)$. We introduce the following functions up and $down$. Given $(v, m) \in V \times M$, let $up(m, \alpha(v)) = \max\{m, \alpha(v)\}$. Recall that such an update from m to $m' = up(m, \alpha(v))$ is used in the edges $((v, m), (v', m'))$ of $G \times M$. Function $down$ is the inverse reasoning of function up: given $(v', m') \in V \times M$ and $(v, v') \in E$, the value $down(m', \alpha(v))$ yields the maximal value m such that $(v', up(m, \alpha(v))) \preceq (v', m')$.

Definition 8. *Given* $(v', m') \in V \times M$ *and* $p = \alpha(v)$, *we define* $m = down(m', p)$ *as follows: (i)* **Case** p **even**: *if* $p < m'$ *then* $m = m'$, *else* $m = \max\{p - 1, 0\}$; *(ii)* **Case** p **odd**: *if* $p \leq m'$ *then* $m = m'$, *else* $m = p + 1$ *except if* $p = d$ *in which case* $down(m', p)$ *is not defined.*

The next proposition states that if a set $U \subseteq V \times M$ is closed, then the set $\mathsf{Cpre}_0(G \times M, U)$ is also closed. It also indicates how to design an antichain-based algorithm for computing $\mathsf{Cpre}_0(G \times M, U)$.

Proposition 9. *If* $U \subseteq V \times M$ *is a closed set, then* $\mathsf{Cpre}_0(G \times M, U)$ *is closed. Let* $A = \lceil U \rceil$ *be the antichain representing* U. *Then* $\lceil \mathsf{Cpre}_0(G \times M, U) \rceil = \lceil B_0 \cup B_1 \rceil$ *where* B_0, B_1 *are the following antichains:*

$$B_0 = \bigcup\nolimits_{v \in V_0} \Big[\{(v, down(m', \alpha(v))) \mid (v, v') \in E, (v', m') \in A\} \Big],$$

$$B_1 = \bigcup\nolimits_{v \in V_1} \Big[\bigcap\nolimits_{(v, v') \in E} \lceil \{(v, down(m', \alpha(v))) \mid (v', m') \in A\} \rceil \Big].$$

Let us come back to the computation of $\mathsf{GoodEp}_0(G, \alpha)$, which depends on Eqs. (3–5) and Eqs. (1–2). As each T_j of Equation (3) is a closed set represented by the antichain $\lceil T_j \rceil = \{(v, 0) \mid v \in F_j\}$, by Propositions 6 and 9, we get an antichain-based algorithm for computing $\mathsf{GoodEp}_0(G, \alpha)$ as announced.

This antichain approach can be extended to generalized parity games and their extended game $G \times M_1 \times \ldots \times M_k$. The approach is similar and works *dimension by dimension* as we did before for parity games. First, we define a partial order on $V \times M_1 \times \ldots \times M_k$ such that the partial order of Definition 7 is used on each dimension. More precisely, $(v', m'_1, \ldots, m'_k) \prec (v, m_1, \ldots, m_k)$ if and only if $v = v'$ and for all ℓ, both m_ℓ and m'_ℓ respect Definition 7. Second, functions up_ℓ and $down_\ell$, with $\ell \in \{1, \ldots, k\}$, are defined exactly as previous functions up and $down$ (see Definition 8), for each dimension ℓ and with respect to priority function α_ℓ. Third, we adapt (as expected) Proposition 9 for the computation of $\mathsf{Cpre}_0(G \times M_1 \times \ldots \times M_k, U)$ for a closed set $U \subseteq V \times M_1 \times \ldots \times M_k$. Finally, we obtain an antichain-based algorithm for $\mathsf{GoodEp}_0(G, \alpha_1, \ldots, \alpha_k)$ as each set T_j in (3) is a closed set represented by the antichain $\lceil T_j \rceil = \{(v, 0, \ldots, 0) \mid v \in F_j\}$.

6 Algorithms LaySolver and GenLaySolver

In [13], the authors study another polynomial time partial solver for parity games, called psolQ, that has similarities with the GoodEpSolver approach of Sect. 5. It also generalizes the BüchiSolver approach of Sect. 4. It is a more complex partial solver that we present on an example. We then explain how to modify it for generalized parity games.

Partial Solver for Parity Games. The new partial solver works on the initial game structure G, focuses on a *subset* $P_{\geq q}$ of i-*priorities* and computes a set similar to $\mathsf{GoodEp}_i(G, \alpha)$ such that the positive attractor $\mathsf{PAttr}_i(G \times M, T_j)$ of Eq. (4) is replaced by a *layered attractor* $\mathsf{LayAttr}_i(G, \alpha, P_{\geq q}, U)$ (one layer per priority $p \in P_{\geq q}$).

Let us explain on an example. First, we denote by $\mathsf{PSafeAttr}_i(G, U, U')$ the *positive safe attractor* composed of vertices from which player i can ensure to visit U in any positive number of steps while *not visiting* U'. Now, take the example of a parity game with $d = 9$ and fix a 0-priority $q = 4$. Let $P_{\geq q}$ be the set of all 0-priorities $p \geq q$, that is, $P_{\geq q} = \{4, 6, 8\}$. Given some set $U \subseteq \{v \in V \mid \alpha(v) \in P_{\geq q}\}$, we consider U_8 (resp. U_6, U_4) being the set of vertices of U with priority 8 (resp. priorities in $\{6, 8\}$, in $\{4, 6, 8\}$). Notice that $U_8 \subseteq U_6 \subseteq U_4 = U$. We also consider U_8' (resp. U_6', U_4') the set of vertices with priority 9 (resp. priorities in $\{7, 9\}$, in $\{5, 7, 9\}$).

We compute the following sequence of positive safe attractors (see Fig. 1): Initially $B_{10} = \emptyset$ and for all $p \in P_{\geq q}$, B_p is computed from B_{p+2} as follows:

$$B_p = B_{p+2} \cup \mathsf{PSafeAttr}_0(G, U_p \cup B_{p+2}, U_p' \setminus B_{p+2}). \tag{6}$$

The last computed set B_4 is the layered attractor $\mathsf{LayAttr}_0(G, \alpha, P_{\geq q}, U)$. Notice that $B_8 \subseteq B_6 \subseteq B_4$. Let us give some intuition. From a vertex in $B_4 \setminus B_6$ (lowest layer 4), player 0 can ensure to visit $U_4 \cup B_6$ without visiting $U_4' \setminus B_6$. In case of a visit to U_4, this is a good episode for himself (in the sense of Sect. 5) since the maximum visited priority is a 0-priority ≥ 4. In case of a visit to some $v \in B_6 \setminus B_8$, player 0 can now ensure to visit $U_6 \cup B_8$ without visiting $U_6' \setminus B_8$. In case of a visit to U_6, this is again a good episode for him, otherwise it is a visit to B_8 in the highest layer from which player 0 can ensure to visit U_8 without visiting U_8' (since B_{10} is empty). Thus from all vertices of B_4, player 0 can ensure a good episode for himself.

The new partial solver, called LaySolver (see Algorithm 5), is the same as Algorithm GoodEpSolver of Sect. 5 except that *(i)* in Eq. (4), the layered attractor $\mathsf{LayAttr}_i(G, \alpha, P_{\geq q}, U)$ in the game G replaces $\mathsf{PAttr}_i(G \times M, T)$ in the extended game $G \times M$, and *(ii)* the subset $P_{\geq q} = \{q, q+2, q+4, \ldots\}$ of i-priorities replaces the set of all i-priorities.

Partial Solver for Generalized Parity Games. We explain how to adapt the LaySolver approach to generalized parity games, only for player 0. Indeed for player 1 we can apply the previous LaySolver approach on each α_ℓ separately.

For player 0, take the example of a generalized parity game $d_\ell = 9$ for all ℓ. We fix $q = (4, \ldots, 4)$ and $P_{\geq q} = \{(4, \ldots, 4), (6, \ldots, 6), (8, \ldots, 8)\}$. Take a vector $p \in P_{\geq q}$, (say with $p_\ell = 6, \forall \ell$) and a subset $U \subseteq V$. In a first step, let us focus on how player 0 can ensure to visit U such that along the history, for all ℓ, a 0-priority $\geq p_\ell$ is visited and no 1-priority $> p_\ell$ is visited (in a way to extend Eq. (6) temporarily without set B_{p+2}). Such a generalized reachability can be reduced to reachability by working with an extended game G_p such that in vertex (v, N), the memory $N \subseteq \{1, \ldots, k\}$ records the dimensions ℓ for which a vertex with 0-priority $\geq p_\ell$ is already visited. We then work with the positive safe attractor $\mathsf{PSafeAttr}_0(G_p, T_p, T_p')$ such that $T_p = \{(v, N) \mid v \in U, N = \{1, \ldots, k\}\}$ and $T_p' = \{(v, N) \mid \exists \ell, \alpha_\ell(v) \text{ is a 1-priority} > p_\ell\}$. In a second step, let us show how to manage the set B_{p+2} in Eq. (6). For parity games we explained how player 0 has to adapt his attractor strategy when he shifts from layer p to some higher layer $p' > p$ due to the visit to some $v \in B_{p+2}$.

Here when player 0 visits some vertex (v, N) in layer p for which he has to shift to layer p', he stops applying his current strategy, and begins applying his strategy for layer p' from the vertex $(v, N_{p'}(v))$ belonging to layer p' with initial memory $N_{p'}(v) = \{\ell \mid \alpha_\ell(v) \text{ is a 0-priority} \geq p_\ell\}$. All these modifications lead to the layered attractor $\mathsf{LayAttr}_0(G, \alpha_1, \ldots, \alpha_k, P_{\geq q}, F_j)$ used in place of $\mathsf{LayAttr}_0(G, \alpha, P_{\geq q}, F_j)$. In this way, we derive an algorithm called GenLaySolver with $O((\max_{\ell=1}^k \frac{d_\ell}{2})^2 \cdot |V|^2 \cdot |E| \cdot 2^k)$ time complexity.

7   Empirical Evaluation

For parity games, the polynomial time partial solvers BüchiSolver[5], GoodEp-Solver, and LaySolver are theoretically compared in [13,14]. It is proved that the partial solutions computed by Algorithm BüchiSolver are included in those computed by Algorithm LaySolver themselves included in those computed by Algorithm GoodEpSolver. Examples of parity games are also given that distinguish the three partial solvers (strict inclusion of partial solutions), as well as an example that is not completely solved by the most powerful of these partial solvers. This behavior also holds for the partial solvers proposed here for generalized parity games. Moreover, their time complexity is exponential in the number k of priority functions while the classical algorithm for generalized parity games is exponential in both k and all d_ℓ [6].

For both parity games and generalized parity games, we implemented in Python 2.7 the three partial solvers (with the antichain approach for Algorithm GenGoodEpSolver), Algorithm Zielonka (resp. GenZielonka) and its combination Ziel&PSolver (resp. GenZiel&PSolver) with each partial solver, and we executed all these algorithms on a large set of benchmarks. Our benchmarks were generated from TLSF specifications used for the Reactive Synthesis Competition (SYNTCOMP [15]) using the compositional translation explained in the introduction.[6] The source code for our prototype tool along with all the information about our benchmarks is made publicly available at https://github.com/Skar0/generalizedparity. Our experiments have been carried out on a server with Mac OS X 10.13.4 (build 17E199). As hardware, the server had as CPU one 6-Core Intel Xeon; as processor speed, 3.33 GHz; as L2 Cache (per Core), 256 KB; as L3 Cache, 12 MB; as memory, 32 GB; and as processor interconnect speed, 6.4 GT/s.

Experiments. We considered 240 benchmarks for parity games. Those games have a mean size $|V|$ around 46K with a maximal size of 3157K, and a mean number d of priorities of 4.1 with a maximal number $d = 15$. The statistics about the behaviors of the different algorithms are summarized in Table 1 and divided into two parts: the first part concerns all the 240 benchmarks and the second part the 20 most difficult benchmarks for Zielonka's algorithm. Column 1 indicates

---

[5] The variant with safety objectives.

[6] The tool we implemented to realize this translation can be fetched from https://github.com/gaperez64/tlsf2gpg.

the name of the solver, Columns 2 and 6 count the number of benchmarks completely solved (for the partial solvers, the second number is the number of incomplete solutions), Columns 3 and 7 count the number of timeouts (fixed at 60000 ms), and Columns 4 and 8 count how many times the solver was the fastest (excluding examples with timeout). For the 233 (resp. 13) benchmarks without timeout for Zielonka's algorithm and all its combinations with a partial solver, Column 5 (resp. 9) indicates the mean execution time in milliseconds.

We considered 152 benchmarks for generalized parity games. Those games have a mean size $|V|$ around 207K with a maximal size of 7009K. The mean number of priority functions is equal to 4.53 with a maximum number of 17. The statistics about the behaviors of the different algorithms are summarized in Table 2. The columns have the same meaning as before and the last column concerns the 87 benchmarks without timeout for all Algorithms GenZielonka and GenZiel&PSolver.

Table 1. Statistics on the one dimensional benchmarks.

Solver	Solved	T.O.	Fastest	Mean	Solved	T.O.	Fastest	Mean
Zielonka	240 (100%)	0	150 (63%)	272 ms	20 (100%)	0	11 (55%)	451 ms
Ziel&BüchiSolver	240 (100%)	0	89 (37%)	480 ms	20 (100%)	0	8 (40%)	7746 ms
Ziel&GoodEpSolver	233 (97%)	7	0 (0%)	1272 ms	13 (65%)	7	0 (0%)	20025 ms
Ziel&LaySolver	238 (99%)	2	1 (0%)	587 ms	18 (99%)	2	1 (5%)	9079 ms
BüchiSolver	203 (84%) - 37	0	-	-	15 (75%) - 5	0	-	-
GoodEpSolver	233 (97%) - 0	7	-	-	13 (65%) - 0	7	-	-
LaySolver	232 (97%) - 6	2	-	-	18 (90%) - 0	2	-	-

Table 2. Statistics on the multi-dimensional benchmarks.

Solver	Solved	T.O.	Fastest	Mean
GenZielonka	128 (84%)	24	33 (25%)	66 ms
GenZiel&GenBüchiSolver	130 (86%)	22	72 (55%)	56 ms
GenZiel&GenGoodEpSolver	112 (74%)	40	24 (18%)	644 ms
GenZiel&GenLaySolver	110 (72%)	42	3 (2%)	1133 ms
GenBüchiSolver	110 (72%) - 20	22	-	-
GenGoodEpSolver	112 (74%) - 0	40	-	-
GenLaySolver	104 (68%) - 6	42	-	-

Observations. Our experiments show that for parity games, Zielonka's algorithm is faster than partial solvers on average which was not observed on random graphs in [14]. For generalized parity games, they show that 4 benchmarks can be solved only by the combination of GenZielonka with a partial solver. Our experiments also show that the combination with a partial solver improves the performances of Zielonka's algorithm (resp. GenZielonka): for 90 cases over 240 (38%) (resp. for 99 cases over 132 (75%)). For generalized parity games, they suggest that it is interesting to launch in parallel all three Algorithms Gen-Ziel&PSolver, as none appears to dominate the other ones. Nevertheless, the combination of GenZielonka with GenBüchiSolver is a good compromise.

References

1. Ah-Fat, P., Huth, M.: Partial solvers for parity games: effective polynomial-time composition. In: GandALF Proceedings. EPTCS, vol. 226, pp. 1–15 (2016). https://doi.org/10.4204/EPTCS.226.1
2. Bloem, R., Chatterjee, K., Greimel, K., Henzinger, T.A., Jobstmann, B.: Robustness in the presence of liveness. In: Touili, T., Cook, B., Jackson, P. (eds.) CAV 2010. LNCS, vol. 6174, pp. 410–424. Springer, Heidelberg (2010). https://doi.org/10.1007/978-3-642-14295-6_36
3. Calude, C.S., Jain, S., Khoussainov, B., Li, W., Stephan, F.: Deciding parity games in quasipolynomial time. In: STOC Proceedings, pp. 252–263. ACM (2017). https://doi.org/10.1145/3055399.3055409
4. Chatterjee, K., Dvorák, W., Henzinger, M., Loitzenbauer, V.: Conditionally optimal algorithms for generalized Büchi games. In: MFCS Proceedings. LIPIcs, vol. 58, pp. 25:1–25:15. Schloss Dagstuhl - Leibniz-Zentrum fuer Informatik (2016). https://doi.org/10.4230/LIPIcs.MFCS.2016.25
5. Chatterjee, K., Henzinger, M.: Efficient and dynamic algorithms for alternating büchi games and maximal end-component decomposition. J. ACM **61**(3), 15:1–15:40 (2014). https://doi.org/10.1145/2597631
6. Chatterjee, K., Henzinger, T.A., Piterman, N.: Generalized parity games. In: Seidl, H. (ed.) FoSSaCS 2007. LNCS, vol. 4423, pp. 153–167. Springer, Heidelberg (2007). https://doi.org/10.1007/978-3-540-71389-0_12
7. Dijk, T.: Oink: an implementation and evaluation of modern parity game solvers. In: Beyer, D., Huisman, M. (eds.) TACAS 2018. LNCS, vol. 10805, pp. 291–308. Springer, Cham (2018). https://doi.org/10.1007/978-3-319-89960-2_16
8. Doyen, L., Raskin, J.-F.: Antichain algorithms for finite automata. In: Esparza, J., Majumdar, R. (eds.) TACAS 2010. LNCS, vol. 6015, pp. 2–22. Springer, Heidelberg (2010). https://doi.org/10.1007/978-3-642-12002-2_2
9. Emerson, E.A., Jutla, C.S.: Tree automata, mu-calculus and determinacy (extended abstract). In: FOCS Proceedings, pp. 368–377. IEEE Computer Society (1991). https://doi.org/10.1109/SFCS.1991.185392
10. Filiot, E., Jin, N., Raskin, J.: Exploiting structure in LTL synthesis. STTT **15**(5-6), 541–561 (2013). https://doi.org/10.1007/s10009-012-0222-5
11. Friedmann, O., Lange, M.: Solving parity games in practice. In: Liu, Z., Ravn, A.P. (eds.) ATVA 2009. LNCS, vol. 5799, pp. 182–196. Springer, Heidelberg (2009). https://doi.org/10.1007/978-3-642-04761-9_15
12. Grädel, E., Thomas, W., Wilke, T. (eds.): Automata Logics, and Infinite Games. LNCS, vol. 2500. Springer, Heidelberg (2002). https://doi.org/10.1007/3-540-36387-4
13. Huth, M., Kuo, J.H.-P., Piterman, N.: Fatal attractors in parity games. In: Pfenning, F. (ed.) FoSSaCS 2013. LNCS, vol. 7794, pp. 34–49. Springer, Heidelberg (2013). https://doi.org/10.1007/978-3-642-37075-5_3
14. Huth, M., Kuo, J.H.-P., Piterman, N.: Static analysis of parity games: alternating reachability under parity. In: Probst, C.W., Hankin, C., Hansen, R.R. (eds.) Semantics, Logics, and Calculi. LNCS, vol. 9560, pp. 159–177. Springer, Cham (2016). https://doi.org/10.1007/978-3-319-27810-0_8

15. Jacobs, S., et al.: The 5th reactive synthesis competition (SYNTCOMP 2018): Benchmarks, participants & results. CoRR abs/1904.07736 (2019). http://arxiv.org/abs/1904.07736
16. Jurdzinski, M.: Deciding the winner in parity games is in UP ∩ co-UP. Inf. Process. Lett. **68**(3), 119–124 (1998). https://doi.org/10.1016/S0020-0190(98)00150-1
17. Martin, D.A.: Borel determinacy. Ann. Math. **102**, 363–371 (1975)
18. Pnueli, A., Rosner, R.: On the synthesis of a reactive module. In: POPL Proceedings, pp. 179–190. ACM Press (1989). https://doi.org/10.1145/75277.75293
19. Zielonka, W.: Infinite games on finitely coloured graphs with applications to automata on infinite trees. Theor. Comput. Sci. **200**(1–2), 135–183 (1998). https://doi.org/10.1016/S0304-3975(98)00009-7

Reachability in Augmented Interval Markov Chains

Ventsislav Chonev[✉]

MPI-SWS, Saarbrücken, Germany
v.chonev@gmail.com

Abstract. In this paper we propose augmented interval Markov chains (AIMCs): a generalisation of the familiar interval Markov chains (IMCs) where uncertain transition probabilities are in addition allowed to depend on one another. This new model preserves the flexibility afforded by IMCs for describing stochastic systems where the parameters are unclear, for example due to measurement error, but also allows us to specify transitions with probabilities known to be identical, thereby lending further expressivity.

The focus of this paper is reachability in AIMCs. We study the qualitative, exact quantitative and approximate reachability problem, as well as natural subproblems thereof, and establish several upper and lower bounds for their complexity. We prove the exact reachability problem is at least as hard as the well-known square-root sum problem, but, encouragingly, the approximate version lies in **NP** if the underlying graph is known, whilst the restriction of the exact problem to a constant number of uncertain edges is in **P**. Finally, we show that uncertainty in the graph structure affects complexity by proving **NP**-completeness for the qualitative subproblem, in contrast with an easily-obtained upper bound of **P** for the same subproblem with known graph structure.

Keywords: Interval Markov decision processes · Reachability

1 Introduction

Discrete-time Markov chains are a well-known stochastic model, one which has been used extensively to reason about software systems [7,14,21]. They comprise a finite set of states and a set of transitions labelled with probabilities in such a way that the outgoing transitions from each state form a distribution. They are useful for modelling systems with inherently probabilistic behaviour, as well as for abstracting complexity away from deterministic ones. Thus, it is a long-standing interest of the verification community to develop logics for describing properties concerning realiability of software systems and to devise verification algorithms for these properties on Markov chains and their related generalisations, such as Markov decision processes [2,17].

One well-known such generalisation is motivated by how the assumption of precise knowledge of a Markov chain's transition relation often fails to hold.

© Springer Nature Switzerland AG 2019
E. Filiot et al. (Eds.): RP 2019, LNCS 11674, pp. 79–92, 2019.
https://doi.org/10.1007/978-3-030-30806-3_7

Indeed, a real-world system's dynamics are rarely known exactly, due to incomplete information or measurement error. The need to model this uncertainty and to reason about robustness under perturbations in stochastic systems naturally gives rise to *interval Markov chains (IMCs)*. In this model, uncertain transition probabilities are constrained to intervals, with two different semantic interpretations. Under the *once-and-for-all* interpretation, the given interval Markov chain is seen as representing an uncountably infinite collection of Markov chains refining it, and the goal is to determine whether some (or alternatively, all) refinements satisfy a given property. In contrast, the *at-every-step* interpretation exhibits a more game-theoretic flavour by allowing a choice over the outgoing transition probabilities prior to every move. The goal is then to determine strategies which optimise the probability of some property being satisfied. Originally introduced in [15], interval Markov chains have recently elicited considerable attention: see for example references [23], [6] and [3], which study the complexity of model checking branching- and linear-time properties, as well as [9], where the focus is on consistency and refinement.

While IMCs are very natural for modelling uncertainty in stochastic dynamics, they lack the expressiveness necessary to capture dependencies between transition probabilities arising out of domain-specific knowledge of the underlying real-world system. Such a dependency could state e.g. that, although the probabilities of some transitions are only known to lie within a given interval, they are all identical. Disregarding this information and studying only a dependence-free IMC is impractical, as allowing these transitions to vary independently of one another results in a vastly over-approximated space of possible behaviours.

Therefore, in the present paper we study *augmented interval Markov chains (AIMCs)*, a generalisation of IMCs which allows for dependencies of this type to be described. We study the effect of this added expressivity through the prism of the existentially quantified reachability problem, exclusively under the once-and-for-all interpretation. Our results are the following. First, we show that the full problem is hard for both the square-root sum problem (Theorem 6) and for the class **NP** (Theorem 3). The former hardness is present even when the underlying graph structure is known and acyclic, whilst the latter arises even in the qualitative subproblem when transition intervals are allowed to include zero, rendering the structure uncertain. Second, assuming known structure, we show the approximate reachability problem to be in **NP** (Theorem 7). Third, we show that the restriction of the reachability problem to a constant number of uncertain (i.e. interval-valued) transitions is in **P** (Theorem 4). The problem in full generality is in **PSPACE** via a straightforward reduction to the existential theory of the reals (Theorem 5).

The model studied here can be viewed as a simple variant of *parametric Markov chains*. These have an established presence in the literature, typically with practical and scalable synthesis procedures as the main focus, rather than complexity classification. See for example references [8, 10, 13, 18].

2   Preliminaries

Markov Chains. A *discrete-time Markov chain* or simply *Markov chain (MC)* is a tuple $M = (V, \delta)$ which consists of a finite set of *vertices* or *states* V and a *one-step transition function* $\delta : V^2 \to [0, 1]$ such that for all $v \in V$, we have $\sum_{u \in V} \delta(v, u) = 1$. For the purposes of specifying Markov chains as inputs to decision problems, we will assume δ is given by a square matrix of rational numbers. The transition function gives rise to a probability measure on V^ω in the usual way. We denote the probability of reaching a vertex t starting from a vertex s in M by $\mathbb{P}^M(s \twoheadrightarrow t)$. The *structure* of M is its underlying directed graph, with vertex set V and edge set $E = \{(u, v) \in V^2 : \delta(u, v) \neq 0\}$. Two Markov chains with the same vertex set are said to be *structurally equivalent* if their edge sets are identical.

An *interval Markov chain (IMC)* generalises the notion of a Markov chain. Formally, it is a pair (V, Δ) comprising a vertex set V and a transition function Δ from V^2 to the set $Int_{[0,1]}$ of intervals contained in $[0, 1]$. For the purposes of representing an input IMC, we will assume that each transition is given by a lower and an upper bound, together with two Boolean flags indicating the strictness of the inequalities. A Markov chain $M = (V, \delta)$ is said to *refine* an interval Markov chain $\mathcal{M} = (V, \Delta)$ with the same vertex set if $\delta(u, v) \in \Delta(u, v)$ for all $u, v \in V$. We denote by $[\mathcal{M}]$ the set of Markov chains which refine $\mathcal{M}$. An IMC's structure is said to be *known* if all elements of $[\mathcal{M}]$ are structurally equivalent. Moreover, if there exists some $\epsilon > 0$ such that for all $M = (V, \delta) \in [\mathcal{M}]$ and all $u, v \in V$, either $\delta(u, v) = 0$ or $\delta(u, v) > \epsilon$, then the IMC's structure is *ϵ-known*. An IMC can have known structure but not ϵ-known structure for example by having an edge labelled with an open interval whose lower bound is 0.

An *augmented interval Markov chain (AIMC)* generalises the notion of an IMC further by equipping it with pairs of edges whose transition probabilities are required to be identical. Formally, an AIMC is a tuple (V, Δ, C), where (V, Δ) is an IMC and $C \subseteq V^4$ is a set of *edge equality constraints*. A Markov chain (V, δ) is said to refine an AIMC (V, Δ, C) if it refines the IMC (V, Δ) and for each $(u, v, x, y) \in C$, we have $\delta(u, v) = \delta(x, y)$. We extend the notation $[\mathcal{M}]$ to AIMCs for the set of Markov chains refining $\mathcal{M}$.

The *reachability problem* for AIMCs is the problem of deciding, given an AIMC $\mathcal{M} = (V, \Delta, C)$, an initial vertex $s \in V$, a target vertex $t \in V$, a threshold $\tau \in [0, 1]$ and a relation $\sim \in \{\leq, \geq\}$, whether there exists $M \in [\mathcal{M}]$ such that $\mathbb{P}^M(s \twoheadrightarrow t) \sim \tau$. The *qualitative* subproblem is the restriction of the reachability problem to inputs where $\tau \in \{0, 1\}$.

Finally, in the *approximate reachability problem*, we are given a (small) rational number θ and a reachability problem instance. If $\sim$ is $\geq$, our procedure is required to accept if there exists some refining Markov chain with reachability probability greater than $\tau + \theta/2$, it is required to reject if all refining Markov chains have reachability probability less than $\tau - \theta/2$, and otherwise it is allowed to do anything. Similarly if $\sim$ is $\leq$. Intuitively, this is a promise problem: in the given instance the optimal reachability probability is guaranteed to be outside the interval $[\tau - \theta/2, \tau + \theta/2]$. A similar type of problem was studied in [19].

First-Order Theory of the Reals. We write $\mathcal{L}$ to denote the first-order language $\mathbb{R}\langle +, \times, 0, 1, <, =\rangle$. Atomic formulas in $\mathcal{L}$ are of the form $P(x_1, \ldots, x_n) = 0$ and $P(x_1, \ldots, x_n) > 0$ for $P \in \mathbb{Z}[x_1, \ldots, x_n]$ a polynomial with integer coefficients. We denote by $Th(\mathbb{R})$ *the first-order theory of the reals*, that is, the set of all valid sentences in the language $\mathcal{L}$. Let $Th^\exists(\mathbb{R})$ be the *existential first-order theory of the reals*, that is, the set of all valid sentences in the existential fragment of $\mathcal{L}$. A celebrated result [24] is that $\mathcal{L}$ admits quantifier elimination: each formula $\phi_1(\bar{x})$ in $\mathcal{L}$ is equivalent to some effectively computable formula $\phi_2(\bar{x})$ which uses no quantifiers. This immediately entails the decidability of $Th(\mathbb{R})$. Tarski's original result had non-elementary complexity, but improvements followed, culminating in the detailed analysis of [20]:

Theorem 1. *(i) $Th(\mathbb{R})$ is complete for* **2-EXPTIME**. *(ii) $Th^\exists(\mathbb{R})$ is decidable in* **PSPACE**. *(iii) If $m \in \mathbb{N}$ is a fixed constant and we consider only existential sentences where the number of variables is bounded above by m, then validity is decidable in* **P**.

We denote by $\exists\mathbb{R}$ the class, introduced in [22], which lies between **NP** and **PSPACE** and comprises all problems reducible in polynomial time to the problem of deciding membership in $Th^\exists(\mathbb{R})$.

Square-Root Sum Problem. The *square-root sum* problem is the decision problem where, given $r_1, \ldots, r_m, k \in \mathbb{N}$, one must determine whether $\sqrt{r_1} + \cdots + \sqrt{r_m} \geq k$. Originally posed in [16], this problem arises naturally in computational geometry and other contexts involving Euclidean distance. Its exact complexity is open. Membership in **PSPACE** is straightforward via a reduction to the existential theory of the reals. Later this was sharpened in [1] to **PosSLP**, the complexity class whose complete problem is deciding whether a division-free arithmetic circuit represents a positive number. This class was introduced and bounded above by the fourth level of the counting hierarchy **CH** in the same paper. However, containment of the square-root sum problem in **NP** is a long-standing open question, originally posed in [12], and the only obstacle to proving membership in **NP** for the exact Euclidean travelling salesman problem. This highlights a difference between the familiar integer model of computation and the Blum-Shub-Smale Real RAM model [4], under which the square-root sum is decidable in polynomial time [25]. See also [11] for more background.

3   Qualitative Case

In this section, we will focus on the qualitative reachability problem for AIMCs. We show that, whilst membership in **P** is straightforward when the underlying graph is known, uncertainty in the structure renders the qualitative problem **NP**-complete.

Theorem 2. *The qualitative reachability problem for AIMCs with known structure is in* **P**.

Proof. Let the given AIMC be $\mathcal{M}$ and s, t the initial and target vertices, respectively. Since the structure $G = (V, E)$ of $\mathcal{M}$ is known, the qualitative reachability problem can be solved simply using standard graph analysis techniques on G. More precisely, for any $M \in [\mathcal{M}]$, $\mathbb{P}^M(s \twoheadrightarrow t) = 1$ if and only if there is no path in G which starts in s, does not enter t and ends in a bottom strongly connected component which does not contain t. Similarly, $\mathbb{P}^M(s \twoheadrightarrow t) = 0$ if and only if there is no path from s to t in G. $\qquad\square$

Theorem 3. *The qualitative reachability problem for AIMCs is* **NP**-*complete.*

Proof. Membership in **NP** is straightforward. The equivalence classes of $[\mathcal{M}]$ under structure equivalence are at most 2^{n^2}, where n is the number of vertices, since for each pair (u, v) of vertices, either an edge (u, v) is present in the structure or not. This upper bound is exponential in the size of the input. Thus, we can guess the structure of the Markov chain in nondeterministic polynomial time and then proceed to solve an instance of the qualitative reachability problem on an AIMC with known structure in polynomial time by Theorem 2.

We now proceed to show **NP**-hardness using a reduction from 3-SAT. Suppose we are given a propositional formula φ in 3-CNF: $\varphi \equiv \varphi_1 \wedge \varphi_2 \wedge \cdots \wedge \varphi_k$, where each clause is a disjunction of three literals: $\varphi_i \equiv l_{i,1} \vee l_{i,2} \vee l_{i,3}$. Let the variables in φ be $x_1, \ldots, x_m$.

Let $\mathcal{M} = (V, \Delta, C)$ be the following AIMC, also depicted in Fig. 1. The vertex set has $3m + k + 3$ vertices:

$$V = \{x_1, \ldots, x_m, \overline{x_1}, \ldots, \overline{x_m}\} \cup \{\varphi_1, \ldots, \varphi_k\} \cup \{S, F\}, \cup \{v_0, \ldots, v_m\}$$

that is, one vertex for each possible literal over the given variables, one vertex for each clause, two special sink vertices S, F (*success* and *failure*) and $m + 1$ auxiliary vertices. Through a slight abuse of notation, we use $x_i, \overline{x_i}$ to refer both to the literals over the variable x_i and to their corresponding vertices in $\mathcal{M}$, and similarly, φ_i denotes both the clause in the formula and its corresponding vertex.

The transitions are the following. For all $i \in \{1, \ldots, m\}$, we have:

$$\Delta(v_{i-1}, x_i) = \Delta(v_{i-1}, \overline{x_i}) = \Delta(x_i, v_i) =$$
$$\Delta(x_i, F) = \Delta(\overline{x_i}, F) = \Delta(\overline{x_i}, v_i) = [0, 1].$$

For all $i \in \{1, \ldots, k\}$ and $j \in \{1, \ldots, 3\}$, we have: $\Delta(\varphi_i, l_{i,j}) = [0, 1]$. For all $i \in \{1, \ldots, k\}$,

$$\Delta(v_m, S) = \Delta(v_m, \varphi_i) = \left[\frac{1}{k+1}, \frac{1}{k+1}\right].$$

Finally, $\Delta(S, S) = \Delta(F, F) = [1, 1]$. For all other pairs of vertices u, v, we have $\Delta(u, v) = [0, 0]$.

The edge equality constraints are:

$$C = \bigcup_{i=1, \ldots, m} \{(v_{i-1}, x_i, x_i, v_i), (v_{i-1}, x_i, \overline{x_i}, F)\}.$$

Intuitively, the sequence of 'diamonds' comprised by $v_0, \ldots, v_m$ and the vertices corresponding to literals are a *variable setting gadget*. Choosing transition probabilities $\delta(v_{i-1}, x_i) = \delta(x_i, v_i) = 1$, and hence necessarily $\delta(x_i, F) = 0$, corresponds to setting x_i to true, whereas $\delta(v_{i-1}, \overline{x_i}) = \delta(\overline{x_i}, v_i) = 1$ and $\delta(\overline{x_i}, F) = 0$ corresponds to setting x_i to false. On the other hand, the branching from v_m into $\varphi_1, \ldots, \varphi_k$ and the edges from clauses to their literals makes up the *assignment testing gadget*. Assigning non-zero probability to the edge $(\varphi_i, l_{i,j})$ corresponds to selecting the literal $l_{i,j}$ as witness that the clause φ_i is satisfied.

Formally, we claim that there exists a Markov chain $M \in [\mathcal{M}]$ such that $\mathbb{P}^M(v_0 \twoheadrightarrow S) = 1$ if and only if φ is satisfiable.

Suppose first that φ is satisfiable and choose some satisfying assignment $\sigma : \{x_1, \ldots, x_m\} \to \{0, 1\}$. Let $M = (V, \delta) \in [\mathcal{M}]$ be the refining Markov chain which assigns the following transition probabilities to the interval-valued edges of $\mathcal{M}$. First, let $\delta(v_{i-1}, x_i) = \delta(x_i, v_i) = \delta(\overline{x_i}, F) = \sigma(x_i)$, and $\delta(v_{i-1}, \overline{x_i}) = \delta(\overline{x_i}, v_i) = \delta(x_i, F) = 1 - \sigma(x_i)$ for all $i \in \{1, \ldots, m\}$. Second, for each clause φ_i, choose some literal $l_{i,j}$ which is true under σ and set $\delta(\varphi_i, l_{i,j}) = 1$ and consequently $\delta(\varphi_i, l) = 0$ for the other literals l. Now we can observe that the structure of M has two bottom strongly-connected components, namely $\{S\}$ and $\{F\}$, and moreover, F is unreachable from v_0. Therefore, $\mathbb{P}^M(v_0 \twoheadrightarrow S) = 1$.

Conversely, suppose there exists some $M = (V, \delta) \in [\mathcal{M}]$ such that $\mathbb{P}^M(v_0 \twoheadrightarrow S) = 1$. We will prove that φ has a satisfying assignment. For each $i \in \{1, \ldots, m\}$, write $p_i = \delta(v_{i-1}, x_i) = \delta(x_i, v_i) = \delta(\overline{x_i}, F)$, and $1 - p_i = \delta(v_{i-1}, \overline{x_i}) = \delta(\overline{x_i}, v_i) = \delta(x_i, F)$. Notice that $\mathbb{P}^M(v_0 x_1 F^\omega) = \mathbb{P}^M(v_0 \overline{x_1} F^\omega) = p_1(1 - p_1)$, so we can conclude $p_1 \in \{0, 1\}$, otherwise $\mathbb{P}^M(v_0 \twoheadrightarrow S) \neq 1$, a contradiction. If $p_1 = 1$, then

$$\mathbb{P}^M(v_0 x_1 v_1 x_2 F^\omega) = \mathbb{P}^M(v_0 x_1 v_1 \overline{x_2} F^\omega) = p_2(1 - p_2),$$

whereas if $p_1 = 0$, then

$$\mathbb{P}^M(v_0 \overline{x_1} v_1 x_2 F^\omega) = \mathbb{P}^M(v_0 \overline{x_1} v_1 \overline{x_2} F^\omega) = p_2(1 - p_2).$$

Either way, we must have $p_2 \in \{0, 1\}$ to ensure $\mathbb{P}^M(v_0 \twoheadrightarrow S) = 1$. Unrolling this argument further shows $p_i \in \{0, 1\}$ for all i. In particular, there is exactly one path from v_0 to v_m and it has probability 1. Let σ be the truth assignment $x_i \to p_i$, we show that σ satisfies φ. Indeed, if some clause φ_i is unsatisfied under σ, then its three literals $l_{i,1}, \ldots, l_{i,3}$ are all unsatisfied, so $\delta(l_{i,j}, F) > 0$ for all $j = 1, \ldots, 3$. Moreover, for at least one of these three literals, say $l_{i,1}$, we will have $\delta(\varphi_i, l_{i,1}) > 0$, so the path $v_0 \ldots v_m \varphi_i l_{i,1} F^\omega$ will have non-zero probability:

$$\mathbb{P}^M(v_0 \ldots v_m \varphi_i l_{i,1} F^\omega) = \frac{1}{k+1} \delta(\varphi_i, l_{i,1}) \delta(l_{i,1}, F) \neq 0,$$

which contradicts $\mathbb{P}^M(v_0 \twoheadrightarrow S) = 1$. Therefore, σ satisfies φ, which completes the proof of **NP**-hardness and of the Theorem. $\qquad\square$

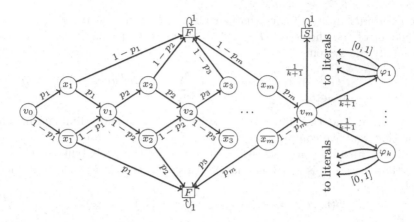

Fig. 1. Construction used in Theorem 3 for showing **NP**-hardness of the qualitative AIMC reachability problem. The sink F is duplicated to avoid clutter.

4 Quantitative Case: Upper Bound

We now shift our attention to the subproblem of AIMC reachability which arises when the number of interval-valued transitions is fixed, that is, bounded above by some absolute constant. Our result is the following.

Theorem 4. *Fix a constant $N \in \mathbb{N}$. The restriction of the reachability problem for AIMCs to inputs with at most N interval-valued transitions lies in $\mathbf{P}$. Hence, the approximate reachability problem under the same restriction is also in $\mathbf{P}$.*

Proof. Let $\mathcal{M} = (V, \Delta, C)$ be the given AIMC and suppose we wish to decide whether there exists $M \in [\mathcal{M}]$ such that $\mathbb{P}^M(s \twoheadrightarrow t) \sim \tau$. Let $U \subseteq V$ be the set of vertices which have at least one interval-valued outgoing transition, together with s and t: $U = \{s, t\} \cup \{u \in V : \exists v \in V. \Delta(u, v) \text{ is not a singleton}\}$. Notice that $|U| \leq N + 2 = const$. Write $W = V \setminus U$, so that $\{U, W\}$ is a partition of V.

Let $\mathbf{x}$ be a vector of variables, one for each interval-valued transition of $\mathcal{M}$. For vertices v_1, v_2, let $\delta(v_1, v_2)$ denote the corresponding variable in $\mathbf{x}$ if the transition (v_1, v_2) is interval-valued, and the only element of the singleton set $\Delta(v_1, v_2)$ otherwise. Let φ_1 be the following propositional formula over the variables $\mathbf{x}$ which captures the set of 'sensible' assignments:

$$\varphi_1 \equiv \bigwedge_{v_1 \in V} \sum_{v_2 \in V} \delta(v_1, v_2) = 1$$
$$\wedge \bigwedge_{v_1, v_2 \in V} \delta(v_1, v_2) \in \Delta(v_1, v_2) \cap [0, 1] \wedge \bigwedge_{(a,b,c,d) \in C} \delta(a, b) = \delta(c, d).$$

There is clearly a bijection between $[\mathcal{M}]$ and assignments of $\mathbf{x}$ which satisfy φ_1.

For vertices v_1, v_2, use the notation $v_1 \rightsquigarrow v_2$ to denote the event 'v_2 is reached from v_1 along a path consisting only of vertices in W, with the possible exception *of the endpoints* v_1, v_2'. Notice that for all $u \in U$ and $w \in W$, $\mathbb{P}^M(w \rightsquigarrow u)$ is independent of the choice of $M \in [\mathcal{M}]$. Denote these probabilities by $\alpha(w, u)$. They satisfy the system

$$\bigwedge_{w \in W, u \in U} \alpha(w, u) = \delta(w, u) + \sum_{w' \in W} \delta(w, w')\alpha(w', u),$$

which is linear and therefore easy to solve with Gaussian elimination. Thus, assume that we have computed $\alpha(w, u) \in \mathbb{Q}$ for all $w \in W$ and $u \in U$.

Similarly, for all $u_1, u_2 \in U$, write $\beta(u_1, u_2)$ for the probability of $u_1 \rightsquigarrow u_2$. Notice that $\beta(u_1, u_2)$ is a polynomial of degree at most 1 over the variables $\mathbf{x}$, given by

$$\beta(u_1, u_2) = \delta(u_1, u_2) + \sum_{w \in W} \delta(u_1, w)\alpha(w, u_2).$$

Thus, assume we have computed symbolically $\beta(u_1, u_2) \in \mathbb{Q}[\mathbf{x}]$ for all $u_1, u_2 \in U$.

Finally, for each $u \in U$, let $y(u)$ be a variable and write $\mathbf{y}$ for the vector of variables $y(u)$ in some order. Consider the following formula in the existential first-order language of the real field: $\varphi \equiv \exists \mathbf{x} \exists \mathbf{y} . \varphi_1 \wedge \varphi_2 \wedge \varphi_3$, where

$$\varphi_2 \equiv y(t) = 1 \wedge \bigwedge_{u \in U \setminus \{t\}} y(u) = \sum_{u' \in U} \beta(u, u')y(u'),$$

$\varphi_3 \equiv y(s) \sim \tau$, and φ_1 is as above. Intuitively, φ_1 states that the variables $\mathbf{x}$ describe a Markov chain in $[\mathcal{M}]$, φ_2 states that $\mathbf{y}$ gives the reachability probabilities from U to t, and φ_3 states that the reachability probability from s to t meets the required threshold τ. The problem instance is positive if and only if φ is a valid sentence in the existential theory of the reals, which is decidable. Moreover, the formula uses exactly $2|U| \leq 2(N + 2) = const$ variables, so by Theorem 1, the problem is decidable in polynomial time, as required. $\quad\square$

Notice that removing the assumption of a constant number of interval-valued transitions only degrades the complexity upper bound, but not the described reduction to the problem of checking membership in $Th^\exists(\mathbb{R})$. As an immediate corollary, we have:

Theorem 5. *The reachability problem and the approximate reachability problem for AIMCs are in* $\exists \mathbb{R}$.

Note that Theorem 5 can be shown much more easily, without the need to consider separately U-vertices and W-vertices as in the proof of Theorem 4. It is sufficient to use one variable per interval-valued transition to capture its transition probability as above and one variable per vertex to express its reachability probability to the target. Then write down an existentially quantified formula with the usual system of equations for reachability in a Markov chain obtained by conditioning on the first step from each vertex. While this easily gives the $\exists \mathbb{R}$ upper bound, it uses at least $|V|$ variables, so it is insufficient for showing membership in **P** for the restriction to a constant number of interval-valued transitions.

5 Quantitative Case: Lower Bound

In this section, we give a lower bound for the AIMC reachability problem. This bound remains in place even when the structure of the AIMC is ϵ-known and acyclic, except for the self-loops on two sink vertices.

Theorem 6. *The AIMC reachability problem is hard for the square-root sum problem, even when the structure of the AIMC is ϵ-known and is acyclic, except for the self-loops on two sink vertices.*

Proof. The reduction is based on the gadget depicted in Fig. 2. It is an AIMC with two sinks, S and F (*success* and *failure*), each with a self-loop with probability 1, and 12 vertices: $\{a, b_1, \ldots, b_4, c_1, \ldots, c_4, d_1, d_4, e\}$. The structure is acyclic and comprises four chains leading to S, namely, $ab_1c_1d_1eS$, ab_2c_2S, ab_3c_3S and $ab_4c_4d_4S$. From each vertex other than a and S there is also a transition to F.

The probabilities are as follows. The transition (b_3, c_3) has probability α, whilst (b_1, c_1), (b_2, c_2), (b_4, c_4) have probability β, for rationals α, β to be specified later. Consequently, the remaining outgoing transition to F out of each b_i has probability $1 - \alpha$ or $1 - \beta$. The transitions (a, b_i) for $i = 1, \ldots, 4$ all have probability $1/4$. Finally, the transitions (c_1, F), (c_2, F), (c_3, S), (c_4, F), (d_1, e), (d_4, S) and (e, S) are interval-valued and must all have equal probability in any refining Markov chain. Assign the variable x to the probability of these transitions. The interval to which these transition probabilities are restricted (i.e. the range of x) is to be specified later. Consequently, the remaining transitions (c_1, d_1), (d_1, F), (e, F), (c_2, S), (c_3, F), (c_4, d_4), (d_4, F) are also interval-valued, with probability $1 - x$.

The gadget is parameterised by an input $r \in \mathbb{N}$, on which the transition probabilities depend. Let M be a positive integer large enough to ensure

$$x^* := \frac{3\sqrt{r}}{2M} \in (0, 1).$$

Then choose a positive integer N large enough, so that

$$\alpha := \frac{4M}{N} \in (0, 1), \beta := \frac{16M^3}{27rN} \in (0, 1), \text{ and } p_{opt} := \frac{\sqrt{r}}{N} + \frac{\beta}{4} \in (0, 1).$$

Now, a straightforward calculation shows

$$\mathbb{P}(a \twoheadrightarrow S) = \mathbb{P}(ab_1c_1d_1eS) + \mathbb{P}(ab_2c_2S) + \mathbb{P}(ab_3c_3S) + \mathbb{P}(ab_4c_4d_4S)$$
$$= \frac{\beta x^2(1 - x)}{4} + \frac{\beta(1 - x)}{4} + \frac{\alpha x}{4} + \frac{\beta x(1 - x)}{4}$$
$$= \frac{\alpha x - \beta x^3 + \beta}{4}.$$

Analysing the derivative of this cubic, we see that $\mathbb{P}(a \twoheadrightarrow S)$ increases on $[0, x^*)$, has its maximum at $x = x^*$ and then decreases on $(x^*, 1]$. This maximum is

$$\frac{\alpha x^* - \beta(x^*)^3 + \beta}{4} = \frac{\sqrt{r}}{N} + \frac{\beta}{4} = p_{opt}.$$

Thus, if we choose some closed interval which contains x^* but not 0 and 1 to be the range of x, then the gadget described thus far will have ϵ-known structure and maximum reachability probability from a to S given by $\sqrt{r}$ scaled by a constant and offset by another constant.

Now, suppose we wish to decide whether $\sqrt{r_1} + \cdots + \sqrt{r_m} \geq k$ for given positive integers $r_1, \ldots, r_m$ and k. Construct m gadgets as above, with values of the parameter r given by $r_1, \ldots, r_m$, respectively. The constants α, N, M are shared across the gadgets, as are the sinks S, F, but each gadget has its own constant β_i in place of β, and its own copy of each non-sink vertex. The edge equality constraints are the same as above within each gadget, and there are no equality constraints across gadgets. Assign a variable x_i to those edges in the i-th gadget which in the description above were labelled x, and choose a range for x_i as described above for x. Finally, add a new initial vertex v_0, with m equiprobable outgoing transitions to the a-vertices of the gadgets.

In this AIMC, the probability of $v_0 \twoheadrightarrow S$ is given by the multivariate polynomial

$$\frac{1}{m} \sum_{i=1}^{m} \frac{\alpha x_i - \beta_i x_i^3 + \beta_i}{4},$$

whose maximum value on $[0,1]^m$ is

$$\frac{1}{m} \sum_{i=1}^{m} \left(\frac{\sqrt{r_i}}{N} + \frac{\beta_i}{4} \right).$$

Therefore, if we denote

$$\tau = \frac{k}{mN} + \frac{1}{m} \sum_{i=1}^{m} \frac{\beta_i}{4},$$

then we have $\sqrt{r_1} + \cdots + \sqrt{r_m} \geq k$ if and only if there exists a refining Markov chain of this AIMC with $\mathbb{P}(v_0 \twoheadrightarrow S) \geq \tau$, so the reduction is complete. Note that if we represent rational numbers as usual as pairs of integers in binary, the bit-length of τ and all intermediate constants is bounded above by a polynomial in the bit-lengths of the inputs $r_1, \ldots, r_m, k$, so the reduction can be carried out in polynomial time.   □

Remark 1. It is easy to see that if we are given an acyclic AIMC with the interval-valued edges labelled with variables, the reachability probabilities from all vertices to a single target vertex are multivariate polynomials and can be computed symbolically with a backwards breadth-first search from the target. Then optimising reachability probabilities reduces to optimising the value of a polynomial over given ranges for its variables.

It is interesting to observe that a reduction holds in the other direction as well. Suppose we wish to decide whether there exist values of $x_1 \in I_1, \ldots, x_n \in I_n$ such that $P(x_1, \ldots, x_n) \geq \tau$ for a given multivariate polynomial P, intervals

$I_1, \ldots, I_n \subseteq [0,1]$ and $\tau \in \mathbb{Q}$. Notice that P can easily be written in the form $P(x_1, \ldots, x_n) = \beta + N \sum_{i=1}^{m} \alpha_i Q_i(x_1, \ldots, x_n)$, where $N > 0$, $\alpha_1, \ldots, \alpha_m \in (0,1)$ are constants such that $\sum_{i=1}^{m} \alpha_i \leq 1$, each Q_i is a non-empty product of terms drawn from $\bigcup_{j=1}^{n} \{x_j, (1-x_j)\}$, and β is a (possibly negative) constant term. For example, we rewrite the monomial $-2x_1 x_2 x_3$ as $2(1 - x_1)x_2 x_3 + 2(1 - x_2)x_3 + 2(1 - x_3) - 2$. Do this to all monomials with a negative coefficient, then choose an appropriately large N to obtain the desired form.

Now it is easy to construct an AIMC with two sinks S, F and a designated initial vertex v_0 where the probability of $v_0 \twoheadrightarrow S$ is $\sum_{i=1}^{m} \alpha_i Q_i$. We use a chain to represent each Q_i, and then branch from v_0 into the first vertices of the chains with distribution given by the α_i. There exist values of the x_i in their appropriate intervals such that $P(x_1, \ldots, x_n) \geq \tau$ if and only if there exists a refining Markov chain such that $\mathbb{P}(v_0 \twoheadrightarrow S) \geq (\tau - \beta)/N$.

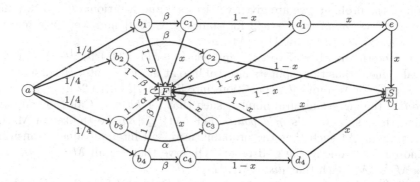

Fig. 2. Gadget for reduction from square-root sum problem to AIMC reachability.

6   Approximate Case

In this section, we focus on the approximate reachability problem for AIMCs. To obtain our upper bound, we will use a result from [5].

Definition 1. *If $M_1 = (V, \delta_1)$ and $M_2 = (V, \delta_2)$ are Markov chains with the same vertex set, then their* absolute distance *is*

$$dist_A(M_1, M_2) = \max_{u,v \in V} \{|\delta_1(u,v) - \delta_2(u,v)|\}.$$

Lemma 1 (Theorem 5 in [5]). *Let $M_1 = (V, \delta_1)$ and $M_2 = (V, \delta_2)$ be structurally equivalent Markov chains, where $n = |V|$ and for all $u, v \in V$, we have either $\delta_1(u,v) = 0$ or $\delta_1(u,v) \geq \epsilon$. Then for any two vertices $s, t \in V$, we have*

$$\left| \mathbb{P}^{M_1}(s \twoheadrightarrow t) - \mathbb{P}^{M_2}(s \twoheadrightarrow t) \right| \leq \left(1 + \frac{dist_A(M_1, M_2)}{\epsilon} \right)^{2n} - 1.$$

In particular, if $dist_A(M_1, M_2) \leq d < \epsilon$ for some d, then

$$\left| \mathbb{P}^{M_1}(s \twoheadrightarrow t) - \mathbb{P}^{M_2}(s \twoheadrightarrow t) \right| \leq \left(1 + \frac{d}{\epsilon - d} \right)^{2n} - 1.$$

We will also need a version of Bernoulli's inequality:

Lemma 2. *For all $x \geq -1$ and $r \in [0, 1]$, we have $(1 + x)^r \leq 1 + rx$.*

Now we proceed to prove our upper bound.

Theorem 7. *The approximate reachability problem for AIMCs with ϵ-known structure is in* **NP**.

Proof. Let $\mathcal{M}$ be the given AIMC and let $\epsilon > 0$ be a lower bound on all non-zero transitions across all $M \in [\mathcal{M}]$. Suppose we are solving the maximisation version of the problem: we are given vertices s, t and a rational $\theta > 0$, we must accept if $\mathbb{P}^M(s \twoheadrightarrow t) > \tau + \theta/2$ for some $M \in [\mathcal{M}]$ and we must reject if $\mathbb{P}^M(s \twoheadrightarrow t) < \tau - \theta/2$ for all $M \in [\mathcal{M}]$.

Let n be the number of vertices and let $d := \epsilon \left(1 - (1 + \theta)^{-1/2n} \right)$. For each interval-valued transition, split its interval into at most $1/d$ intervals of length at most d each. For example, $[l, r]$ partitions into $[l, l+d), [l+d, l+2d), \ldots, [l+kd, r]$, where k is the largest natural number such that $l + kd \leq r$. Call the endpoints defining these subintervals *grid points*. Let $\langle \mathcal{M} \rangle \subseteq [\mathcal{M}]$ be the set of Markov chains refining $\mathcal{M}$ such that the probabilities of all interval-valued transitions are chosen from among the grid points. Observe that for all $M_1 \in [\mathcal{M}]$, there exists $M_2 \in \langle \mathcal{M} \rangle$ such that $dist_A(M_1, M_2) \leq d$.

Our algorithm showing membership in **NP** is the following. We choose $M \in \langle \mathcal{M} \rangle$ nondeterministically and compute $p := \mathbb{P}^M(s \twoheadrightarrow t)$ using Gaussian elimination. Then if $p \geq \tau - \theta/2$, we accept, and otherwise we reject.

To complete the proof, we need to argue two points. First, that $\langle \mathcal{M} \rangle$ is at most exponentially large in the size of the input, so that M can indeed be guessed in nondeterministic polynomial time. Second, that if for all $M \in \langle \mathcal{M} \rangle$ we have $\mathbb{P}^M(s \twoheadrightarrow t) < \tau - \theta/2$, then it is safe to reject, i.e. there is no M' with $\mathbb{P}^{M'}(s \twoheadrightarrow t) \geq \tau + \theta/2$. (Note that the procedure is obviously correct when it accepts.)

To the first point, we apply Lemma 2 with $x = -\theta/(\theta + 1)$ and $r = 1/2n$:

$$(1 + \theta)^{-1/2n} = \left(1 - \frac{\theta}{1 + \theta} \right)^{1/2n} \leq 1 - \frac{1}{2n} \frac{\theta}{1 + \theta}$$

$$\implies d^{-1} = \epsilon^{-1} \frac{1}{1 - (1 + \theta)^{-1/2n}} \leq \frac{1}{\epsilon} 2n \frac{1 + \theta}{\theta} = \frac{1}{\epsilon} 2n \left(1 + \frac{1}{\theta} \right).$$

This upper bound is a polynomial in n, $1/\theta$ and $1/\epsilon$, and hence at most exponential in the length of the input data. Therefore, for each interval-valued transition, we can write down using only polynomially many bits which grid point we wish to use for the probability of that transition. Since the number of transitions is polynomial in the length of the input, it follows that an element of $\langle \mathcal{M} \rangle$ may be specified using only polynomially many bits, as required.

To the second point, consider $M_1, M_2 \in [\mathcal{M}]$ such that $dist_A(M_1, M_2) \leq d$. Noting that $d < \epsilon$, by Lemma 1, we have

$$
\left| \mathbb{P}^{M_1}(s \twoheadrightarrow t) - \mathbb{P}^{M_2}(s \twoheadrightarrow t) \right| \leq \left(1 + \frac{d}{\epsilon - d} \right)^{2n} - 1
$$

$$
= \left(\frac{\epsilon}{\epsilon(1 + \theta)^{-1/2n}} \right)^{2n} - 1 = \theta.
$$

In other words, changing the transition probabilities by at most d does not alter the reachability probability from s to t by more than θ. However, recall that we chose $\langle \mathcal{M} \rangle$ in such a way that for all $M_1 \in [\mathcal{M}]$, there is some $M_2 \in \langle \mathcal{M} \rangle$ with $dist_A(M_1, M_2) \leq d$. In particular, if $\mathbb{P}^{M_2}(s \twoheadrightarrow t) < \tau - \theta/2$ for all $M_2 \in \langle \mathcal{M} \rangle$, then certainly $\mathbb{P}^{M_1}(s \twoheadrightarrow t) < \tau + \theta/2$ for all $M_1 \in [\mathcal{M}]$, so it is safe to reject. This completes the proof.   □

References

1. Allender, E., Bürgisser, P., Kjeldgaard-Pedersen, J., Miltersen, P.B.: On the complexity of numerical analysis. SIAM J. Comput. **38**(5), 1987–2006 (2009)
2. Bellman, R.: A Markovian decision process. Technical report, DTIC Document (1957)
3. Benedikt, M., Lenhardt, R., Worrell, J.: LTL model checking of interval Markov chains. In: Piterman, N., Smolka, S.A. (eds.) TACAS 2013. LNCS, vol. 7795, pp. 32–46. Springer, Heidelberg (2013). https://doi.org/10.1007/978-3-642-36742-7_3
4. Blum, L., Shub, M., Smale, S.: On a theory of computation and complexity over the real numbers: *NP*-completeness, recursive functions and universal machines. Bull. (New Ser.) Am. Math. Soc. **21**(1), 1–46 (1989)
5. Chatterjee, K.: Robustness of structurally equivalent concurrent parity games. In: Birkedal, L. (ed.) FoSSaCS 2012. LNCS, vol. 7213, pp. 270–285. Springer, Heidelberg (2012). https://doi.org/10.1007/978-3-642-28729-9_18
6. Chatterjee, K., Sen, K., Henzinger, T.A.: Model-checking ω-regular properties of interval Markov chains. In: Amadio, R. (ed.) FoSSaCS 2008. LNCS, vol. 4962, pp. 302–317. Springer, Heidelberg (2008). https://doi.org/10.1007/978-3-540-78499-9_22
7. Courcoubetis, C., Yannakakis, M.: The complexity of probabilistic verification. J. ACM (JACM) **42**(4), 857–907 (1995)
8. Cubuktepe, M., Jansen, N., Junges, S., Katoen, J.-P., Topcu, U.: Synthesis in pMDPs: a tale of 1001 parameters. In: Lahiri, S.K., Wang, C. (eds.) ATVA 2018. LNCS, vol. 11138, pp. 160–176. Springer, Cham (2018). https://doi.org/10.1007/978-3-030-01090-4_10
9. Delahaye, B., Larsen, K.G., Legay, A., Pedersen, M.L., Wąsowski, A.: Decision problems for interval Markov chains. In: Dediu, A.-H., Inenaga, S., Martín-Vide, C. (eds.) LATA 2011. LNCS, vol. 6638, pp. 274–285. Springer, Heidelberg (2011). https://doi.org/10.1007/978-3-642-21254-3_21

10. Delahaye, B., Lime, D., Petrucci, L.: Parameter synthesis for parametric interval Markov chains. In: Jobstmann, B., Leino, K.R.M. (eds.) VMCAI 2016. LNCS, vol. 9583, pp. 372–390. Springer, Heidelberg (2016). https://doi.org/10.1007/978-3-662-49122-5_18

11. Etessami, K., Yannakakis, M.: Recursive Markov chains, stochastic grammars, and monotone systems of nonlinear equations. J. ACM (JACM) **56**(1), 1 (2009)

12. Garey, M.R., Graham, R.L., Johnson, D.S.: Some NP-complete geometric problems. In: Proceedings of the Eighth Annual ACM Symposium on Theory of Computing, STOC 1976, pp. 10–22. ACM, New York (1976). https://doi.org/10.1145/800113.803626

13. Hahn, E.M., Hermanns, H., Zhang, L.: Probabilistic reachability for parametric Markov models. Int. J. Softw. Tools Technol. Transf. **13**(1), 3–19 (2011)

14. Hansson, H., Jonsson, B.: A logic for reasoning about time and reliability. Formal Aspects Comput. **6**(5), 512–535 (1994)

15. Jonsson, B., Larsen, K.G.: Specification and refinement of probabilistic processes. In: Proceedings of Sixth Annual IEEE Symposium on Logic in Computer Science. LICS 1991, pp. 266–277. IEEE (1991)

16. O'Rourke, J.: Advanced problem 6369. Am. Math. Monthly **88**(10), 769 (1981)

17. Puterman, M.L.: Markov Decision Processes: Discrete Stochastic Dynamic Programming. Wiley, Hoboken (2014)

18. Quatmann, T., Dehnert, C., Jansen, N., Junges, S., Katoen, J.-P.: Parameter synthesis for Markov models: faster than ever. In: Artho, C., Legay, A., Peled, D. (eds.) ATVA 2016. LNCS, vol. 9938, pp. 50–67. Springer, Cham (2016). https://doi.org/10.1007/978-3-319-46520-3_4

19. Randour, M., Raskin, J.F., Sankur, O.: Percentile queries in multi-dimensional Markov decision processes. Formal Methods Syst. Des. **50**(2–3), 207–248 (2017)

20. Renegar, J.: On the computational complexity and geometry of the first-order theory of the reals. Part I: Introduction. Preliminaries. The geometry of semi-algebraic sets. The decision problem for the existential theory of the reals. J. Symb. Comput. **13**(3), 255–299 (1992). https://doi.org/10.1016/S0747-7171(10)80003-3

21. Rutten, J.J.M.M., Kwiatkowska, M., Norman, G., Parker, D.: Mathematical Techniques for Analyzing Concurrent and Probabilistic Systems. American Mathematical Society, Providence (2004)

22. Schaefer, M., Štefankovič, D.: Fixed points, Nash equilibria, and the existential theory of the reals. Theory Comput. Syst. **60**, 172–193 (2011)

23. Sen, K., Viswanathan, M., Agha, G.: Model-checking Markov chains in the presence of uncertainties. In: Hermanns, H., Palsberg, J. (eds.) TACAS 2006. LNCS, vol. 3920, pp. 394–410. Springer, Heidelberg (2006). https://doi.org/10.1007/11691372_26

24. Tarski, A.: A Decision Method for Elementary Algebra and Geometry (1951)

25. Tiwari, P.: A problem that is easier to solve on the unit-cost algebraic RAM. J. Complex. **8**(4), 393–397 (1992). https://doi.org/10.1016/0885-064X(92)90003-T

On Solving Word Equations Using SAT

Joel D. Day[1], Thorsten Ehlers[2], Mitja Kulczynski[3(✉)], Florin Manea[3],
Dirk Nowotka[3], and Danny Bøgsted Poulsen[3]

[1] Department of Computer Science, Loughborough University, Loughborough, UK
j.day@lboro.ac.uk
[2] German Aerospace Center (DLR), Helmholtz Association, Hamburg, Germany
thorsten.ehlers@dlr.de
[3] Department of Computer Science, Kiel University, Kiel, Germany
{mku,flm,dn,dbp}@informatik.uni-kiel.de

Abstract. We present WOORPJE, a string solver for bounded word
equations (i.e., equations where the length of each variable is upper
bounded by a given integer). Our algorithm works by reformulating the
satisfiability of bounded word equations as a reachability problem for
nondeterministic finite automata, and then carefully encoding this as
a propositional satisfiability problem, which we then solve using the
well-known Glucose SAT-solver. This approach has the advantage of
allowing for the natural inclusion of additional linear length constraints.
Our solver obtains reliable and competitive results and, remarkably,
discovered several cases where state-of-the-art solvers exhibit a faulty
behaviour.

1 Introduction

Over the past twenty years, applications of software verification have scaled from
small academic programs to finding errors in the GNU Coreutils [7]. In principle, the employed verification strategies involve exploring the control-flow-graph
of the program, gathering constraints over program variables and passing these
constraints to a constraint solver. The primary worker of software verification is
thus the constraint solver, and the scalability of software verification achieved
by improving the efficiency of constraint solvers. The theories supported by constraint solvers are likewise highly influenced by the needs of software verification
tools (e.g. array theory and bitvector arithmetic). A recent need of software verification tools is the ability to cope with equations involving string constraints,
i.e. equations over string variables composing equality between concatenation of
strings and string variables. This need arose from the desire to do software verification of languages with string manipulation as a core part of the language (e.g.
JavaScript and Java) [9,19]. To accomplish this goal, we have seen the advent of

Florin Manea's work was supported by the DFG grant MA 5725/2-1.
Danny Bøgsted Poulsen's work was supported by the BMBF through the ARAMiS2
(01IS160253) project.

E. Filiot et al. (Eds.): RP 2019, LNCS 11674, pp. 93–106, 2019.
https://doi.org/10.1007/978-3-030-30806-3_8

dedicated string solvers as well as constraint solvers implementing string solving techniques. As an incomplete list we mention HAMPI [15], CVC4 [4], Ostrich [8], Sloth [11], Norn [1], S3P [20] and Z3str3 [5].

Although the need for string solving only recently surfaced in the software verification community, the problem is in fact older and known as *Word Equations* (a term that we will use from now on). The word equation satisfiability problem is to determine whether we can unify the two strings, i.e., transform them into two equal strings containing constant letters only, by substituting the variables consistently by strings of constants. For example, consider the equation defined by the two strings $XabY$ and $aXYb$, denoted $XabY \doteq aXYb$, with variables X, Y and constants a and b. It is satisfiable because X can be substituted by a and Y by b, which produces the equality $aabb = aabb$. In fact, substituting X by an arbitrary amount of a's and Y by an arbitrary amount of b's unifies the two sides of the equation.

The word equation problem is decidable [16] and NP-hard. In a series of works, Jeż [12, 13] showed that word equations can be solved in non-deterministic linear space. It has been shown by Plandowski [18] that there exists an upper bound of $2^{2^{O(n^4)}}$ for the smallest solution to a word equation of length n. Having this in mind, a standard method for solving word equations is known as *filling the positions* [14,17]. In this method a length for each of the string variables is non-deterministically selected. Having a fixed length of the variables reduces the problem to lining up the positions of the two sides of the equation, and filling the unknown positions of the variables with characters, making the two sides equal.

In this paper we present a new solver for word equation with linear length constraints, WOORPJE. In particular, it guesses the maximal length of variables and encodes a variation of *filling the positions* method into an automata-construction, thereby reducing the search for a solution to a reachability question of this automata. Preliminary experiments with a pure automata-reachability-based approach revealed however, that this would not scale for even small word equations. WOORPJE therefore encodes the automata into SAT and uses the tool Glucose [3] as a backend. Unlike other approaches based on the filling the positions method (e.g. [6,19]), WOORPJE does not need an exact bound for each variable, but only an upper bound. Experiments indicate that WOORPJE is not only reliable but also competitive with the more mature CVC4 and Z3. Results indicate that WOORPJE is quicker on pure word equations (no linear length constraints), and that CVC4 and Z3 mainly have an edge on word equations with linear constraints. This may be due to our naive solution for solving linear length constraints.

2   Preliminaries

Let $\mathbb{N}$ be the set of natural numbers, let $[n]$ be the set $\{0, 1, 2, \ldots, n-1\}$ and $[n]_0$ the set $[n] \setminus \{0\}$. For a finite set Δ of symbols, we let Δ^* be the set of all words over Δ and ε be the empty word. For an alphabet Δ and $a \notin \Delta$, we let

Δ_a denote the set $\Delta \cup \{a\}$. For a word $w = x_0 x_1 \ldots x_{n-1}$ we let $|w| = n$ refer to its length. For $i \in [[w]]$ we denote by $w[i]$ the symbol on the ith position of w i.e. $w[i] = x_i$. For $a \in \Delta$ and $w \in \Delta^*$ we let $|w|_a$ denote the number of as in w. If $w = v_1 v_2$ for some words $v_1, v_2 \in \Delta^*$, then v_1 is called a *prefix* of w and v_2 is a *suffix* of w. In the remainder of the paper, we let $\Xi = \Sigma \cup \Gamma$ where Σ (Γ) is a set of symbols called letters (variables) and $\Sigma \cap \Gamma = \emptyset$. We call a word $w \in \Xi^*$ a *pattern* over Ξ. For a pattern $w \in \Xi^*$ we let $\mathrm{var}(w) \subseteq \Gamma$ denote the set of variables from Γ occurring in w. A *substitution* for Ξ is a morphism $S : \Xi^* \to \Sigma^*$ with $S(a) = a$ for every $a \in \Sigma$ and $S(\varepsilon) = \varepsilon$. Note, that to define a substitution S, it suffices to define $S(X)$ for all $X \in \Gamma$.

A *word equation* over Ξ is a tuple $(u, v) \in \Xi^* \times \Xi^*$ written $u \doteq v$. A substitution S over Ξ is a *solution* to a word equation $u \doteq v$ (denoted $S \models u \doteq v$) if $S(u) = S(v)$. A word equation $u \doteq v$ is *satisfiable* if there exists a substitution S such that $S \models u \doteq v$. A *system of word equations* is a set of word equations $P \subseteq \Xi^* \times \Xi^*$. A system of word equations P is satisfiable if there exists a substitution S that is a solution to all word equations (denoted $S \models E$). Karhumäki et al. [14] showed that for every system of word equations, a single equation can be constructed which is satisfiable if and only if the initial formula was satisfiable. The solution to the constructed word equation can be directly transferred to a solution of the original word equation system.

Bounded Word Equations. A natural sub-problem of solving word equations is that of *Bounded Word Equations*. In this problem we are not only given a word equation $u \doteq v$ but also a set of length constraints $\{|X| \leq b_X \mid X \in \Gamma \wedge b_X \in \mathbb{N}\}$. The bounded word equation is satisfiable if there exists a substitution S such $S \models u \doteq v$ and $|S(X)| \leq b_X$ for each $X \in \Gamma$. For convenience, we shall sometimes refer to the set of bounds b_X as a function $B : \Gamma \to \mathbb{N}$ such that $b_X = B(X)$.

Word Equations with Linear Constraints. A word equation with linear constraints is a word equation $u \doteq v$ accompanied by a system θ of linear Diophantine equations, where the unknowns correspond to the lengths of possible substitutions of the variables in Γ. A word equation with linear constraints is satisfiable if there exists a substitution S such that $S \models u \doteq v$ and S satisfies θ. Note that the bounded word equation problem is in fact a special case of word equations with linear constraints.

SAT Solving. A Boolean formula φ with finitely many Boolean variables $\mathrm{var}(\varphi) = \{x_1, \ldots, x_n\}$ is usually given in conjunctive normal form. This is a conjunction over a set of disjunctions (called clauses), i.e. $\varphi = \bigwedge_i \bigvee_j l_{i,j}$, where $l_{i,j} \in \bigcup_{i \in [n]} \{x_i, \neg x_i\}$ is a literal. A mapping $\beta : \mathrm{var}(\varphi) \to \{0, 1\}$ is called an *assignment*; for such an assignment, the literal l evaluates to true if and only if $l = x_i$ and $\beta(x_i) = 1$, or $l = \neg x_i$ and $\beta(x_i) = 0$. A clause inside a formula in conjunctive normal form is evaluated to true if at least one of its literals evaluates true. We call a formula φ *satisfied* (under an assignment) if all clauses are evaluated to true. If there does not exists a satisfying assignment, φ is unsatisfiable.

3   Word Equation Solving

In this section we focus on solving *Bounded Word Equations* and *Word Equations with Linear Constraints*. We proceed by first solving bounded word equations, and secondly, we discuss a minor change, that allows solving word equations with linear constraints.

3.1   Solving Bounded Word Equation

Recall that a bounded word equation consists of a word equation $u \doteq v$ along with a set of equations $\{|X| \leq b_X\}$ providing upper bounds for the solution of each variable X. In our approach we use these bounds to create a finite automaton which has an accepting run if and only if the bounded word equation is satisfiable.

Before the actual automata construction, we need some convenient transformations of the word equation itself. For a variable X with length bound b_X, we replace X with a sequence of new *'filled variables'* $X^{(0)} \cdots X^{(b_X-1)}$ which we restrict to only be substituted by either a single letter or the empty word. A pattern containing only filled variables, as well as letters, is called a *filled pattern*. For a pattern $w \in \Xi^*$ we denote its corresponding filled pattern by w_ξ. In the following, we refer to the alphabet of filled variables by Γ_ξ and by $\Xi_\xi = \Sigma \cup \Gamma_\xi$ the alphabet of the filled patterns. Let $S : \Xi^* \to \Sigma^*$ be a substitution for $w \in \Xi^*$. We can canonically define the induced substitution for filled patterns as $S_\xi : (\Sigma \cup \Gamma_\xi) \to \Sigma_\lambda$ with $S_\xi(a) = S(a)$ for all $a \in \Sigma$, $S_\xi(X^{(i)}) = S(X)[i]$ for all $X^{(i)} \in \Gamma_\xi$ and $i < |S(X)|$, and $S_\xi(X^{(j)}) = \lambda$ for all $X^{(j)} \in \Gamma_\xi$ and $|S(X)| \leq j < b_X$. Here, λ is a new symbol ($\lambda \notin \Xi_\xi$) to indicate an unused position at the end of a filled variable. Note that the substitution of a single filled variable always maps to exactly one character from Σ_λ, and, as soon as we discover $S_\xi(X^{(j)}) = \lambda$ for $j \in [b_X]$ it also holds that $S_\xi(X^{(i)}) = \lambda$ for all $j \leq i < b_X$. In a sense, the new element λ behaves in the same way as the neutral element of the word monoid Σ^*, being actually a place holder for this element ε. In the other direction, if we have found a satisfying filled substitution to our word equation, the two filled patterns obtained from the left hand side and the right hand side of an equation, respectively, we can transform it to a substitution for our original word equation by defining $S(X)$ as the concatenation $S_\xi(X^{(0)}) \dots S_\xi(X^{(i)})$ in which each occurrence of λ is replaced by the empty word ε, for all $X \in \Gamma$ and $i \in [b_X]$.

Our goal is now to build an automaton which calculates a suitable substitution for a given equation. During the calculation there are situations where a substitution does not form a total function. To extend a partial substitution $S : \Xi \rightharpoonup \Sigma^*$ we define for $X \in \Xi$ and $b \in \Sigma^*$ the notation $S\left[\frac{X}{b}\right] = S \cup \{X \mapsto b\}$ whenever $S(X)$ is undefined and otherwise $S\left[\frac{X}{b}\right] = S$. This definition can be naturally applied to filled substitutions. We define a congruence relation which sets variables and letters in relation whenever their substitution with respect to a partial substitution S_ξ is equal or undefined. For all $a, b \in \Xi_\xi \cup \{\lambda\}$ we define

$$a \overset{S_\xi}{\sim} b \text{ iff } S_\xi(a) = S_\xi(b) \text{ or } S_\xi(b) \notin \Sigma_\lambda^* \text{ or } S_\xi(a) \notin \Sigma_\lambda^*.$$

Definition 1. *For a word equation $u \doteq v$ for $u, v \in \Xi^*$ and a mapping $B : \Gamma \to \mathbb{N}$ defining the bounds $B(X) = b_X$, we define the* equation automaton *$A(u \doteq v, B) = (Q, \delta, I, F)$ where $Q = ([|u_\xi| + 1] \times [|v_\xi| + 1]) \times (\Xi_{\xi_\lambda} \twoheadrightarrow \Sigma_\lambda)$ is a set of states consisting of two integers which indicate the position inside the two words u_ξ and v_ξ and a partial substitution, the transition function $\delta : Q \times \Sigma_\lambda \to Q$ defined by*

$$\delta(((i,j),S),a) = \begin{cases} \left((i+1,j+1), S\left[\frac{u_\xi[i]}{a}\right]\left[\frac{v_\xi[j]}{a}\right]\right) & \text{if } u_\xi[i] \overset{S_\xi}{\sim} v_\xi[j] \overset{S_\xi}{\sim} a, \\ \left((i+1,j), S\left[\frac{u_\xi[i]}{\lambda}\right]\right) & \text{if } u_\xi[i] \overset{S_\xi}{\sim} \lambda = a, \\ \left((i,j+1), S\left[\frac{v_\xi[j]}{\lambda}\right]\right) & \text{if } v_\xi[j] \overset{S_\xi}{\sim} \lambda = a. \end{cases}$$

an initial state $I = ((0,0), \{a \mapsto a \mid a \in \Sigma_\lambda\})$ and the set of final states $F = \{((i,j), S_\xi) \mid i = |u_\xi|, j = |v_\xi|\}$.

The state space of our automaton is finite since the filled substitution S_ξ maps each input to exactly one character in Σ. The automaton is nondeterministic, as the three choices we have for a transition are not necessarily mutually exclusive.

As an addition to the above definition, we introduce the notion of *location* as a pair of integers (i, j) corresponding to two positions inside the two words u_ξ and v_ξ. A location (i, j) can also be seen as the set of states of the form $((i, j), S)$ for all possible partial substitutions S.

A run of the above nondeterministic automaton constructs a partial substitution for the given equation which is extended with each change of state. The equation has a solution if one of the accepting states $(|u_\xi|, |v_\xi|, S)$, where S is a total substitution, is reachable, because the automaton simulates a walk through our input equation left to right, with all its positions filled in a coherent way.

Example 1. Consider the equation $u \doteq v$ for $u = aZXb, v = aXaY \in \Xi^*$. We choose the bounds $b_X = b_Y = b_Z = 1$. This will give us the words $u_\xi = aZ^{(0)}X^{(0)}b$ and $v_\xi = aX^{(0)}aY^{(0)}$. Figure 1 visualizes the corresponding automaton. A run starting with the initial substitution $S_i = \{a \mapsto a \mid a \in \Sigma_\lambda\}$ reaching one of the final states gives us a solution to the equation. In this example we get the substitutions $Z \mapsto a, X \mapsto a, Y \mapsto b$ and $Z \mapsto a, X \mapsto \varepsilon, Y \mapsto b$.

Theorem 1. *Given a bounded word equation $u \doteq v$ for $u, v \in \Xi^*$, with bounds B, then the automaton $A(u \doteq v, B)$ reaches an accepting state if and only if there exists S such that $S \models u \doteq v$ and $|S(X)| \le B(X)$ for all $X \in \Gamma$.*

SAT Encoding. We now encode the solving process into propositional logic. For that we impose an ordering on the finite alphabets $\Sigma = \{a_0, \ldots, a_{n-1}\}$ and $\Gamma = \{X_0, \ldots, X_{m-1}\}$ for $n, m \in \mathbb{N}$. Using the upper bounds given for all variables $X \in \Gamma$, we create the filled variables alphabet Γ_ξ. Further, we create

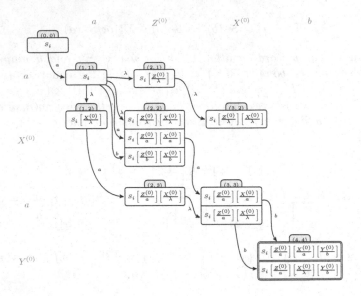

Fig. 1. Automaton for the word equation $aZXb \doteq aXaY$, with the states grouped according to their locations. Only reachable states are shown.

the Boolean variables $\mathsf{K}^a_{X^{(i)}}$, for all $X^{(i)} \in \Gamma_\xi$, $a \in \Sigma_\lambda$ and $i \in [b_X]$. Intuitively, we want to construct our formula such that an assignment β sets $\mathsf{K}^a_{X^{(i)}}$ to 1, if the solution of the word equation, which corresponds to the assignment β, is such that at position i of the variable X an a is found, meaning $S_\xi(X^{(i)}) = a$. To make sure $\mathsf{K}^a_{X^{(i)}}$ is set to 1 for exactly one $a \in \Sigma_\lambda$ we define the clause $\bigvee_{a \in \Sigma_\lambda} \mathsf{K}^a_{X^{(i)}}$ which needs to be assigned true, as well the constraints $\mathsf{K}^a_{X^{(i)}} \to \neg \mathsf{K}^b_{X^{(i)}}$, for all $a, b \in \Sigma_\lambda, X \in \Gamma, i \in [b_X]$ where $a \neq b$, which also need to be all true.

To match letters we add the variables $\mathsf{C}_{a,a} \leftrightarrow \top$ and $\mathsf{C}_{a,b} \leftrightarrow \bot$ for all $a, b \in \Sigma_\lambda$ with $a \neq b$. As such, the actual encoding of our equation can be defined as follows: for $w \in \{u_\xi, v_\xi\}$ and each position i of w and letter $a \in \Sigma_\lambda$ we introduce a variable which is true if and only if $w[i]$ will correspond to an a in the solution of the word equation. More precisely, we make a distinction between constant letters and variable positions and define: $\mathsf{word}^a_{w[i]} \leftrightarrow \mathsf{C}_{w[i],a}$ if $w[i] \in \Sigma_\lambda$ and $\mathsf{word}^a_{w[i]} \leftrightarrow \mathsf{K}^a_{w[i]}$ if $w[i] \in \Gamma_\xi$. The equality of two characters, corresponding to position i in u and, respectively, j in v, is encoded by introducing a Boolean variable $\mathsf{wm}_{i,j} \leftrightarrow \bigvee_{a \in \Sigma_\lambda} \mathsf{word}^a_{u[i]} \wedge \mathsf{word}^a_{v[j]}$ for appropriate $i \in [|u_\xi|], j \in [|v_\xi|]$.

Based on this setup, each location of the automaton is assigned a Boolean variable. As seen in Definition 1 we process both sides of the equation simultaneously, from left to right. As such, for a given equation $u \doteq v$ we create $n \cdot m = (|u_\xi| + 1) \cdot (|v_\xi| + 1)$ many Boolean variables $\mathsf{S}_{i,j}$ for $i \in [n]$ and $j \in [m]$. Each variable corresponds to a location in our automaton. The location $(0,0)$ is our initial location and $(|u_\xi|, |v_\xi|)$ our accepting location. The goal is to find a path between those two locations, or, alternatively, a satisfying assignment β, which sets the variables corresponding to these locations to 1. Every path between the location $(0,0)$ and another location corresponds to matching prefixes of u and v, under proper substitutions. We will call locations where an assignment β sets a variable $S_{i,j}$ to 1, active locations. Our transitions are now defined by a set of constraints. We fix $i \in [n]$ and $j \in [m]$ in the following. The constraints are given as follows: The first constraint (1) ensures that every active location has at least one active successor. The next three constraints (2)–(4) ensure the validity of the paths we follow: from a location we can only proceed to exactly one other location, in order to find a satisfying assignment; therefore we disallow simultaneous steps in multiple directions. In (5), (6) we forbid using an λ-transition whenever there is another possibility of moving forward. In the same manner we proceed in the case of two matching λ in (7); this part is especially important for finding substitutions which are smaller than the given bounds. The idea applies in the same way for matching letters, whose encoding is given in (8). The actual transitions which are possible from one state to another are encoded in (9) by using our Boolean variables $\mathsf{wm}_{i,j}$ which are true for matching positions in the two sides of the equation. This constraint allows us to move forward in both words if there was a match of two letters in the previous location. When the transitions are pictured as movements in the plane, this corresponds to a diagonal move. A horizontal or vertical move corresponds to a match with the empty word. The last constraint (10) ensures a valid predecessor. This is supposed to help the solver in deciding the satisfiability of the obtained formula, i.e., to guide the search in an efficient way. It can be seen as a local optimization step.

$$S_{i,j} \rightarrow S_{i+1,j} \vee S_{i,j+1} \vee S_{i+1,j+1} \tag{1}$$

$$(S_{i,j} \wedge S_{i,j+1}) \rightarrow (\neg S_{i+1,j+1} \wedge \neg S_{i+1,j}) \tag{2}$$

$$(S_{i,j} \wedge S_{i+1,j}) \rightarrow (\neg S_{i+1,j+1} \wedge \neg S_{i,j+1}) \tag{3}$$

$$(S_{i,j} \wedge S_{i+1,j+1}) \rightarrow (\neg S_{i,j+1} \wedge \neg S_{i+1,j}) \tag{4}$$

$$S_{i,j} \wedge \neg \mathsf{word}^{\lambda}_{u[i]} \rightarrow \neg S_{i+1,j} \text{ and } S_{i,j} \wedge \mathsf{word}^{\lambda}_{u[i]} \wedge \neg \mathsf{word}^{\lambda}_{v[j]} \rightarrow S_{i+1,j} \tag{5}$$

$$S_{i,j} \wedge \neg \mathsf{word}^{\lambda}_{v[j]} \rightarrow \neg S_{i,j+1} \text{ and } S_{i,j} \wedge \neg \mathsf{word}^{\lambda}_{u[i]} \wedge \mathsf{word}^{\lambda}_{v[j]} \rightarrow S_{i,j+1} \tag{6}$$

$$S_{i,j} \wedge \mathsf{word}^{\lambda}_{u[i]} \wedge \mathsf{word}^{\lambda}_{v[j]} \rightarrow S_{i+1,j+1} \tag{7}$$

$$S_{i,j} \wedge S_{i+1,j+1} \rightarrow \mathsf{wm}_{i,j} \tag{8}$$

$$S_{i,j} \leftrightarrow (S_{i-1,j-1} \wedge \mathsf{wm}_{i-1,j-1}) \vee (S_{i,j-1} \wedge \neg \mathsf{wm}_{i,j-1}) \vee (S_{i-1,j} \wedge \neg \mathsf{wm}_{i-1,j}) \tag{9}$$

$$S_{i+1,j+1} \rightarrow S_{i,j} \vee S_{i+1,j} \vee S_{i,j+1} \tag{10}$$

The final formula is the conjunction of all constraints defined above. This formula is true iff location (n,m) is reachable from location $(0,0)$, and this is true iff the given word equation is satisfiable w.r.t. the given length bounds.

Lemma 1. *Let $u \doteq v$ be a word equation, B be the function giving the bounds for the word equation variable, and φ the corresponding formula consisting of the conjunction (1)–(10) and the earlier defined constraints in this section, then $\varphi \wedge S_{0,0} \wedge S_{|u_\xi|,|v_\xi|}$ has a satisfying assignment if and only if $A(u \doteq v, B)$ reaches an accepting state.*

Example 2. Consider the word equation $u \doteq v$ where $u = XaXbYbZ$ and $v = aXYYbZZbaa \in \varXi^*$ where $\varSigma = \{\, a \,\}$ and $\varGamma = \{\, X, Y, Z \,\}$. Using the approach discussed above, we find the solution $S(X) = aaaaaaaa$, $S(Y) = aaaa$ and $S(Z) = aa$ using the bounds $b_X = 8$ and $b_Y = b_Z = 6$. We set up an automaton with $32 \cdot 38 = 1216$ states to solve the equation. In Fig. 2 we show the computation of the SAT-Solver. Light grey markers indicate states considered in a run of the automaton. In this case only 261 states are needed. The dark grey markers visualize the actual path in the automaton leading to the substitution. Non-diagonal stretches are λ transitions.

3.2   Refining Bounds and Guiding the Search

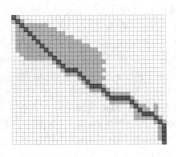

Fig. 2.   Solver computation on $XaXbYbZ$ $\doteq aXYYbZZbaa$

Initial experiments revealed a major inefficiency of our approach: most of the locations were not used during the search and only increased the encoding time. The many white markers in Fig. 2 indicating unused locations visualizes the problem. Since we create all required variables $x \in \varGamma$ and constraints for every position $i < b_X$, we can reduce the automaton size by lowering these upper bounds. Abstracting a word equation by the length of the variables gives us a way to refine the bounds b_X for some of the variables $X \in \varGamma$. By only considering length we obtain a Diophantine equation in the following manner. We assume an ordering on the variable alphabet $\varGamma = \{\, X_0, \ldots, X_{n-1} \,\}$. We associate to each word equation variable X_j an integer variable I_j.

Definition 2. *For a word equation $u \doteq v$ with $\varGamma = \{\, X_0, \ldots, X_{n-1} \,\}$ we define its length abstraction by $\sum_{j \in [n]} \left(|u|_{X_j} - |v|_{X_j} \right) \cdot I_j = \sum_{a \in \varSigma} |v|_a - |u|_a$ for $j \in [n]$.*

If a word equation has a solution S, then so does its length abstraction with variable $I_j = |S(X_j)|$. Our interest is computing upper bounds for each variable $X_k \in \varGamma$ relative to the upper bounds of the bounded word equation problem. To this end consider the following natural deductions

$$\sum_{j \in [n]} \left(|u|_{X_j} - |v|_{X_j} \right) \cdot I_j = \sum_{a \in \varSigma} \left(|v|_a - |u|_a \right)$$

$$\Longleftrightarrow \quad I_k = \frac{\sum_{a \in \varSigma} \left(|v|_a - |u|_a \right)}{|u|_{X_k} - |v|_{X_k}} - \frac{\sum_{j \in [n] \setminus k} \left(|u|_{X_j} - |v|_{X_j} \right) \cdot I_j}{|u|_{X_k} - |v|_{X_k}}$$

$$\Longrightarrow \quad I_k \leq \frac{\sum_{a \in \varSigma} \left(|v|_a - |u|_a \right)}{\left(|u|_{X_k} - |v|_{X_k} \right)} - \frac{\sum_{j \in \kappa} \left(|u|_{X_j} - |v|_{X_j} \right) \cdot b_{X_j}}{\left(|u|_{X_k} - |v|_{X_k} \right)} = b_{X_k}^{\mathsf{S}},$$

where $\kappa = \{\, m \in [n] \setminus k \mid (\, |u|_{X_k} - |v|_{X_k}) \cdot (|u|_{X_m} - |v|_{X_m}) < 0 \,\}$. Whenever $0 < b_{X_k}^{\mathsf{S}} < b_{X_k}$ holds, we use $b_{X_k}^{\mathsf{S}}$ instead of b_{x_k} to prune the search space.

The length abstraction is also useful because it might give information about the unsatisfiability of an equation: if there is no solution to the Diophantine equation, there is no solution to the word equation. We use this acquired knowledge and directly report this fact. Unfortunately whenever $|u|_X - |v|_X = 0$ holds for a variable X we cannot refine the bounds, as they are not influenced by the above Diophantine equation.

Guiding the Search. The length abstraction used to refine upper bounds can also be used to guide the search in the automaton. In particular it can restrict allowed length of one variable based on the length of others. We refer to the coefficient of variable I_j in Definition 2 by $\mathsf{Co}_{u,v}(I_j) = \left(|u|_{X_j} - |v|_{X_j} \right)$.

To benefit from the abstraction of the word equation inside our propositional logic encoding we use Reduced Ordered Multi-Decision Diagrams (MDD) [2]. An MDD is a directed acyclic graph, with two nodes having no outgoing edges (called **true** and **false** terminal nodes). A Node in the MDD is associated to exactly one variable I_j, and has an outgoing edge for each element of I_j's domain. In the MDD, a node labelled I_j is only connected to nodes labelled I_{j+1}. A row $(\mathsf{r}(I_j))$ in an MDD is a subset of nodes corresponding to a certain variable I_j.

We create the MDD following Definition 2. The following definition creates the rows of the MDD recursively. An MDD node is a tuple consisting of a variable I_j and an integer corresponding the partial sum which can be obtained using the coefficients and position information of all previous variables I_k for $k < j$. We introduce a new variable I_{-1} labelling the initial node of the MDD. The computation is done as follows:

$$\mathsf{r}(I_i) = \{ (I_i, s + k \cdot \mathsf{Co}_{u,v}(X_i)) \mid s \in \{ s' \mid (I_{i-1}, s') \in \mathsf{r}(I_{i-1}) \}, k \in [b_{X_i}] \} \quad (11)$$

and $\mathsf{r}(I_{-1}) = \{ (I_{-1}, 0) \}$. Since I_j is associated to the word equations variable X_j, we let $\mathsf{r}(X_j) = \mathsf{r}(I_j)$. We denote the whole set of nodes in the MDD by $M^C = \bigcup_{X \in \Gamma \cup \{ I_{-1} \}} \mathsf{r}(X)$. The **true** node of the MDD is $(I_{n-1}, \mathsf{s}_\#)$, where $\mathsf{s}_\# = \sum_{a \in \Sigma} |v|_a - |u|_a$. If the initial creation of nodes did not add this node, the given equation (Definition 2) is not satisfiable hence the word equation has no solution given the set bounds. Furthermore there is no need to encode the full MDD, when only a subset of its nodes can reach $(I_{n-1}, \mathsf{s}_\#)$. For reducing the MDD nodes to this subset, we calculate all predecessors of a given node $(I_i, s) \in M^C$ as follows

$$\mathsf{pred}((I_i, s)) = \{ (I_{i-1}, s') \mid s' = s - k \cdot \mathsf{Co}_{u,v}(X_{i-1}), k \in [b_{X_{i-1}}] \}.$$

The minimized set $M = F(T)$ of reachable nodes starting at the only accepting node $T = \{ (I_n, \mathsf{s}_\#) \}$ is afterwards defined through a fixed point by

$$T \subseteq F(T) \land \left(\forall\, p \in F(T) : q \in \mathsf{pred}(p) \land q \in M^C \Rightarrow q \in F(T) \right) \quad (12)$$

We continue by encoding this into a Boolean formula. For that we need information on the actual length of a possible substitution. We reuse the Boolean variables of our filled variables $X \in \Gamma_\xi$. The idea is to introduce $b_X + 1$ many Boolean variables $(\mathsf{OH}_i(0)\ldots\mathsf{OH}_i(b_X+1))$ for each $X_i \in \Gamma$, where $\mathsf{OH}_i(j)$ is true if and only if X_i has length j in the actual substitution. This kind of encoding is known as a *one-hot encoding*. To achieve this we add a constraint forcing substitutions to have all λ in the end. We force our solver to adapt to this by adding clauses $\mathsf{K}^\lambda_{X^{(j)}} \to \mathsf{K}^\lambda_{X^{(j+1)}}$ for all $j \in [b_{X_i} - 1]$ and $X_i^{(j)} \in \Gamma_\xi$. The actual encoding is done by adding the constraints $\mathsf{OH}_i(0) \leftrightarrow \mathsf{K}^\lambda X_i^{(0)}$ and $\mathsf{OH}_i(b_{X_i}) \leftrightarrow \neg\mathsf{K}^\lambda_{X_i^{(b_{X_i}-1)}}$, which fixes the edge cases for the substitution by the empty word and no λ inside it. For all $j \in [b_{X_i}]_0$, we add the constraints $\mathsf{OH}_i(j) \leftrightarrow \mathsf{K}^\lambda_{X_i^{(j)}} \wedge \neg\mathsf{K}^\lambda_{X_i^{(j-1)}}$, which marks the first occurrence of λ. The encoding of the MDD is done nodewise by associating a Boolean variable $\mathsf{M}_{i,j}$ for each $i \in [|\Gamma|]$, where $(I_i, j) \in M$. Our goal is now to find a path inside the MDD from node $(I_{-1}, 0)$ to $(I_{n-1}, \mathsf{s}_\#)$. Therefore we enforce a true assignment for the corresponding variables $\mathsf{M}_{-1,0}$ and $\mathsf{M}_{n-1,\mathsf{s}_\#}$. A valid path is encoded by the constraint $\mathsf{M}_{i-1,j} \wedge \mathsf{OH}_i(k) \to \mathsf{M}_{i,s}$ for each variable $X_i \in \Gamma$, $k \in [b_{X_i}]_0$, where $s = j + k \cdot \mathsf{Co}_{u,v}(X_i)$ and $(I_i, s) \in M$. This encodes the fact that whenever we are at a node $(I_{i-1}, s) \in M$ and the substitution for a variable X_i has length k $(|S(X_i)| = k)$, we move on to the next node, which corresponds to X_i and an integer obtained by taking the coefficient of the variable X_i, multiplying it by the substitution length, and adding it to the previous partial sum s. Whenever there is only one successor to a node (I_i, j) within our MDD, we directly force its corresponding one hot encoding to be true by adding $\mathsf{M}_{i-1,j} \to \mathsf{OH}_i(j)$. This reduces the amount of guesses on variables.

Example 3. Consider the equation $u \doteq v$ for $u = aX_2X_0b$, $v = aX_0aX_1 \in \Xi^*$, where $\Sigma = \{a,b\}$ and $\Gamma = \{X_0, X_1, X_2\}$. The corresponding linear equation therefore has the form $0 \cdot I_0 + (-1) \cdot I_1 + 1 \cdot I_2 = 0$ which gives us the coefficients $\mathsf{Co}_{u,v}(X_0) = 0$, $\mathsf{Co}_{u,v}(X_1) = -1$ and $\mathsf{Co}_{u,v}(X_2) = 1$. For given

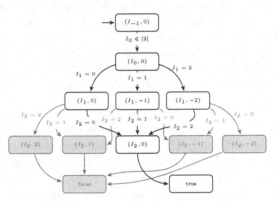

Fig. 3. The MDD for $aX_2X_0b \doteq aX_0aX_1$

bounds $b_{X_0} = b_{X_1} = b_{X_2} = 2$ the induced MDD has the form shown in Fig. 3. In this example $\mathsf{s}_\# = 0$, and therefore $(I_2, 0)$ is the only node connected to the **true** node. The minimization of the MDD by using the fixed point decribed in (12) removes all grey nodes, since they are not reachable starting at the **true** node. The solver returns the substitution $S(X_0) = \varepsilon$, $S(X_0) = b$ and $S(X_0) = a$. It took the centred path consisting of the nodes $(I_{-1}, 0), (I_0, 0), (I_1, -1), (I_2, 0),$ **true** inside the MDD.

Adding Linear Length Constraints. Until now we have only concerned ourselves with bounded word equations. As mentioned in the introduction however, bounded equations with linear constraints are of interest as well. In particular, without

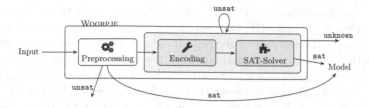

Fig. 4. Architecture of WOORPJE

loss of generality we restrict to linear constraints of the form [2] $c_0I_0 + \cdots + c_{n-1}I_{n-1} \leq c$ where $c, c_i \in \mathbb{Z}$ are integer coefficients and I_i are integer variables with a domain $D_i = \{ m \in \mathbb{N} \mid 0 \leq m \leq d_i \}$ and a corresponding $d_i \in \mathbb{N}$. Each I_i corresponds to the length of a substitution to a variable of the given word equation.

Notice that the structure of the linear length constraint is similar to that of Definition 2. For handling linear constraints we can adapt the generation of MDD nodes to keep track the partial sum of the linear constraint, and define the accepting node of the MDD to one where all rows have been visited and the inequality is true. We simply extend the set T which was used in the fix point iteration in (12) to the set $T = \{ (I_n, s) \mid (I_n, s) \in M^C \wedge s \leq \mathsf{s}_\# \}$.

4 Experiments

The approach described in the previous sections has been implemented in the tool WOORPJE. The inner workings of WOORPJE is visualised in Fig. 4. WOORPJE first has a preprocessing step where obviously satisfiable/unsatisfiable word equations are immediately reported.

After the preprocessing step, WOORPJE iteratively encodes the word equation into a propositional logic formula and solves it with Glucose [3] for increasing maximal variable lengths (i^2, where i is the current iteration). If a solution is found, it is reported. The maximal value of i is user definable, and by default set to 2^n where n is the length of the given equation. If WOORPJE reaches the given bound without a verdict, it returns unknown.

We have run WOORPJE and state of the art word equation solvers (CVC4 1.6, Norn 1.0.1, Sloth 1.0, Z3 4.8.4) on several word equation benchmarks with linear length constraints. The benchmarks range from theoretically-interesting cases to variations of the real-world application set Kaluza [19]. All tests were performed on Ubuntu Linux 18.04 with an Intel Xeon E5-2698 v4 @ 2.20 GHz CPU and 512 GB of memory with a timeout of 30 s.

We used five different kind of benchmarks. The first track (I) was produced by generating random strings, and replacing factors with variables at random, in a coherent fashion. This guarantees the existence of a solution. The generated word equations have at most 15 variables, 10 letters, and length 300. The second track (II) is based on the idea in Proposition 1 of [10], where the equation $X_naX_nbX_{n-1}bX_{n-2}\cdots bX_1 \doteq aX_nX_{n-1}X_{n-1}bX_{n-2}X_{n-2}b\cdots bX_1X_1baa$ is shown to have a minimal solution of exponential length w.r.t. the length of the equation. The third track (III) is based on the second track, but each letter b is replaced by the left hand side or the right hand side of some randomly generated word equation (e.g., with the methods from track (I)). In the fourth track (IV) each benchmark consists of a system of 100 small random word equations with at most 6 letters, 10 variables and length 60. The hard aspect of this track is solving multiple equations at the same time. Within the fifth track (V) each benchmark enriches a system of 30 word equations by suitable linear constraints, as presented in this paper. This track is inspired by the Kaluza benchmark set in terms of having many small equations enriched by linear length constraints. All tracks, except track II which holds 9 instances, consist of 200 benchmarks. The full benchmark set is available at https://www.informatik.uni-kiel.de/~mku/woorpje. Table 1 is read as follows: ✓ is the count of instances classified as correctly, where ● marks the incorrect classified cases. For instances marked with ● the solver returned no answer but terminated before the timeout of 30 s was reached, where in ⑩ marked cases the solver was killed after 30 s. The row marked by ⊘ states the overall solving time. The produced substitutions were checked regarding their correctness afterwards. The classification of ⊗ was done by ad-hoc case inspection whenever not all solvers agreed on a result. In the cases one solver produced a valid solution, and others did not, we validated the substitutions manually. For the cases where one solver determined an equation is unsatisfied and all others timed out, we treated the unsat result as correct. This means that we only report errors if a solver reports unsat and we know the equation was satisfiable. During our evaluation of track I CVC4 crashed with a null-pointer exception regarding the word equation $dbebgddbecfcbbAadeeaecAgebegeecafegebdbagddaadbddcaeeebfabfef\text{-}$ $abfacdgAgaabgegagf \doteq dbebgddbeAfcbbAaIegeeAaDegagf$, where lowercase symbols are letters and uppercase symbols are variables. Worth mentioning is the reporting of 14 satisfiable benchmarks by the tool Sloth without being able to produce a valid model, while at least two other tools classified them as unsatisfiable. We treated this as an erroneous behaviour.

Table 1. Benchmark results (✓: correct classified, ●: reported unknown, ⑩: timed out after 30 s, ⊗: incorrectly classified, ⊘: total Time in seconds)

	TRACK I					TRACK II					TRACK III					TRACK IV					TRACK V				
	✓	●	⑩	⊗	⊘	✓	●	⑩	⊗	⊘	✓	●	⑩	⊗	⊘	✓	●	⑩	⊗	⊘	✓	●	⑩	⊗	⊘
WOORPJE	200	0	0	0	8.10	5	0	4	0	123.85	189	0	11	0	341.74	196	2	2	0	136.20	178	9	13	0	399.22
CVC4	182	0	17	0	543.32	1	0	8	0	240.03	165	0	35	0	1055.24	172	0	28	0	925.13	179	0	21	0	635.97
Z3STR3	197	1	2	0	105.07	0	9	0	0	0.33	93	43	58	6	2089.92	175	10	12	3	490.23	198	1	1	0	41.52
Z3SEQ	183	0	17	0	545.24	9	0	0	0	1.81	126	0	73	1	2199.99	193	0	6	1	200.09	193	0	7	0	217.35
NORN	176	0	20	4	1037.63	0	0	9	0	270.00	71	0	128	1	4038.83	60	0	72	68	3216.95	112	0	9	79	742.34
SLOTH	101	0	99	0	3658.34	7	0	2	0	124.56	121	0	67	12	2808.48	16	0	184	0	5615.81	9	2	187	2	5750.79

The result shows that WOORPJE produces reliable results (0 errors) in competitive time. It outperforms the competitors in track I, III and IV and sticks relatively tight to the leaders Z3str3, Z3Seq and CVC4 on track V. On track II WOORPJE trails CVC4 and Z3Str3. The major inefficiency of WOORPJE is related to multiple equations with large alphabets and linear length constraints.

It is worth emphasising, that the benchmarks developed here seem of intrinsic interest, as they challenge even established solvers.

5 Conclusion

In this paper we present a method for solving word equations by using a SAT-Solver. The method is implemented in our new tool WOORPJE and experiments show it is competitive with state-of-the-art string solvers. WOORPJE solves word equations instances that other solvers fail to solve. This indicates that our technique can complement existing techniques in a portfolio approach.

In the future, we aim to extend our approach to include regular constraints. As our approach relies on automata theory, it is expected that this could be achievable. Another step is the enrichment of our linear constraint solving. We currently do a basic analysis by using the MDDS. There are a few refinement steps described in [2] which seem applicable. A next major step is to develop a more efficient encoding of the alphabet of constants. Currently the state space explodes due to the massive branching caused by the usage of large alphabets.

References

1. Abdulla, P.A., et al.: Norn: an SMT solver for string constraints. In: Kroening, D., Pǎsǎreanu, C.S. (eds.) CAV 2015. LNCS, vol. 9206, pp. 462–469. Springer, Cham (2015). https://doi.org/10.1007/978-3-319-21690-4_29
2. Abío, I., Stuckey, P.J.: Encoding linear constraints into SAT. In: O'Sullivan, B. (ed.) CP 2014. LNCS, vol. 8656, pp. 75–91. Springer, Cham (2014). https://doi.org/10.1007/978-3-319-10428-7_9
3. Audemard, G., Simon, L.: On the glucose SAT solver. Int. J. Artif. Intell. Tools 27(01), 1840001 (2018)
4. Barrett, C., et al.: CVC4. In: Gopalakrishnan, G., Qadeer, S. (eds.) CAV 2011. LNCS, vol. 6806, pp. 171–177. Springer, Heidelberg (2011). https://doi.org/10.1007/978-3-642-22110-1_14
5. Berzish, M., Ganesh, V., Zheng, Y.: Z3str3: a string solver with theory-aware heuristics. In: 2017 Formal Methods in Computer Aided Design (FMCAD), pp. 55–59, October 2017
6. Bjørner, N., Tillmann, N., Voronkov, A.: Path feasibility analysis for string-manipulating programs. In: Kowalewski, S., Philippou, A. (eds.) TACAS 2009. LNCS, vol. 5505, pp. 307–321. Springer, Heidelberg (2009). https://doi.org/10.1007/978-3-642-00768-2_27
7. Cadar, C., Dunbar, D., Engler, D.R.: KLEE: unassisted and automatic generation of high-coverage tests for complex systems programs. In: Draves, R., van Renesse, R. (eds.) 8th USENIX Symposium on Operating Systems Design and Implementation, OSDI 2008, 8–10 December 2008, San Diego, California, USA, Proceedings, pp. 209–224. USENIX Association (2008). http://www.usenix.org/events/osdi08/tech/full_papers/cadar/cadar.pdf

8. Chen, T., Hague, M., Lin, A.W., Rümmer, P., Wu, Z.: Decision procedures for path feasibility of string-manipulating programs with complex operations. Proc. ACM Program. Lang. **3**(POPL), 49 (2019)
9. Cordeiro, L., Kesseli, P., Kroening, D., Schrammel, P., Trtik, M.: JBMC: a bounded model checking tool for verifying Java bytecode. In: Chockler, H., Weissenbacher, G. (eds.) CAV 2018. LNCS, vol. 10981, pp. 183–190. Springer, Cham (2018). https://doi.org/10.1007/978-3-319-96145-3_10
10. Day, J.D., Manea, F., Nowotka, D.: The hardness of solving simple word equations. In: Proceedings of MFCS 2017. LIPIcs, vol. 83, pp. 18:1–18:14 (2017)
11. Holík, L., Jank P., Lin, A.W., Rümmer, P., Vojnar, T.: String constraints with concatenation and transducers solved efficiently. Proc. ACM Program. Lang. **2**(POPL), 4 (2017)
12. Jeż, A.: Recompression: a simple and powerful technique for word equations. In: 30th International Symposium on Theoretical Aspects of Computer Science, STACS 2013, 27 February- 2 March 2013, Kiel, Germany, pp. 233–244 (2013). https://doi.org/10.4230/LIPIcs.STACS.2013.233
13. Jeż, A.: Word equations in nondeterministic linear space. In: Proceedings of ICALP 2017. LIPIcs, vol. 80, pp. 95:1–95:13. Schloss Dagstuhl - Leibniz-Zentrum fuer Informatik (2017)
14. Karhumäki, J., Mignosi, F., Plandowski, W.: The expressibility of languages and relations by word equations. J. ACM (JACM) **47**(3), 483–505 (2000)
15. Kiezun, A., Ganesh, V., Guo, P.J., Hooimeijer, P., Ernst, M.D.: Hampi: a solver for string constraints. In: Proceedings of the Eighteenth International Symposium on Software Testing and Analysis, pp. 105–116. ACM (2009)
16. Makanin, G.S.: The problem of solvability of equations in a free semigroup. Sbornik: Math. **32**(2), 129–198 (1977)
17. Plandowski, W., Rytter, W.: Application of Lempel-Ziv encodings to the solution of word equations. In: Larsen, K.G., Skyum, S., Winskel, G. (eds.) ICALP 1998. LNCS, vol. 1443, pp. 731–742. Springer, Heidelberg (1998). https://doi.org/10.1007/BFb0055097
18. Plandowski, W.: Satisfiability of word equations with constants is in PSPACE. In: 40th Annual Symposium on Foundations of Computer Science, pp. 495–500. IEEE (1999)
19. Saxena, P., Akhawe, D., Hanna, S., Mao, F., McCamant, S., Song, D.: A symbolic execution framework for Javascript. In: 2010 IEEE Symposium on Security and Privacy, pp. 513–528. IEEE (2010)
20. Trinh, M.-T., Chu, D.-H., Jaffar, J.: Progressive reasoning over recursively-defined strings. In: Chaudhuri, S., Farzan, A. (eds.) CAV 2016. LNCS, vol. 9779, pp. 218–240. Springer, Cham (2016). https://doi.org/10.1007/978-3-319-41528-4_12

Parameterised Verification
of Publish/Subscribe Networks
with Exception Handling

Giorgio Delzanno[(✉)]

DIBRIS, University of Genova, Genoa, Italy
`giorgio.delzanno@unige.it`

Abstract. We present a formal model of publish/subscribe network architectures in which a central communication broker is in charge of distributing messages to clients subscribed to certain topics. We consider different semantics for the notification phase in order to take into consideration exceptions due to node crashes. For the considered model, we study decidability of verification problems formulated in terms of coverability, a non trivial class of reachability problems well-suited to validate properties of parameterised systems.

1 Introduction

Publish/subscribe protocols such as MQTT are widely used for interconnecting heterogeneous collections of network services and devices, e.g., in Internet of Things applications. In this paper we present a new formal model of publish/-subscribe protocols such as MQTT in which a broker is in charge of distributing messages to clients subscribed to certain topics. In this setting we use a transition system parametric on the specification of individual nodes to provide an operational semantics to basic operations such as (un)subscription and push notifications. Node crashes and connection failures are modelled via state information included in the representation of individual nodes. We then provide a formal specification of different implementations of the broker internal structure. The semantics is inspired to a working prototype of pub/sub broker that we implemented in Java using RMI (Remote Method Invocation) communication. More in detail, we consider two scenarios in which to model the delivery of a published message m to subscribers of a given topic t.

In the first scenario the broker acknowledges the request and, inside a synchronisation block, forwards the message the other clients subscribed to topic t. Communication failures are captured locally via a try-catch statement as in Fig. 1. In the considered example the broker stores client communication data in a shared Map protected by a synchronization region (the Map *topicRelation*). In RMI the communication data are encapsulated in stub objects that act as interfaces when invoking remote methods on each client (in our example the method

© Springer Nature Switzerland AG 2019
E. Filiot et al. (Eds.): RP 2019, LNCS 11674, pp. 107–120, 2019.
https://doi.org/10.1007/978-3-030-30806-3_9

```
boolean publish(String topic, String news, String sender)
throws Exception {
  ClientInterface client;
  synchronized(topicRelation) {
    Map<Integer, ClientInterface>
        subscriberList = topicRelation.get(topic);
    synchronized(subscriberList) {
      Iterator<Map.Entry<Integer, ClientInterface>> entries =
          subscriberList.entrySet().iterator();
        while (entries.hasNext()) {
          Map.Entry<Integer, ClientInterface>
              entry = entries.next();
          client = entry.getValue();
          try {    stub.send(topic,sender,news); }
          catch (RemoteException e) {
                  System.out.println("Notification error);
          }
        }
    }
  }
  return true;
}
```

Fig. 1. Broker in Java: first scenario.

send). Remote methods throw RemoteExceptions. In this example, since exceptions are handled locally, notification always reach all connected nodes leaving the state of all other nodes unchanged.

In the second scenario we consider an implementation in which an exception generated during a push notification sent to a certain client is propagated to the caller of the *publish* method. Going back to our Java example, this scenario corresponds to the implementation of the *publish* method with synchronized regions, an iterator over a Map, and a remote callback on a client stub (method *send*) in Fig. 2. In this scenario we assume that every invocation to the *publish* method is embedded into a try-catch statement, e.g., to propagate error notifications to the server or to modify the current list of active clients.

In the paper we give a formal account of the above mentioned scenarios by introducing a transition system modelling configurations with an arbitrary number of publishers and subscribers and a single broker. The behaviour of the broker is hard-wired in the semantics of the transition system. For the considered model, we focus our attention on decidability properties of verification problems formulated in terms of coverability, a non trivial class of reachability problems well-suited to validate properties of parameterised systems.

The reason why we consider parameterised formulations of verification problems is strictly related to the nature of distributed algorithms and protocols. Indeed, protocols designed to operate in distributed systems are often defined

```
boolean publish(String topic, String news, String sender)
    throws Exception {
 synchronized(topicRelation) {
     Map<Integer, InfoSub> subscriberList =
        topicRelation.get(topic);
    synchronized(subscriberList) {
        Iterator<Map.Entry<Integer, InfoSub>>
            entries = subscriberList.entrySet().iterator();
        while (entries.hasNext()) {
            Map.Entry<Integer, InfoSub>
                entry = entries.next();
            entry.getValue().send(topic,sender,news);
        }
    }
 }
 return true;
}
```

Fig. 2. Broker in Java: second scenario.

for an arbitrary number of components. Formal specification languages like Petri nets and automata are often used to model skeletons of this kind of systems. In this setting the coverability decision problem [1] is typically used to formulate reachability of bad configurations independently from the number of components of a system. Furthermore, to express safety properties of distributed systems we can lift the coverability decision problem, in which the initial configuration is fixed a priory, to a formulation in which the initial configuration is picked up from an infinite set of initial configurations [3,4]. This formulation of the coverability problem has been considered in [5,7–12] in order to reason on Broadcast Protocols. Falsification of this decision problem provides a characterisation of initial configurations from which it is possible to reach a bad configuration.

Plan of the Paper. In Sect. 2 we introduce our formal model of Pub/Sub Networks, inspired to extensions of Petri nets with data with a first formulation of the notification phase. In Sect. 4 we study decidability properties for coverability in parameterised formulations of Pub/Sub Networks. In Sect. 5 we consider an extensions with retained messages inspired to the MQTT protocol. In Sect. 6 we introduce a variant of the notification phase in which we model exception handling using a global conditions on the operating status of client nodes and reconsider decidability properties for coverability in parameterised formulations of the proposed variant of Pub/Sub Networks. In Sect. 7 we address some conclusions, consider other extensions and proposed some open problems.

2   Formal Model of Pub/Sub Architectures

In this section we introduce a formal model that will help us to analyse the interactions between publishers and subscribers via a server with synchronized operations on internal data structures. In the rest of the paper we will use the following terminology. Multisets over elements $e_1, e_2, \ldots$ in a support set D will be indicated as $\{\!\!\{e_1, e_2, \ldots\}\!\!\}$, multiset union as $\oplus$, multiset difference as $\ominus$, $\sqsubseteq$, $\sqsubset$ as multiset inclusion, and $\in$ as membership. Furthermore, we will use the standard notation such as $\cup, \cap, \setminus, \subseteq, \subset$ and $\in$ for operations over sets. We will use 2^A to indicate the powerset of A. Finally, we will use $\langle e_1, \ldots, e_n \rangle$ to denote tuples of elements in D.

Topics, States, Messages and Actions

We define T to be a finite set of, fixed a priori, labels representing topics names. We define Q to be a finite set of labels of client states. Furthermore, we define M to be a finite set of message labels. Finally, we consider a finite set of action labels having the following form:

- *local*, that denotes a local transition,
- *subscribe(s)* for $s \subseteq T$ that denotes subscription to a subset of topics,
- *unsubscribe(s)* for $s \subseteq T$ that denotes unsubscription from a subset of topics,
- *publish(m, t)* with $m \in M$ and $t \in T$, that denotes publishing of message m on topic t.

The above listed type of actions are strictly related to the communication model typical of publish/subscribe architecture in which every message is delivered to all subscribers via a shared broker.

Client Specification

In the rest of the section we will first introduce the static specification of individual clients. The dynamic semantics of a client will be described only after having introduced the notion network configuration. In this setting we will consider systems composed by a single server and an arbitrary number of client instances, each one defined by the same client specification. A client configuration c is a tuple $\langle q, s, b, f \rangle$, where $q \in Q$ is the current client state, $s \in 2^T$ is the set of topics for which the client is a subscriber, $b \in 2^M$ is the set of messages received so far, and $f \in \{\top, \bot\}$ is a flag that defines the connection status of the client with respect to the global network, namely $\top$ corresponds to the normal operating status, whereas $\bot$ corresponds to a disconnection event.

A client specification P is a tuple $\langle Q, q_0, R \rangle$, where Q is a finite set of states, $q_0 \in Q$ is the initial state, and $R \subseteq Q \times A \times 2^M \times Q$ defines state transitions induced by action labels. In other words a client specification can be viewed as a finite state automata with labelled transitions which statically defines its behavior. The tuple $\langle q, a, s, q' \rangle$ denotes a transition from q to q' associated to action a whose firing requires the presence of at least the set of messages s in the local message list. For instance, $\langle q_1, local, \{m, n\}, q_2 \rangle$ can be fired in $\langle q_1, s, b, f \rangle$ only if $\{m, n\} \subseteq b$.

We assume here that disconnected clients cannot roll back to a normal status, i.e., when they restart they will be assigned a new identity, their internal state being completely reset. In other words we will simulate restart by using client creation. The model can naturally be extended in order to consider a richer set of operations for the manipulation of local message sets (e.g. remove messages, reset buffers, etc.). For the sake of brevity, we will keep the model simple and discuss how these extension affect the decidability of coverability when necessary.

2.1 Pub/Sub Networks

We are ready now to define a model for Pub/Sub Networks. In this paper we will consider a single Pub/Sub broker. The client-server architecture of such a server will be implicitly defined via the semantics of *publish* operations.

A Pub/Sub Network S consists of fixed sets A, Q, M, and a client specification $P = \langle Q, q_0, R \rangle$. In this setting a network configuration is defined as a multiset $\gamma = \{c_1, \ldots, c_k\}$ of client configurations, i.e., $c_i = \langle q_i, s_i, b_i, f_i \rangle$ with $q_i \in Q$, $s_i \in 2^T$, $b \in 2^M$ and $f_i \in \{\bot, \top\}$ for $i : 1, \ldots, k$. We use N to denote the set of Network Configurations of finite but arbitrary size. The set N_0 of Initial Network Configurations is the subset of N is which client configurations are restricted to those with form $c_0 = \langle q_0, \emptyset, \emptyset, \top \rangle$.

Operational Semantics

The operational semantics of a Pub/Sub Network η is defined via a transitions system defined through a binary relation $\rightarrow$ over Network Configurations. To specify the semantics of this operation we first introduce some auxiliary definitions. Let $t \in T$ and γ be a configuration, $E_f(t, \gamma)$ is the multiset containing all client configurations of the form $\langle q, s, b, f \rangle$ occurring in γ such that $t \in s$. Furthermore, we use $Add_m(t, \gamma)$ to denote the multiset obtained from γ by adding message m in all local message sets of configurations of subscribers of topic t with flag f. More precisely, $Add_m(t, \{\}) = \{\}$, $Add_m(t, \{\langle q, s, b, \top \rangle\} \oplus \gamma) = \{\langle q, s, \{m\} \cup b, \top \rangle\} \oplus Add_m(t, \gamma)$, if $t \in s$; $Add_m(t, \{c\} \oplus \gamma) = \{c\} \oplus Add_m(t, \gamma)$ otherwise. We also use $Add_m(t, \gamma)$ to denote the multiset obtained from γ by adding message m in all local message sets of configurations of subscribers of topic t with flag f. The relation $\rightarrow \subseteq N \times N$ is the least relation satisfying one of the conditions listed below.

Local Operations. We first consider actions with local effect.

$$Local \quad \{\langle q, s, b, \top \rangle\} \oplus C \;\rightarrow\; \{\langle q', s, b, \top \rangle\} \oplus C$$

under the assumption $\langle q, local, q' \rangle \in R$. With this rule a client instance updates its local state after firing a local action.

$$Subscription \quad \{\langle q, s, b, \top \rangle\} \oplus C \;\rightarrow\; \{\langle q', s \cup s_1, b, \top \rangle\} \oplus C$$

under the assumption $\langle q, subscribe(s_1), q' \rangle \in R$. With this rule a client instance subscribes to the set of topics s_1.

$$Unsubscription \quad \{\langle q, s, b, \top \rangle\} \oplus C \;\rightarrow\; \{\langle q', s \setminus s_1, b, \top \rangle\} \oplus C$$

under the assumption $\langle q, unsubscribe(s_1), q' \rangle \in R$. With this rule a client unsubscribes from the set of topics s_1.

$$Disconnection \quad \{\!\{\langle q, s, b, \top \rangle\}\!\} \oplus C \;\to\; \{\!\{\langle q, s, b, \bot \rangle\}\!\} \oplus C$$

With this rule we can non-deterministically turn an active client instance into a disconnected instance. This rule models a fairly realistic scenario in which the broker is equipped with a background service for sending heartbeat messages to each client in order to check their connection status.

Global Operations. We now turn our attention to global operations that model the publish action. As discussed in the introduction, we first consider a semantics based on a broadcast message embedded into synchronisation blocks in which possible errors during individual notifications are handled locally, e.g., using try-catch statements. The semantics of *publish* is defined as follows.

$$Publish \quad \{\!\{\langle q, s, m, \top \rangle\}\!\} \oplus \gamma \to \{\!\{\langle q', s, m, \top \rangle\}\!\} \oplus \gamma'$$

under the following assumptions:

- $\langle q, publish(m, t), q' \rangle \in R$,
- $\xi = E_\top(t, \gamma)$,
- $\mu = \gamma \ominus \xi$,
- $\gamma' = Add_m(t, \xi) \oplus \mu$.

With this rule a client sends a publish request for message m on topic t to the server. The server acknowledges the request and, inside a synchronisation block, forwards the message m to all other active clients subscribed (at least) to topic t. We assume that m is not sent to the sender client. Since disconnected clients are not selected in $E_\top(t, \gamma)$, the server forwards the message only to active clients. The state of all other clients (disconnected ones and clients that are not subscribed to topic t) remain unchanged. Since our semantics does not keep track of exact time information on node failures and message receptions, the message m is not added to disconnected nodes in order to avoid confusion when inspecting a final configuration.

Computations. A computation σ is a (possibly infinite) sequence of network configurations $\sigma = \gamma_0 \dots \gamma_i \dots$ such that $\gamma_i \to \gamma_{i+1}$ for $i \geq 0$. Using a standard notation, we will use $\to^*$ to denote the transitive closure of $\to$.

3   Example: Specification of an IoT System

Let us consider an example inspired to the standard workflow of an IoT application based on MQTT. MQTT is often used for both device discovery and data acquisition. In the discovery phase a subscriber registers to a topic exposing a list of available sensors. After receiving access details for a specific sensor, a subscriber can start listening to data coming from the sensor.

$$r_1 = \langle init, subscribe(sensors), \emptyset, listen \rangle$$
$$r_2 = \langle listen, local, \{s_i\}, aux_i \rangle,$$
$$r_3 = \langle aux_i, unsubscribe(sensors), \emptyset, sub_i \rangle$$
$$r_4 = \langle sub_i, subscribe(s_i), \emptyset, acquire_i \rangle$$
$$r_5 = \langle acquire_i, \{d_{i1}, \ldots, d_{ik}\}, \emptyset, ok_i \rangle$$
$$r_6 = \langle ok_i, unsubscribe(s_i), \emptyset, init \rangle$$
$$r_7 = \langle ok_i, unsubscribe(s_i), \emptyset, end_i \rangle$$

Fig. 3. IoT subscriber

Subscriber. The workflow of a subscriber can be described as follows. Consider the set of topics $T = \{sensors, s_1, \ldots, s_n\}$ and messages $M = \{s_1, \ldots, s_n\} \cup \bigcup_{i=1}^{n} \{d_{i1}, \ldots, d_{ik}\}$ in which $d_{i,j}$ represents a piece of data coming from sensor s_i. A subscriber can then be described by the specification in Fig. 3. In this model the subscriber first register on topic $sensors$. Then he waits for message s_i for some i and use the message label to register to topic s_i. After registration the subscriber waits for messages $d_{i1}, \ldots, d_{ik}$. Rule r_6 specifies the completion the reception phase, i.e., it simulates the reception of all data sent by sensor s_i. Notice that communication is asynchronous, i.e., the subscriber accumulates individual messages in its message set and then moves to state ok_i only when all messages have been received. The subscriber then unsubscribes from s_i and moves either to state $init$ or to the halting state end_i.

Discovery Service. A discovery service is in charge of generating from time to time the list of available sensors on the $sensor$ topic. This service can be described via the following specification.

$$\langle sinit, publish(s_i, sensors), send_i \rangle \ for \ i : 1, \ldots, n$$

Sensors. Each sensor s_i acts as a publisher that is in charge of sending data along the corresponding topic s_i. The publisher associate to sensor s_i can be described via the following model.

$$\langle init_i, publish(d_{ij}, s_i), init_i \rangle \ for \ i : 1, \ldots, n, \ j : 1, \ldots, k$$

In our model the broker is implicitly defined via the semantics of the $subscribe$, $unsubscribe$, and $publish$ operations. In Fig. 4 we present an example of computation in the above defined Pub/Sub Network in which for clarity states are labeled with process indexes. Notice that messages associated along with a certain topic are delivered only to the current set of subscribed clients. For instance, in our example the client in state $init_2$ never receives message s_1 since it is not subscribed to topic $sensors$ when the publisher sends the message. In other words clients do not read messages from a shared global memory. Subscriber groups are formed dynamically and messages are delivered to the current set of subscribers. In our example when sensor s_i is added to the public registry via

$\langle\langle sinit, \emptyset, \emptyset, \top\rangle, \langle init_1, \emptyset, \emptyset, \top\rangle, \langle init_2, \emptyset, \emptyset, \top\rangle, \langle init_3, \emptyset, \emptyset, \top\rangle,$
$\langle init_4, \emptyset, \emptyset, \top\rangle\rangle \rightarrow$
$\langle\langle sinit, \emptyset, \emptyset, \top\rangle, \langle init_1, \emptyset, \emptyset, \top\rangle, \langle init_2, \emptyset, \emptyset, \top\rangle, \langle listen, \{sensors\}, \emptyset, \top\rangle,$
$\langle init_4, \emptyset, \emptyset, \top\rangle\rangle \rightarrow$
$\langle\langle sinit, \emptyset, \emptyset, \top\rangle, \langle init_1, \emptyset, \emptyset, \top\rangle, \langle init_2, \emptyset, \emptyset, \top\rangle, \langle listen, \{sensors\}, \{s_1\}, \top\rangle,$
$\langle init_4, \emptyset, \emptyset, \top\rangle\rangle \rightarrow$
$\langle\langle sinit, \emptyset, \emptyset, \top\rangle, \langle init_1, \emptyset, \emptyset, \top\rangle, \langle init_2, \emptyset, \emptyset, \top\rangle, \langle aux_1, \emptyset, \{s_1\}, \top\rangle,$
$\langle init_4, \emptyset, \emptyset, \top\rangle\rangle \rightarrow$
$\langle\langle sinit, \emptyset, \emptyset, \top\rangle, \langle init_1, \emptyset, \emptyset, \top\rangle, \langle init_2, \emptyset, \emptyset, \top\rangle, \langle sub_1, \{s_1\}, \{s_1\}, \top\rangle,$
$\langle init_4, \emptyset, \emptyset, \top\rangle\rangle \rightarrow$
$\langle\langle sinit, \emptyset, \emptyset, \top\rangle, \langle init_1, \emptyset, \emptyset, \top\rangle, \langle init_2, \emptyset, \emptyset, \top\rangle, \langle acquire_1, \{s_1\}, \{s_1\}, \top\rangle,$
$\langle init_4, \emptyset, \emptyset, \top\rangle\rangle \rightarrow$
$\langle\langle send_1, \emptyset, \emptyset, \top\rangle, \langle init_1, \emptyset, \emptyset, \top\rangle, \langle init_2, \emptyset, \emptyset, \top\rangle, \langle acquire_1, \{s_1\}, \{s_1, d_{1,3}\}, \top\rangle,$
$\langle listen, \{sensors\}, \{s_1\}, \top\rangle\rangle \rightarrow \dots \rightarrow$
$\langle\langle send_1, \emptyset, \emptyset, \top\rangle, \langle init_1, \emptyset, \emptyset, \top\rangle, \langle init_2, \emptyset, \emptyset, \top\rangle, \langle ok_1, \{s_1\}, \{s_1, d_{1,1}, \dots, d_{1,k}\}, \top\rangle,$
$\langle listen, \{sensors\}, \{s_1\}, \bot\rangle\rangle \rightarrow$
$\langle\langle send_1, \emptyset, \emptyset, \top\rangle, \langle init_1, \emptyset, \emptyset, \top\rangle, \langle init_2, \emptyset, \emptyset, \top\rangle, \langle end_1, \emptyset, \{s_1, d_{1,1}, \dots, d_{1,k}\}, \top\rangle,$
$\langle listen, \{sensors\}, \{s_1\}, \bot\rangle\rangle$

Fig. 4. Computations

the discovery service, each subscriber can receive its data and possibly reach its halting state. For instance, it should not be possible to reach a configuration in which a subscriber is in state end_i from initial configurations in which there are no discovery service nodes in state $init_i$. This kind of property should hold for any number of nodes. In the next section we will discuss how this kind of properties can be stated formally on parameterised families of transition systems of Pub/Sub Networks.

4   The Coverability Decision Problem

In this paper we will focus our attention on safety properties for Pub/Sub Networks with a finite but arbitrary number of clients (parameterised system) described via a decision problem called Coverability considered in other formal models of concurrent and distributed systems such as Petri nets, Broadcast Protocols, Lossy FIFO systems, and Ad Hoc Networks (see e.g. [2,14]).

Reachability Sets. In order to formally define the problem we introduce some auxiliary definitions. Given a set of configurations $C \subseteq N$, the Pre and $Post$ operators are defined as follows:

$$Post(C) = \{\gamma' | \exists \gamma \in C \ s.t. \ \gamma \rightarrow \gamma'\}$$
$$Pre(C) = \{\gamma | \exists \gamma' \in C \ s.t. \ \gamma \rightarrow \gamma'\}$$

$Post^*(C)$ [resp. $Pre^*(C)$] is defined as $\bigcup_{i=0}^{\infty} Post^i(C)$ [resp. $\bigcup_{i=0}^{\infty} Pre^i(C)$]. The set of configurations reachable from $C \subseteq N$ is defined as $Post^*(C)$. For instance,

$Post^*(N_0)$ is the set of configurations reachable from initial configurations of arbitrary size. Similarly, given a set of target configurations T, $Pre^*(T)$ is the set of predecessor configurations that can reach configurations in T after finitely many steps.

Coverability. The Coverability Decision problem is strictly related to the above mentioned correctness criterion. Let $\langle N, \leq \rangle$ be a total ordering on Network configurations. Furthermore, for a set of configurations S, let $uc_\leq(S) = \{\gamma' | \gamma \leq \gamma', \gamma \in S\}$.

Definition 1. *Let $\langle \eta, \rightarrow \rangle$ be a Pub/Sub Network defined over the sets A, M, Q and the client specification P, with an associated predecessor operator Pre, with an ordering $\leq$ on Network Configurations, and with a set N_0 of Initial Network Configuration. Given a finite set of configuration $F \subseteq N$, the Coverability Decision Problem consists in checking whether $N_0 \cap Pre^*(uc_\leq(F)) = \emptyset$ or, alternatively, $Post^*(N_0) \cap uc_\leq(F) = \emptyset$.*

The rationale behind this definition is as follows. Assume that $T = uc_\leq(F)$ represents a set of bad configurations of arbitrary size (e.g. violations of a given safety property) that can be finitely generated via the upward closure of F (if γ represents a violations, then any γ' larger than γ represents a violation). The condition $N_0 \cap Pre^*(T) = \emptyset$ [resp. $Post^*(N_0) \cap T = \emptyset$] holds if and only if there exist no finite computations that starting from some initial configuration (of any size) can reach a bad configuration in T.

4.1   Decision Procedure for Coverability in Pub/Sub Networks

In this section we will study instances of the coverability problem that can be applied to verify properties by considering both local states and received messages.

Definition 2. *Given two client configurations c_1, c_2, the ordering $\leq_c$ is defined as follows: $c_1 = \langle q_1, s_1, b_1, f_1 \rangle \leq_c c_2 = \langle q_2, s_2, b_2, f_2 \rangle$ if and only if $q_1 = q_2$, $s_1 = s_2$, $b_1 \subseteq b_2$, and $f_1 = f_2$.*

The ordering on configurations can be lifted to Network configurations as follows.

Definition 3. *Given two Network Configurations γ_1, γ_2, the ordering $\leq_c$ is defined as follows: $\gamma_1 \leq_n \gamma_2$ if and only if there exists an injective map h from the configurations in $\gamma_1 = \{c_1, \dots, c_k\}$ to configurations in $\gamma_2 = \{d_1, \dots, d_n\}$ such that $c_i \leq_c h(c_i)$ for $i : 1, \dots, k$.*

Theorem 1. *The Coverability Decision Problem is decidable for Pub/Sub Networks.*

Proof. We apply the methodology introduced in [2,14] to prove that $\langle \rightarrow, \leq_n \rangle$ is a well-structured transition systems.

We first observe that the ordering $\leq_n$ is obtained embedding equality over finite sets and finite set inclusion into multiset inclusion. By Higman Lemma's [16], the resulting ordering is a well-quasi-ordering, i.e., for any sequence $\gamma_1 \gamma_2 \ldots$ there exist indexes i, j s.t. $\gamma_i \leq_c \gamma_j$.

The transition relation $\rightarrow$ induced by a client specification P is monotone w.r.t. $\leq_n$, i.e., if $\gamma_1 \leq_n \gamma_2$ and $\gamma_1 \rightarrow \gamma_3$, then there exists γ_4 s.t. $\gamma_2 \rightarrow \gamma_4$ and $\gamma_3 \leq_n \gamma_4$. The proof is based on the observation that enabling conditions for a transition rely only on the occurrence of a certain control state and on the presence of at least a certain sets of messages in the local message list. Thus, augmenting the number of client configurations or the size of local message lists cannot prevent the firing of a rule. In particular, this property holds for the *Publish* rule of the operational semantics.

Given a finite set of configuration C it is possible to compute a finite representation of $Pre(uc_{\leq_n}(C))$. An indirect proof can be given via the observation that the semantics of *Publish* can be encoded using a transfer arc operation on Petri Nets. The encoding is based on a preliminary flattening step in which topics set, message lists and connection flag are hardwired into the control state of individual components. In other words we can generate a flatten specification in which control states have the form $\langle q, s, b, f \rangle$ and in which $\leq_n$ is multiset inclusion (over finitely many labels).

The combination of all above properties proves that Pub/Sub Networks are a well-structured transition systems w.r.t. $\leq_n$. Decidability of coverability follows then from the general results in [2,14].   □

5   Notification with Retained Messages

In Pub/Sub protocols such as MQTT the broker can be instructed in order to retained the last published message for every topic. Retained messages are then distributed to new subscribers right after their first connection. The semantics of this kind of operations requires the introduction of a global state to bookkeep published messages. For brevity, we assume here that all messages (a finite set) are maintained in the broker. More specifically, a network configuration with retained messages is defined as a multiset $\gamma = \langle g, \{\!\!\{ c_1, \ldots, c_k \}\!\!\} \rangle$ where $g : T \rightarrow 2^M$ is a mapping from topics to published messages, and c_i is a client configuration. We use $g(s)$ to denote $\bigcup_{t \in s} \{ g(t) \}$.

The semantics of *publish* is redefined in order to update the global configuration.

$$Publish \qquad \langle g, \{\!\!\{ \langle q, s, m, \top \rangle \}\!\!\} \oplus \gamma \rangle \rightarrow \langle g', \{\!\!\{ \langle q', s, m, \top \rangle \}\!\!\} \oplus \gamma' \rangle$$

under the following assumptions:

- $\langle q, publish(m, t), q' \rangle \in R$,
- $\xi = E_\top(t, \gamma)$,

- $\mu = \gamma \ominus \xi$,
- $\gamma' = Add_m(t, \xi) \oplus \mu$,
- $g'(t) = g(t) \cup \{m\}$, $g'(r) = g(r)$ for $r \neq t$.

The semantics of *subscribe* is redefined in order to update the message set of a node upon subscription to a given topic.

$$Subscription \quad \langle g, \{\!\{\langle q, s, b, \top\rangle\}\!\} \oplus \gamma \rightarrow \{\!\{\langle q', s \cup s_1, b \cup g(s_1), \top\rangle\}\!\} \oplus \gamma\rangle$$

under the assumption $\langle q, subscribe(s_1), q'\rangle \in R$. With this rule a client instance subscribes to the set of topics s_1.

For the extended model, the following property then holds.

Theorem 2. *The Coverability Decision Problem is decidable for Pub/Sub Networks with retained messages.*

Proof. The proof of Theorem 1 can be extended in order to deal with the semantics with retained messages. Indeed, we observe that (1) the extended transition system is still monotone and that (2) it is still possible to compute a finite representation of predecessor states passing through a flattening of the transition system that reduces configurations to multisets of control states. The resulting system can then be viewed as a Petri net with transfer arc and a control unit (the global state) for which coverability is known to be decidable [2,14]. □

We notice that synchronisation steps with control unit can also be encoded via simpler models such a Process Rewrite Systems with weak unit or finite state constraints [17–19].

6   Handling Exceptions During Notifications

In this section we consider a semantics of the *publish* operation in which the broker does not handle node failures locally to individual notification messages. In this scenario the failure during a notification for a specific client, e.g. the client is disconnected and the notification generates and exception, can lead to a failure of the entire notification phase. As a consequence, the message might be delivered to a strict subset of the active destination nodes. We assume that the sender proceed with its execution without forcing the broker to roll-back to a previous state. Let $publish_e$ denote the operation with the above described implementation of the publish operation.

Operational Semantics for $publish_e$. Let us first define $Up_\perp(\gamma)$ as the multiset obtained by setting all connection flags occurring in γ to $\perp$, namely $Up_\perp(\{\!\{\}\!\}) = \{\!\{\}\!\}$, $Up_\perp(\{\!\{\langle q, s, b, \top\rangle\}\!\} \oplus \gamma) = \{\!\{\langle q, s, b, \perp\rangle\}\!\} \oplus Up_\perp(\gamma)$, $Up_\perp(t, \{\!\{c\}\!\} \oplus \gamma) = \{\!\{c\}\!\} \oplus Up_\perp(\gamma)$ otherwise. The operational semantics of $publish_e$ is defined as follows.

$$Publish_e \quad \{\!\{\langle q, s, m, \top\rangle\}\!\} \oplus \gamma \rightarrow \{\!\{\langle q', s, m, \top\rangle\}\!\} \oplus \gamma'$$

with the following assumptions:

- $\langle q, publish_e(m,t), q' \rangle \in R$,
  - $E_T(t,\gamma) = \xi \oplus \eta$,
  - $\mu = \gamma \ominus (\xi \oplus \eta)$,
  - $\gamma' = Add_m(\xi) \oplus Up_\perp(\eta) \oplus \mu$.

With this rule a client sends a publish request for message m on topic t to the broker. The server acknowledges the request and, inside a synchronisation block, forwards the message m to all other active clients subscribed (at least) to topic t. If during the notification phase (typically a scanning of an internal data structure) a disconnected client is detected (i.e. the corresponding notification operation fails) the procedure exits. This effect is modelled using a non-deterministically chosen subset of active destination nodes ξ with $\xi \sqsubseteq E_T(t,\gamma)$ that represents subscribers ready to receive message m before failure detection. The remaining potential receivers η are marked as disconnected. In this semantics, among all possible executions, we consider the case in which, during the notification phase, no disconnected clients are detected as well the case in which none or a strict subset of clients receive the notification.

The semantics of the new operation is slightly different from the typical broadcast operations adopted in Petri Nets that we took as target operation to prove decidability of coverability in the first part of the paper. Indeed this operation applies a non-deterministic split during the transfer phase in which instances are transferred from one state to another. The non-deterministic split redistributes all instances in a given state to a finite set of different states without cancellations or duplications. Despite of the use of a non-standard transfer operation, coverability is still decidable as proved in the following theorem.

Theorem 3. *Coverability is decidable for Pub/Sub Networks with the publish$_e$ operation.*

Proof. The proof is based on a reduction of the considered decision problem to coverability for parameterised systems composed by many finite-state components with a single monitor in which each component reacts in a non deterministic way to broadcast messages sent by the monitor. This kind of systems has been introduced in [6] to model the behaviour of synchronous systems. The decision procedure is based on a symbolic reachability algorithm based on a constraint solver for linear integer (in)equalities. The reduction requires the following steps. For the *publish* operation the encoding requires a preliminary flattening step in which topics set, message lists and connection flag are hardwired into the control state of individual clients. The flattening can then be used to associate finitely many counters (to keep track of occurrences of states in network configurations) to each control state in accord with the counting abstraction used e.g. to model Petri nets as vector addition systems. The flattening and the counter representation of control states provides a way to represent transition rule using linear integer inequalities over variables ranging over natural numbers. For instance, enabling conditions of the *publish$_e$* operation can be expressed via lower bounds constraints of the form $X \leq 1$ for the counter X that denotes a given control state, e.g., a publisher state. The effect of a transition can be expressed as an affine transformation, i.e., a linear combination defined on the current number of client instances in different counters associated to control states. Differently from other models such as transfer nets and affine well-structured Nets [15] the semantics of $publish_e$ requires inequalities in special form in which the left hand side may consists of an expression $X'_1 + \ldots + X'_n = X_1 + \ldots + X_n + Exp(Y_1, \ldots, Y_m)$ for variables X'_i denoting the value of the counter in the next state and variables X_i and Y_j denoting the current values of the counters for $i : 1, \ldots, n$ and $j : 1, \ldots, m$. In addition we need to insert the side conditions $X'_1 \geq X_i$ to ensure that variables occurring in $X'_1 + \ldots + X'_n$ will be incremented. As an example, the transitions on counters $X' + Y' = X + Y + Z, X' \geq X, Y' \geq Y, Z' = 0$ can be used to model the transfer of all instances in Z in X and Y. The effect of the transfer is to distribute the instances in Z non-deterministically between X and Y. Decidability of coverability in the resulting counter representation of the flattened transition system follows by observing that the problem can be solved by applying the symbolic backward reachability algorithm based on constraint solvers for inequalities over natural numbers proposed in [6]. The algorithm maintains constraint-based representations of infinite set of configurations via unions of constraints of the form $X_1 \geq c_1, X_n \geq c_n$. To apply termination results based on Dickson's lemma [13] on the $\leq_v$ ordering to the resulting procedure, transitions of the form $X' + Y' = X + Y + Z, X' \geq X, Y' \geq Y, \ldots$ require a further normalization steps in order to eliminate constraints of the form $X' \geq X$. The idea here is to associate an auxiliary variable $AuxZ$ to each variable Z whose value must be split between several variables. Before firing the transfer action, the transition system enters a special state in which instances in Z are moved to $AuxZ$. This phase is non-deterministically terminated in order to start the transfer arc from Z and $AuxZ$ to variables X and Y respectively.   □

7   Conclusions

We have studied coverability problems for a formal model of Publish/Subscribe Networks inspired to extensions of Petri nets with broadcast and transfer arcs. Our model combines asynchronous communication with global operations and non-deterministic actions to model the effect of exceptions generated during communication between broker and individual clients. The proposed model, extensions and variants seem to be different from other infinite-state models proposed in the literature, see e.g., [1,3] for a survey on extensions of Petri nets used to model distributed systems.

For the considered model, we prove preliminary results for the coverability decision problem. The model discussed in this paper can be extended in different directions. One possible extension consists of a new operation $publish_r(m, t, q)$ in which m is a message, t is a topic name, and $q \in Q$ denotes the state in which the sender is redirected when an exception is generated during the notification step. The operational semantics requires to detect absence/presence of crashed nodes, i.e., global, universally quantified, condition for firing a transition. Finding suitable semantics for this operation for which coverability remains decidable is still an open problem.

References

1. Abdulla, P.A., Delzanno, G.: Parameterized verification. STTT **18**(5), 469–473 (2016)
2. Abdulla, P.A., Jonsson, B.: Ensuring completeness of symbolic verification methods for infinite-state systems. Theor. Comput. Sci. **256**(1–2), 145–167 (2001)
3. Bloem, R., et al.: Decidability of Parameterized Verification. Synthesis Lectures on Distributed Computing Theory. Morgan & Claypool Publishers, San Rafael (2015)
4. Bloem, R., et al.: Decidability in parameterized verification. SIGACT News **47**(2), 53–64 (2016)
5. Conchon, S., Delzanno, G., Ferrando, A.: Parameterized verification of topology-sensitive distributed protocols goes declarative. In: NETYS 2018, pp. 209–224 (2018)
6. Delzanno, G.: Constraint-based model checking for parameterized synchronous systems. In: Armando, A. (ed.) FroCoS 2002. LNCS (LNAI), vol. 2309, pp. 72–86. Springer, Heidelberg (2002). https://doi.org/10.1007/3-540-45988-X_7
7. Delzanno, G.: A logic-based approach to verify distributed protocols. In: CILC 2016, pp. 86–101 (2016)
8. Delzanno, G.: A unified view of parameterized verification of abstract models of broadcast communication. STTT **18**(5), 475–493 (2016)
9. Delzanno, G.: Formal verification of internet of things protocols. In: FRIDA@FLOC (2018)
10. Delzanno, G., Sangnier, A., Zavattaro, G.: Parameterized verification of ad hoc networks. In: Gastin, P., Laroussinie, F. (eds.) CONCUR 2010. LNCS, vol. 6269, pp. 313–327. Springer, Heidelberg (2010). https://doi.org/10.1007/978-3-642-15375-4_22
11. Delzanno, G., Sangnier, A., Zavattaro, G.: On the power of cliques in the parameterized verification of ad hoc networks. In: Hofmann, M. (ed.) FoSSaCS 2011. LNCS, vol. 6604, pp. 441–455. Springer, Heidelberg (2011). https://doi.org/10.1007/978-3-642-19805-2_30
12. Delzanno, G., Sangnier, A., Zavattaro, G.: Verification of ad hoc networks with node and communication failures. In: Giese, H., Rosu, G. (eds.) FMOODS/FORTE-2012. LNCS, vol. 7273, pp. 235–250. Springer, Heidelberg (2012). https://doi.org/10.1007/978-3-642-30793-5_15
13. Dickson, L.E.: Finiteness of the odd perfect and primitive abundant numbers with n distinct prime factors. Am. J. Math. **35**(4), 413–422 (1913)
14. Finkel, A., Schnoebelen, P.: Well-structured transition systems everywhere! Theor. Comput. Sci. **256**(1–2), 63–92 (2001)
15. Finkel, A., McKenzie, P., Picaronny, C.: A well-structured framework for analysing petri net extensions. Inf. Comput. **195**(1–2), 1–29 (2004)
16. Higman, G.: Ordering by divisibility in abstract algebras. Proc Lond. Math. Soc. **2**(7), 326–336 (1952)
17. Kretínský, M., Rehák, V., Strejcek, J.: On extensions of process rewrite systems: rewrite systems with weak finite-state unit. Electr. Notes Theor. Comput. Sci. **98**, 75–88 (2004)
18. Kretínský, M., Rehák, V., Strejcek, J.: Extended process rewrite systems: expressiveness and reachability. In: CONCUR 2004, pp. 355–370 (2004)
19. Mayr, R.: Process rewrite systems. Inf. Comput. **156**(1–2), 264–286 (2000)

Cellular Automata for the Self-stabilisation of Colourings and Tilings

Nazim Fatès[1]([⊠]), Irène Marcovici[2], and Siamak Taati[3]

[1] Université de Lorraine, CNRS, Inria, LORIA, 54000 Nancy, France
nazim.fates@loria.fr
[2] Université de Lorraine, CNRS, Inria, IECL, 54000 Nancy, France
irene.marcovici@univ-lorraine.fr
[3] Bernoulli Institute, University of Groningen, Groningen, The Netherlands
siamak.taati@gmail.com

Abstract. We examine the problem of self-stabilisation, as introduced by Dijkstra in the 1970's, in the context of cellular automata stabilising on k-colourings, that is, on infinite grids which are coloured with k distinct colours in such a way that adjacent cells have different colours. Suppose that for whatever reason (e.g., noise, previous usage, tampering by an adversary), the colours of a finite number of cells in a valid k-colouring are modified, thus introducing errors. Is it possible to reset the system into a valid k-colouring with only the help of a local rule? In other words, is there a cellular automaton which, starting from any finite perturbation of a valid k-colouring, would always reach a valid k-colouring in finitely many steps? We discuss the different cases depending on the number of colours, and propose some deterministic and probabilistic rules which solve the problem for $k \neq 3$. We also explain why the case $k = 3$ is more delicate. Finally, we propose some insights on the more general setting of this problem, passing from k-colourings to other tilings (subshifts of finite type).

Keywords: Cellular automata · Self-stabilisation · Self-correction · k-colourings · Subshifts of finite type

1 Introduction

Self-stabilisation is a property omnipresent in biological systems. Indeed, living cells always need to correct their defects in order to keep their behaviour as stable as possible (see e.g. Ref. [2]). The study of self-stabilisation in computational systems was proposed by Dijkstra [4]. The objective is to incorporate self-stabilisation in discrete parallel models of computation.

In the present article, we explore the phenomenon of self-stabilisation in the context of two-dimensional cellular automata which operate on k-colourings.

S. Taati—The work of ST was partially supported by NWO grant 612.001.409 of Tobias Müller.

E. Filiot et al. (Eds.): RP 2019, LNCS 11674, pp. 121–136, 2019.
https://doi.org/10.1007/978-3-030-30806-3_10

To illustrate the problem, imagine that an artist has a plan to create a two-dimensional tiling with the constraint that two adjacent tiles necessarily bear different colours. When this tiling is realised, the artist realises that (a) some mistakes have occurred during the tiling process and (b) the original tiling plan has been lost. The question is to know whether it is possible to correct the tiling to respect the constraints of non-adjacency of colours only by following local rules. In other words, we can reformulate the question as a reachability problem: given a set of admissible states of the system, under which conditions is this set always reachable from the set of its finite perturbations?

The problem of designing self-correcting or self-stabilising cellular automata has been explored since the 1970's. Two main models of errors have been considered: (a) the errors can happen at each time step and are thus concurrent with the correction process [7,8,11,12] or, (b) the errors are present at the beginning and are then corrected [6]. Pippenger has studied this latter question, for the binary case, where the configurations to correct are those which only contain a unique colour [10]. He has shown that the problems can have positive or negative answers depending on the specification of the problem such as dimension, symmetry constraints, etc.

We re-examine this problem in the setting of k-colourings. The cases $k = 2$ and $k \geq 5$ are the simplest and the case $k = 4$ can be dealt with rather easily. However, the case $k = 3$ is much more delicate. We will also explore the question of symmetries of the rules we use. The case of k-colourings should be considered as a first step towards a wider view of self-stabilisation in cellular automata. We indicate some directions on how to consider more general tiling constraints.

2   Setting of the Problem

Let Σ be a finite set that represents the different colours of the tiling. Given two configurations $x, y \in \Sigma^{\mathbb{Z}^2}$, we write $\Delta(x, y) \triangleq \{i \in \mathbb{Z}^2 : x_i \neq y_i\}$ for the set of sites at which x and y disagree.

A *finite perturbation* of a configuration $x \in \Sigma^{\mathbb{Z}^2}$ is a configuration $y \in \Sigma^{\mathbb{Z}^2}$ such that $\Delta(x, y)$ is finite. Given a set $\Lambda \subseteq \Sigma^{\mathbb{Z}^2}$, representing the set of *valid* configurations, we denote by $\tilde{\Lambda}$ the set of finite perturbations of the elements of Λ, that is:

$$\tilde{\Lambda} \triangleq \{y \in \Sigma^{\mathbb{Z}^2} : \exists x \in \Lambda, \Delta(x, y) \text{ is finite}\}.$$

Our goal is to find a parallel procedure acting in a local way that would, from any element of $\tilde{\Lambda}$, reach and stabilise on an element of Λ in a finite number of steps. The locality of the rule is expressed by the definition of a *neighbourhood*, that is, an ordered list $\mathcal{N} = (n_1, \ldots, n_k)$ of k elements from $\mathbb{Z}^2$, for some $k \in \mathbb{N}$. We use the model of cellular automata to take into account the distributed aspect of the process: each cell $c \in \mathbb{Z}^2$ is updated according to a *local rule* f that depends only on the states of the cells $c + n_1, \ldots, c + n_k$.

Formally, a two-dimensional *cellular automaton (CA)* with neighbourhood $\mathcal{N}$ is a mapping $F : \Sigma^{\mathbb{Z}^2} \to \Sigma^{\mathbb{Z}^2}$ for which there exists a function $f : \Sigma^k \to \Sigma$

satisfying:

$$\forall c \in \mathbb{Z}^2, \; F(x)_c = f(x_{c+n_1}, \ldots, x_{c+n_k}).$$

Now that we have set all the elements, we can define the notion of *self-stabilisation*. We say that a cellular automaton $F : \Sigma^{\mathbb{Z}^2} \to \Sigma^{\mathbb{Z}^2}$ is *self-stabilising* on $\Lambda \subseteq \Sigma^{\mathbb{Z}^2}$ if it satisfies the following conditions:

(i) The configurations of Λ are fixed points of F: $\forall x \in \Lambda, \; F(x) = x$.
(ii) The configurations of $\tilde{\Lambda}$ evolve to Λ in finitely many steps: $\forall y \in \tilde{\Lambda}, \; \exists t \in \mathbb{N}, \; F^t(y) \in \Lambda$.

We will in particular focus on the case where the set Λ is the set of colourings of $\mathbb{Z}^2$ with k distinct colours. Let $k \geq 2$ be an integer, and let the set $\Sigma = \{0, \ldots, k-1\}$ represent the set of possible colours of the cells. We define the set of k-*colourings* of $\mathbb{Z}^2$ by:

$$\Lambda_k \triangleq \{x \in \Sigma^{\mathbb{Z}^2} : c, c' \in \mathbb{Z}^2, \; \|c - c'\|_1 = 1 \implies x_c \neq x_{c'}\}.$$

Our aim is to examine if there exist simple self-stabilising rules, depending on the value of k. We also have a look at other families of *subshifts of finite type*, that is, sets Λ defined by local constraints. More specifically, a nonempty set $\Lambda \subseteq \Sigma^{\mathbb{Z}^2}$ is a *subshift of finite type (SFT)* if there exists a finite set $B \subseteq \mathbb{Z}^2$ and a function $u : \Sigma^B \to \{0, 1\}$ such that:

$$\Lambda = \{x \in S^{\mathbb{Z}^d} : \forall c \in \mathbb{Z}^d, u((x_{c'})_{c' \in c+B}) = 1\}.$$

In the definition above, the function u describes the set of allowed (image 1) and forbidden (image 0) patterns of base B.

We will focus on cases where the SFT can be defined in terms of horizontal and vertical constraints. We will call the elements of such an SFT a *proximity tiling*. Formally, let us denote by (e_1, e_2) the standard basis of $\mathbb{Z}^2$. A nonempty set $\Lambda \subseteq \Sigma^{\mathbb{Z}^2}$ is a *proximity tiling space* if there exist functions $v_1, v_2 : \Sigma^2 \to \{0, 1\}$ such that:

$$\Lambda = \{x \in \Sigma^{\mathbb{Z}^2} : \forall c \in \mathbb{Z}^2, v_1(x_c, x_{c+e_1}) = v_2(x_c, x_{c+e_2}) = 1\}.$$

Note that the notion of proximity tilings we have introduced here is reminiscent of tilings by Wang tiles, but this formalism is more adapted to our context. For example, the set of k-colourings is simply the proximity tiling space defined by the function $v = v_1 = v_2$ where $v(a, b) = 1$ if $a \neq b$, and 0 if $a = b$.

In designing self-stabilising rules and proving their correctness, we will often examine the set of cells where the constraints are not respected. We thus introduce different notions of *error*. For a configuration $x \in \Sigma^{\mathbb{Z}^2}$, a cell $c \in \mathbb{Z}^2$ is said to have an e_i-*error* (with respect to v_i) if $v_i(x_c, x_{c+e_i}) = 0$. It has a $(-e_i)$-*error* if $v_i(x_{c-e_i}, x_c) = 0$. We will also use the terminology *E-error, W-error, N-error, S-error* instead of respectively e_1-error, $(-e_1)$-error, e_2-error, and $(-e_2)$-error. The set of cells having an error is defined by:

$$\mathcal{E}(x) \triangleq \{c \in \mathbb{Z}^2 : \exists e \in \{\pm e_1, \pm e_2\}, \; c \text{ has an } e\text{-error}\}.$$

A cell $c \in \mathbb{Z}^2$ is said to be *error-free* if it does not belong to $\mathcal{E}(x)$, meaning that it obeys the local constraints in the four directions. Note that in some cases, even if $\mathcal{E}(x)$ contains only very few cells, it is necessary to modify a much larger set of cells in order to reach a valid configuration (see Proposition 6).

We will also consider self-stabilising probabilistic CA. For *probabilistic CA*, the outcome of the local rule is a probability distribution on Σ, and the cells of the lattice are updated simultaneously and independently at each time step, according to the distributions prescribed by the local rule. The local rule in this case is given by a function $\varphi : \Sigma^k \to \mathcal{P}(\Sigma)$, where $\mathcal{P}(\Sigma)$ denotes the set of probability distributions on Σ. The probabilistic CA Φ defined by φ maps a configuration x to a probability measure μ, where for each finite set $C \subseteq \mathbb{Z}^2$, we have

$$\mu(\{y : \forall c \in C, y_c = v_c\}) = \prod_{c \in C} \varphi(x_{c+n_1}, x_{c+n_2}, \ldots, x_{c+n_k})(\{v_c\}).$$

The trajectory of a probabilistic CA Φ with initial configuration x is thus a Markov process $X^0, X^1, \ldots$ with $X^0 = x$ such that, for every $t > 0$, conditioned on the value of the configurations $X^0, X^1, \ldots, X^{t-1}$, the configuration X^t is distributed according to the measure $\Phi(X^{t-1})$. We say that a probabilistic CA Φ is *self-stabilising* on Λ if:

(i) The configurations of Λ are left unchanged by Φ: $\forall x \in \Lambda, \Phi(x) = \delta_x$.
(ii) For every $y \in \tilde{\Lambda}$, there exists a finite (random) time T such that $X^T \in \Lambda$ almost surely.

3   The Case of 2-Colourings

In this section, we study the self-stabilisation problem for 2-colourings. We thus set $\Sigma = \{0, 1\}$, and consider the set Λ_2. Note that Λ_2 contains only two elements, corresponding to the two (odd and even) chequerboard configurations.

3.1   Directional Self-stabilisation by a Deterministic CA

Let us define a cellular automaton F on $\Sigma^{\mathbb{Z}^2}$ by:

$$\forall c \in \mathbb{Z}^2, \quad F(x)_c = \begin{cases} 1 - x_c & \text{if } x_c = x_{c+e_1} = x_{c+e_2}, \\ x_c & \text{otherwise.} \end{cases}$$

The rule above is similar to the well-known majority rule of Toom used to correct errors that appear on a uniform background [3, 12].

Proposition 1. *The cellular automaton F defined above is self-stabilising on Λ_2.*

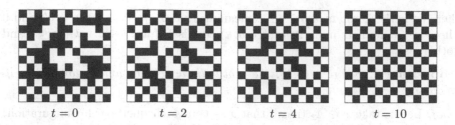

$$t = 0 \qquad t = 2 \qquad t = 4 \qquad t = 10$$

Fig. 1. Evolution of the cellular automaton for self-correction of the 2-colourings. (Color figure online)

Proof. It is clear from the definition that $\forall x \in \Lambda_2$, $F(x) = x$.

For each $n \in \mathbb{N}$, define the triangle $T_n = \{(i,j) \in \mathbb{Z}^2 : i + j \leq n,\ i, j \geq 0\}$ on the grid. Let $x \in \Lambda_2$ (recall that x is thus a chequerboard configuration) and take $y \in \tilde{\Lambda}_2$ such that $\Delta(x, y)$ is finite. By translating x and y if needed, we can assume without loss of generality that the difference set $\Delta(x, y)$ is included in the triangle T_n for some n. It is then easy to verify that $\Delta(x, F(y)) \subseteq T_{n-1}$. Indeed, for every cell outside T_n, the local rule does not modify the state, whereas for the cells (i, j) which are inside T_n and satisfy $i + j = n$, we have $F(y)_{i,j} = x_{i,j}$. Iterating F we obtain $\Delta(x, F^t(y)) \subseteq T_{n-t}$ for each $t \geq 0$. That is, as time goes by, the set of disagreements becomes smaller (see Fig. 1). In particular, for $t = n + 1$, we get $\Delta(x, F^{n+1}(y)) \subseteq T_{-1} = \varnothing$, hence $F^{n+1}(y) = x \in \Lambda_2$. This means that the configuration y has been corrected in $n + 1$ steps. $\qquad\square$

3.2 Isotropic Self-stabilisation by a Probabilistic CA

The cellular automaton F above provides a directional solution: the cells need to distinguish the North and East directions. In the context of deterministic CA, Pippenger has shown that requiring all the symmetries for the local rule can lead to negative results [10]. By contrast, we now propose a probabilistic CA that achieves the self-stabilisation with an isotropic rule, that is, a rule which treats the neighbours "equally" and does not distinguish between the four directions of the grid. This shows that the use of randomness can extend the range of possibilities. More precisely, the rule we propose consists in applying a minority function with probability α, and keeping the state unchanged with probability $1 - \alpha$. Such rules are called α-*asynchronous* and their study has received a continuous attention in the last years [5]; this structure is here used to get out of the potential cyclic behaviours that would prevent the system from reaching the desired stable configurations.

Formally, let $\mathcal{N}$ denote the *von Neumann neighbourhood* $\mathcal{N} = (0, e_1, e_2, -e_1, -e_2)$. We define a probabilistic cellular automaton Φ on $\Sigma^{\mathbb{Z}^2}$ by the local rule $\varphi : \Sigma^5 \to \mathcal{P}(\Sigma)$ given by:

$$\varphi(q_0, q_1, \ldots, q_4) = \alpha \delta_{\mathrm{minority}\,(q_0, q_1, \ldots, q_4)} + (1 - \alpha)\delta_{q_0},$$

where δ_q is the Dirac measure on q, meaning that $\delta_q(\{q'\}) = 1$ if $q = q'$, and 0 otherwise, and where minority (a, b, c, d, e) equals 1 if $a + b + c + d + e \leq 2$ and 0 otherwise.

Proposition 2. *For $\alpha \in (0,1)$, the probabilistic cellular automaton Φ is self-stabilising on Λ_2.*

Proof. Let us take $x \in \Lambda_2$ (recall that x is thus a chequerboard configuration) and $y \in \tilde{\Lambda}_2$ such that $\Delta(x,y)$ is finite. Let $X_0, X_1, \ldots$ be the Markov process described by Φ with initial configuration $X_0 = y$. Let R be a rectangle such that $\Delta(x,y) \subseteq R$. For any $c \notin R$, we have $\varphi(x_c, x_{c+e_1}, x_{c+e_2}, x_{c-e_1}, x_{c-e_2}) = \varphi(y_c, y_{c+e_1}, y_{c+e_2}, y_{c-e_1}, y_{c-e_2}) = \delta_{x_c}$, so that for all $t \geq 0$, $\Delta(X_t, x) \subseteq R$ almost surely. Furthermore, inside R, Φ behaves like an absorbing finite-state Markov chain that eventually reaches the chequerboard configuration $(x_c)_{c \in R}$. Note that from any state, with positive probability the chequerboard configuration can be reached in at most $|R|$ steps. This is because $\alpha < 1$. Otherwise, a monochromatic rectangle could blink between the two states all 0's and all 1's.   □

3.3   Extension to Finite SFT

The methods presented above for the case $k = 2$ can be readily extended to all the cases where Λ is a subshift of finite type that contains only a finite number of configurations (with an arbitrary set of symbols Σ). Indeed, the configurations of such a finite SFT are necessarily spatially periodic.

Let us consider an arbitrary finite SFT Λ on $\Sigma = \{0, \ldots, k-1\}$, for some $k \geq 1$. Then, for each configuration $x \in \Lambda$, there exist integers $m, n \geq 1$ such that $\sigma_{\mathrm{h}}^m(x) = \sigma_{\mathrm{v}}^n(x) = x$, where σ_{h} and σ_{v} denote the horizontal and vertical shift maps. Taking the least common multiple of the collection of integers obtained for the different configurations $x \in \Lambda$ (all these integers are bounded by the cardinality of Λ), we can find horizontal and vertical periods $p_{\mathrm{h}}, p_{\mathrm{v}} \geq 1$ such that $\forall x \in \Lambda$, $\sigma_{\mathrm{v}}^{p_{\mathrm{v}}}(x) = \sigma_{\mathrm{h}}^{p_{\mathrm{h}}}(x) = x$. This means that the elements of Λ are constant on every sublattice $\mathbb{L}_{a,b} = \{(a + p_{\mathrm{h}}i, b + p_{\mathrm{v}}j) : i, j \in \mathbb{Z}^2\}$. Therefore, we can simply use Toom's majority rule on each sublattice, that is, we define a cellular automaton F on $\Sigma^{\mathbb{Z}^2}$ by:

$$F(x)_{a,b} = \mathrm{majority}(x_{a,b}, x_{a+p_{\mathrm{h}},b}, x_{a,b+p_{\mathrm{v}}}),$$

where the majority function associates to three symbols the symbol which is most present in this three symbols, with the convention that when the three symbols are distinct, one can choose arbitrarily the value of the function. Note however that even if Λ is a proximity tiling, the neighbourhood of the cellular automaton F depends on the periods p_{h} and p_{v} and can be much larger than 1.

We can also design an isotropic probabilistic rule that corrects finite SFT. If a state appears strictly more than twice among $x_{a+p_{\mathrm{h}},b}, x_{a,b+p_{\mathrm{v}}}, x_{a-p_{\mathrm{h}},b}, x_{a,b-p_{\mathrm{v}}}$, this state becomes the new value of $x_{a,b}$. Otherwise, we randomly choose a new state in the alphabet Σ. Again, all the errors stay within some enveloping rectangle and are eventually corrected. We can replace both p_{h} and p_{v} by $LCM(p_{\mathrm{h}}, p_{\mathrm{v}})$ in order to have an isotropic rule.

4   The Case of k-Colourings, for $k \geq 5$

We now consider the case $k \geq 5$. Recall that we have $\Sigma = \{0, 1, \ldots, k - 1\}$.

4.1   Directional Self-stabilisation by a Deterministic CA

Let us introduce the following terminology. We say that a cell (i, j) has a *NE-error* if it has either an N-error or an E-error. For $x \in \Sigma^{\mathbb{Z}^2}$, we denote by $\mathcal{E}_{NE}(x)$ the set of cells having a NE-error, that is:

$$\mathcal{E}_{NE}(x) \triangleq \{(i, j) \in \mathbb{Z}^2 : x_c = x_{c+e_1} \text{ or } x_c = x_{c+e_2}\}.$$

Let $\psi : \Sigma^4 \to \Sigma$ be a function which assigns to each quadruplet of colours (a, b, c, d) a colour which is not in the set $\{a, b, c, d\}$, for example $\psi(a, b, c, d) = \min \Sigma \setminus \{a, b, c, d\}$.

We define a cellular automaton F on $\Sigma^{\mathbb{Z}^2}$ by:

$$\forall c \in \mathbb{Z}^2, \quad F(x)_c = \begin{cases} \psi(x_{c-e_1}, x_{c-e_2}, x_{c+e_1}, x_{c+e_2}) \text{ if } c \in \mathcal{E}_{NE}(x) \\ x_c \text{ otherwise.} \end{cases}$$

Proposition 3. *Let $k \geq 5$, the cellular automaton F defined above is self-stabilising on Λ_k.*

Proof. It is clear from the definition that $\forall x \in \Lambda_k$, $F(x) = x$. Let us now take $x \in \tilde{\Lambda}_k$. Without loss of generality, we can assume that there exists an integer $n \geq 0$ such that $\mathcal{E}_{NE}(x) \subseteq T_n$. (Recall that $T_n = \{(i, j) \in \mathbb{Z}^2 : i + j \leq n, \; i, j \geq 0\}$). One can also check that after t steps, we have $\mathcal{E}_{NE}(F^t(x)) \subseteq T_{n-t}$. Indeed, the set of NE-errors can only decrease under the action of F: if $c \notin \mathcal{E}_{NE}(x)$, then $c \notin \mathcal{E}_{NE}(F(x))$, since by definition of F, if $c + e_1$ or $c + e_2$ takes a new colour in $F(x)$, that new colour is different from x_c. Furthermore, if $c \in \mathcal{E}_{NE}(x)$ is such that $c + e_1, c + e_2 \notin \mathcal{E}_{NE}(x)$, then $c \notin \mathcal{E}_{NE}(F(x))$, so that the set of NE-errors is progressively eroded, from the NE to the SW. After $n + 1$ steps, we have: $\mathcal{E}_{NE}(F^{n+1}(x)) = \varnothing$, meaning that the configuration is thus fully corrected: $F^{n+1}(x) \in \Lambda_k$. $\qquad \square$

4.2   Isotropic Self-stabilisation by a Probabilistic CA

Let $\psi : \Sigma^4 \to \Sigma$ be a function as above. For $x \in \Sigma^{\mathbb{Z}^2}$, let us recall that we denote by $\mathcal{E}(x)$ the set of cells having an error, that is:

$$\mathcal{E}(x) \triangleq \{c \in \mathbb{Z}^2 : x_c \in \{x_{c \pm e_1}, x_{c \pm e_2}\}\}$$

We define a probabilistic cellular automaton Φ on $\Sigma^{\mathbb{Z}^2}$ which leaves the state of cell c unchanged if $c \notin \mathcal{E}(x)$ and updates it to a random value with distribution $\alpha \delta_{\psi(x_{c-e_1}, x_{c-e_2}, x_{c+e_1}, x_{c+e_2})} + (1 - \alpha)\delta_{x_c}$ if $c \in \mathcal{E}(x)$. Once again, the use of an α-asynchronous rule is destined to break the potential cycles that could be created by the situations where the value of the update function is *not* deterministic (in the case where only one colour is missing in the neighbourhood).

Proposition 4. *For $k \geq 5$ and $\alpha \in (0,1)$, the probabilistic cellular automaton Φ defined above is self-stabilising on Λ_k.*

Proof. Let $x \in \tilde{\Lambda}_k$ be an initial configuration. Let $X^0, X^1, \ldots$ denote the Markov process described by Φ starting from $X^0 = x$. For any $c \notin \mathcal{E}(X^t)$, the state of cell c remains unchanged, and the neighbouring cells of c cannot take the state X_c^t, so that for each $t' \geq t$, $\mathcal{E}(X^{t'}) \subseteq \mathcal{E}(X^t)$ almost surely. Furthermore, inside $\mathcal{E}(x)$, Φ behaves like an absorbing finite state Markov chain, that eventually reaches an allowed configuration. Indeed, let us consider the cells of $\mathcal{E}(X^t)$ that have at least two correct neighbouring cells (there necessarily exist such cells, since $\mathcal{E}(x)$ is finite). If the function ψ is applied to such a cell c, and if the values of its neighbours remain the same, then $c \notin \mathcal{E}(X^{t+1})$. This happens with probability at least $\alpha(1 - \alpha)^2 > 0$. Consequently, the probability of decreasing the set of errors is strictly larger than this probability at each time step.   □

4.3   Extension to Single-Site Fillable Proximity Tilings

We say that a proximity tiling is *single-site fillable* if there exists a map $\psi : \Sigma^4 \to \Sigma$ such that, for any possible choice $(a, b, c, d) \in \Sigma^4$ of symbols surrounding a cell, assigning the value $\alpha = \psi(a, b, c, d)$ to the central cell ensures that it is error-free [9]. The two constructions above (directional self-stabilisation by a deterministic CA, and isotropic self-stabilisation by a probabilistic cellular automaton) naturally extend to all proximity tiling spaces that are single-site fillable.

5   The Case of 4-Colourings

5.1   Directional Self-stabilisation by a Deterministic CA that Corrects by Blocks

The case of 4-colourings ($\Sigma = \{0, 1, 2, 3\}$) is more delicate. Obviously, it is no longer possible to use a function ψ with the same properties as above. Nevertheless, we propose a solution where we show that the number of errors is decreased by updating 2-squares, that is, 2×2-blocks of cells. We explain the possibility of this update in the next lemma, and the show how to apply this update without generating conflicts.

Lemma 1. *For any possible choice $(a, b, c, d, e, f, g, h) \in \Sigma^8$ of symbols surrounding a 2-square (see right), there exist a choice $(\alpha, \beta, \gamma, \delta) \in \Sigma^4$ for the cells of the 2-square such that the four cells of the 2-square are error-free.*

	c	d	
b	β	γ	e
a	α	δ	f
	h	g	

Proof. If $\{a, d, e, h\} \subsetneq \Sigma$, then we can choose a colour from $\Sigma \setminus \{a, d, e, h\}$ and assign it to both α and γ. We are then sure that we can find suitable colours for the two remaining cells, since each of these two cells is surrounded by at most three different colours. In the same way, if $\{b, c, f, g\} \subsetneq \Sigma$, we can find a valid pattern.

Let us now assume that $\{a, d, e, h\} = \{b, c, f, g\} = \Sigma$. Without loss of generality, we can assume that $a = 0, h = 1, d = 2, e = 3$. The set of allowed colours for α is then $\{2, 3\}$, and the set of allowed colours for γ is $\{0, 1\}$. If the allowed colours for β and δ are $\{0, 1\}$ and $\{2, 3\}$ respectively, then a valid pattern is given by $(\alpha, \beta, \gamma, \delta) = (2, 0, 1, 3)$. If the allowed colours for β and δ are $\{0, 2\}, \{1, 3\}$ respectively, then a valid pattern is given by $(\alpha, \beta, \gamma, \delta) = (2, 0, 1, 3)$. The other cases are analogous. $\square$

We can now design a CA that corrects finite perturbations of Λ_4. Let $\psi : \Sigma^8 \to \Sigma^4$ be a function that maps some $(a, b \ldots, h) \in \Sigma^8$ to a quadruplet $(\alpha, \beta, \gamma, \delta) \in \Sigma^4$ such that the pattern formed by these values as illustrated above is an error-free pattern.

Our aim is to use this function ψ to correct non-overlapping 2-squares, by ensuring that the correcting rule applies without conflicts. In order to do this, we first identify a set of cells that will play the role of the top-right cells of the 2-squares that will be updated.

For a configuration $x \in \Sigma^{\mathbb{Z}^2}$, let us denote again the set of cells having a NE-error by $\mathcal{E}_{NE}(x) = \{c \in \mathbb{Z}^2 : x_c = x_{c+e_1} \text{ or } x_c = x_{c+e_2}\}$. We say that a cell $c \in \mathbb{Z}^2$ is a *NE-corner* if: $c \in \mathcal{E}_{NE}(x)$ and $c + e_1, c - e_1 + e_2, c + e_2, c + e_1 + e_2 \notin \mathcal{E}_{NE}(x)$, see Fig. 2 for an illustration of the definition. We denote by $\mathcal{C}_{NE}(x)$ the set of NE-corners in a configuration $x \in \Sigma^{\mathbb{Z}^2}$, that is:

$$\mathcal{C}_{NE}(x) \triangleq \{c \in \mathcal{E}_{NE}(x); \ c + e_1, c - e_1 + e_2, c + e_2, c + e_1 + e_2 \notin \mathcal{E}_{NE}(x)\}.$$

Note that if $x \in \tilde{\Lambda}_4$, then $\mathcal{E}(x) \neq \varnothing \iff \mathcal{C}_{NE}(x) \neq \varnothing$. Indeed, if $\mathcal{E}(x)$ is a non-empty set, then it contains at least one NE-error. Let us sweep the configuration x by NW-SE diagonals, from the NE to the SW. Since $\mathcal{E}(x)$ is finite, we can consider the first diagonal which contains a NE-error, and on this diagonal, we consider the leftmost NE-error (which is also the uppermost). By definition of a NE-corner, this NE-error is a NE-corner.

We define a CA F by the following rule: if a cell $c = (i, j) \in \mathbb{Z}^2$ is a NE-corner, then apply ψ to the 2-square whose NE-corner is c, that is, we replace the colours of the cells $(i - 1, j - 1), (i - 1, j), (i, j), (i, -j - 1)$ by $\psi(a, b, \ldots, h)$, where $a = x_{i-2,j-1}, b = x_{i-2,j}, \ldots, h = x_{i-1,j-2}$ (see above). Let us first observe that the CA F given by this rule is well-defined. Indeed, by definition of a NE-corner, one can check that there are no two consecutive NE-corners, vertically or horizontally, or in diagonal. Consequently, at each step, the 2-squares that are updated do not overlap (note however that they can share some edges, in which case there can be errors at these edges after applying the CA rule).

Proposition 5. *The cellular automaton F defined above is self-stabilising on Λ_4.*

Proof. Since the initial configuration x is assumed to be a finite perturbation of a valid colouring, the number of NE-corners is finite. We prove that on any

Fig. 2. Illustration of the definition of the cellular automaton used to correct 4-colourings. The central cell is a NE-corner if one of the red/dark gray lines (North or East or both) presents a mistake and all the green/light gray lines are free of errors. The 2-square whose NE-corner is the central cell is then corrected by the cellular automaton. (Color figure online)

configuration in $\tilde{\Lambda}_4 \setminus \Lambda_4$, the number of NE-corners is strictly decreasing. Since every configuration in $\tilde{\Lambda}_4 \setminus \Lambda_4$ has at least one NE-corner, this implies that the self-correction succeeds in finite time.

Let us consider the NE-corners of $F^t(x)$. The rule F consists in updating the 2-squares associated to these NE-corners. At the next time step, one can check that all possible new NE-corners belong to these 2-squares that were updated. Indeed, if a cell does not belong to such a 2-square, then it cannot become a NE-corner at the next time step: if a neighbour of this cell were modified, then its new colour respects the colour constraint. Furthermore, there is at most one new NE-corner in each 2-square that is updated, by definition of a NE-corner.

Now, to end the proof, let us show that there exists at least one of these 2-squares that does not contain a NE-corner any more. This will prove that the number of NE-corners is strictly decreasing. Let us sweep the configuration by NW-SE diagonals, from the NE to the SW.

We consider the first diagonal which contains a NE-corner. After applying F, the 2-squares defined by the NE-corners that are on that diagonal do not contain a NE-corner any more. Indeed, our method of sweeping ensures that the two cells to the North and the two cells to the East of this 2-square were not modified; for an illustration, see this figure: □

5.2   Isotropic Self-stabilisation by a Probabilistic CA

The problem of finding a rule which is isotropic and self-stabilising for four colours is not straightforward. We now propose a rule which we believe answers the problem, but for which we have no formal proof of success yet. Our idea is to modify the method used for the case $k \geq 5$, and make an exception when there is no colour available to directly correct a cell.

So, we now define ψ as a random function which assigns to each quadruplet of colours (a, b, c, d) a colour uniformly chosen in $\Sigma \setminus \{a, b, c, d\}$ if this set is not empty, and a colour uniformly chosen in Σ otherwise. We then consider the probabilistic cellular automaton that, for any configuration $x \in \Sigma^{\mathbb{Z}^2}$, updates to state $\psi(x_{c-e_1}, x_{c-e_2}, x_{c+e_1}, x_{c+e_2})$ the cell c if it has an error ($c \in \mathcal{E}(x)$), and keeps the value x_c otherwise.

Experimentally, we observe that this rule succeeds in correcting rapidly most of the initial perturbations of valid tilings. However, unlike the case $k \geq 5$, for $k = 4$, we cannot ensure with the PCA above that the errors stay in some bounded area.

We conjecture that from any finite perturbation of a valid tiling, this probabilistic cellular automaton almost surely reaches in finite time a valid 4-colouring. To support this claim, one can try to find configurations for which this rule may fail in correcting in finite time for some particular configurations.

Consider the following configuration:

1	2	3	0	1	2	3	0	1	2	3	0	1	2
3	0	1	2	3	0	1	2	3	0	1	2	3	0
2	3	0	1	2	3	0	0	1	2	3	0	1	2
0	1	2	3	0	1	2	3	0	1	2	3	0	1
2	3	0	1	2	3	0	1	2	3	0	1	2	3

It has two cells in error and is such that all cells, even the two that are in error, see the three other colours in their neighbourhood. Consequently, if a cell changes its state alone, it will remain in error. For this specific configuration, some kind of coordination is thus necessary, which cannot here occur by a specific mechanism as for the deterministic case.

It might thus be thought at first that errors may propagate arbitrary far from their origin. However, we experimentally observe that it is not the case: errors have a tendency to stay in the same area, and the correcting process is more rapid than the error-diffusion process. Surprisingly enough, even when the cells are updated successively at random (fully asynchronous case), we also noticed that the rule succeeds in correcting errors. Indeed, when the errors propagate, they modify the configuration in such a way that the property of seeing three different colours in the neighbourhood is lost, which finally enables a correction to take place. By comparison, we believe that for this configuration and this rule, having the possibility to make parallel updates, even if it means using α-asynchronous updates, can only increase the possibilities of correction. It is an open problem to give a formal proof of this self-stabilisation property.

5.3 Extension to ℓ-Fillable Proximity Tilings

We say that a proximity tiling is *strongly ℓ-fillable* if there exists a map ψ : $\Sigma^{4\ell} \to \Sigma^{\ell^2}$ such that, for any possible choice $(a_1, \ldots, a_{2\ell}) \in \Sigma^{4\ell}$ of symbols surrounding an ℓ-square, assigning the values $\psi(a_1, \ldots, a_{2\ell})$ to the inner cells of the ℓ-square ensures that each cell of the ℓ-square is error-free. (Note that here, we do not assume any further condition on $(a_1, \ldots, a_{2\ell}) \in \Sigma^{4\ell}$; we refer again to [1] for a similar but weaker condition of ℓ-fillability). The self-stabilisation by a deterministic CA described above extends to all proximity tilings that are strongly ℓ-fillable. One can indeed extend the notion of NE-corner in that context, see Fig. 3 for an illustration in the case $\ell = 3$. The definition of the CA and the proof that it is self-stabilising can then be easily adapted.

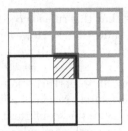

Fig. 3. Illustration of the definition of the cellular automaton used to correct a 3-fillable proximity tiling. The central cell is a NE-corner if one of the red lines (North or East or both) presents a mistake and all the green lines are free of errors. The 3-square whose NE-corner is the central cell is then corrected by the cellular automaton. (Color figure online)

6   The Case of 3-Colourings

6.1   Necessity to Correct Arbitrarily Far from the Locations of Errors

For $k \geq 4$, with the rules defined in the previous sections, one can correct the errors in a local way: if we observe a finite island of errors, then we can always correct the island without modifying the configuration at a distance larger from 1 or 2 from the island. Let us now show that this property no longer holds for $k = 3$. To this end, we will change our representation and associate to each configuration that is a 3-colouring a configuration in the so-called *six-vertex model*.

This model is obtained by associating an arrow to each couple of neighbouring cells (horizontal or vertical), these arrows are represented at the boundary between the two cells according to the following rules. Let q and q' be the colours of the two neighbouring cells. As we have $q' \neq q$, it follows that we either have $q' = q + 1 \mod 3$ or $q' = q - 1 \mod 3$. Depending on this, we draw the arrow in one direction or the other.

- The vertical boundaries which separate q and $q+1$ (resp. $q-1$) have an arrow pointing up (resp. down).
- The horizontal boundaries which separate q and $q + 1$ (resp. $q - 1$) have a right (resp. left) arrow.

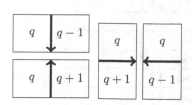 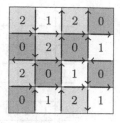

Fig. 4. The convention used for encoding 3-colouring configurations in the six-vertex model and an example of a configuration with its associated six-vertex image. (Color figure online)

These conventions are represented on Fig. 4.

One can then check that starting from a 3-colouring, the resulting arrow configuration is such that at each vertex, there are exactly two incoming arrows and two outgoing arrows. Conversely, from a six-vertex configuration, there are three 3-colourings giving that arrow configuration. (Once we choose the colour of one cell, all the other colours can be deduced).

Figure 4 shows an example of such an encoding of a valid colouring. By contrast, Fig. 5 displays a configuration which holds a finite perturbation of a 3-colouring.

Notice that we have drawn in bold the arrows pointing to the South and the ones pointing to the West. The knowledge of the position of these two types of arrows is sufficient to fully describe the configuration; indeed, the other horizontal or vertical arrows have to be East or North arrows, respectively.

In the example given, let us imagine that we have fixed the value of a set of cells that are located at the boundary of a square. We call this set of cells the *boundary square*, and we want to fill the inner part of that boundary square with an admissible configuration. One can verify that the only way to fill this inner part corresponds to a six-vertex configuration that would have a direct South vertical line: indeed, there is only one bold incoming arrow and one bold outgoing arrow in the boundary square, and we have to connect them. So, we can construct finite perturbations of 3-colourings that present only two cells in error (one single interface with same colours), but for which we need to modify a domain of size arbitrary large in order to recover a valid configuration. This is expressed by the following proposition.

Proposition 6. *For any $m \in \mathbb{N}$, there exists a configuration $y \in \tilde{\Lambda}_3$ such that* card $\mathcal{E}(y) = 2$, *and* $\forall x \in \Lambda_3$, card $\Delta(x, y) \geq m$.

6.2   Deterministic Self-stabilisation by a CA with Additional States

In order to decide if a boundary square is fillable or not, we just need to know if it is possible to associate each incoming arrow with an outgoing arrow. This is easy to do with sequential operations and additional symbols which do not appear in

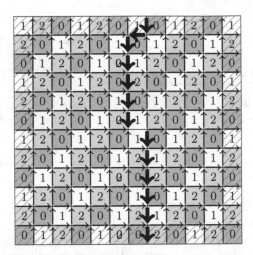

Fig. 5. Example of a finite perturbation of a 3-colouring and its associated six-vertex configuration. South and West arrows are shown in bold, dashed cells indicate the boundary square. (Color figure online)

the initial condition. Starting from the NE-corner, let us enumerate the incoming arrows on the North and then West sides, from 1 to n_i, and the outgoing arrows on the East and then South sides, from 1 to n_o. The boundary square is fillable if we can match each incoming arrow number k with the outgoing arrow number k by a SE-path of arrows (which implies that $n_i = n_o$). In order to know if this can be done, we try to match successively the incoming and outgoing arrows from 1 to $n_i = n_o$ by disjoint paths, by moving E if the edge has not already been selected, and S otherwise. As an additional condition, we need to ensure at each time step that the path does not go beyond the corresponding outgoing arrow or come across another path. This procedure succeeds if and only if there is at least one admissible matching, see the diagram below for an illustration:

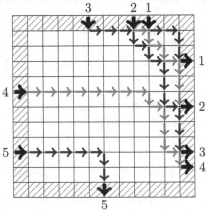

Using this method, let us sketch how to design a deterministic CA that corrects finite perturbations of 3-colourings with additional states. First, we mark error cells and create a boundary square around them. We then use a kind of Turing machine that calculates if this boundary square is fillable or not, with the procedure above. If it is fillable, we fill it with the solution associated to the six-vertex configuration given by the procedure. Otherwise, we consider a new boundary square with a box of size increased by one unit. When different boundary squares meet, they merge and restart their process.

It is an open problem to know if a solution without additional symbols exists.

Moreover, in contrast with the previous the sections, here we cannot use the method of taking an available colour or a random colour when no colour is available. We noticed experimentally that the errors diffuse and we could not find any rule that keeps them confined, even in statistical terms.

7   Conclusion

We presented the study of self-stabilisation problems for k-colourings and for some more general tilings spaces. The easiest cases are $k = 2$ and $k \geq 5$. For $k = 4$, deterministic rules still exist but are not as straightforward to design. In the probabilistic setting, we could propose symmetric rules, which experimentally perform well, but for which no formal proofs are available yet. The three-colour case is the most challenging and it is an open problem to know if efficient deterministic solutions do exist.

In this work, we have searched for solutions that operate in a "reasonable" time scale. However, when no such rules are found, it is still possible to use a kind of "brute-force" process where errors are initially at the centre of a self-correcting zone. The cellular automaton should then test sequentially if there are admissible solutions inside this zone. If the answer is positive, then the part is corrected, if not, then the zone is extended by one cell in each direction. When two such zones meet, there should be some procedures to merge the zones and "restart" the process. It is clear that even though each step can be thought of separately in a clear way, putting all the steps together in a cellular automaton that effectively works is a huge task. Moreover, the time needed for such a rule to operate would be more than exponential in the number of errors.

The question might also be raised for the solutions which make use of additional symbols: can one find rules which also resist the introduction of additional symbols in the initial condition? Another important problem that we are currently addressing is to consider the case where the errors are initially randomly distributed on *all the grid*.

References

1. Alon, N., Briceño, R., Chandgotia, N., Magazinov, A., Spinka, Y.: Mixing properties of colorings of the $\mathbb{Z}^d$ lattice. Preprint arXiv:1903.11685 (2019)

2. Bębenek, A., Ziuzia-Graczyk, I.: Fidelity of DNA replication—a matter of proof-reading. Curr. Genet. **64**, 985–996 (2018)
3. Bušić, A., Fatès, N., Mairesse, J., Marcovici, I.: Density classification on infinite lattices and trees. Electron. J. Probab. **18**(51), 22 (2013)
4. Dijkstra, E.W.: Self-stabilization in spite of distributed control. In: Dijkstra, E.W. (ed.) Selected Writings on Computing: A personal Perspective. Texts and Monographs in Computer Science, pp. 41–46. Springer, New York (1982). https://doi.org/10.1007/978-1-4612-5695-3_7
5. Fatès, N.: Asynchronous cellular automata. In: Adamatzky, A. (ed.) Cellular Automata. ECSSS, pp. 73–92. Springer, New York (2018). https://doi.org/10.1007/978-1-4939-8700-9_671
6. Gach, P., Kurdyumov, G.L., Levin, L.A.: One-dimensional uniform arrays that wash out finite islands. Probl. Inf. Transm. **14**(3), 223–226 (1978)
7. Gács, P.: Reliable computation with cellular automata. J. Comput. Syst. Sci. **32**(1), 15–78 (1986)
8. Gács, P., Reif, J.: A simple three-dimensional real-time reliable cellular array. J. Comput. Syst. Sci. **36**(2), 125–147 (1988)
9. Marcus, B., Pavlov, R.: An integral representation for topological pressure in terms of conditional probabilities. Isr. J. Math. **207**(1), 395–433 (2017)
10. Pippenger, N.: Symmetry in self-correcting cellular automata. J. Comput. Syst. Sci. **49**(1), 83–95 (1994)
11. Toom, A.L.: Nonergodic multidimensional system of automata. Probl. Peredachi Inf. **10**(3), 70–79 (1974)
12. Toom, A.L.: Stable and attractive trajectories in multicomponent systems. In: Multicomponent Random Systems. Advances in Probability. Related Topics, Dekker, New York, vol. 6, pp. 549–575 (1980)

On the Termination Problem for Counter Machines with Incrementing Errors

Christopher Hampson[✉]

Department of Informatics, King's College, London, UK
christopher.hampson@kcl.ac.uk

Abstract. In contrast to their reliable and lossy-error counterparts whose termination problems are either undecidable or non-primitive recursive, the termination problem for counter machines with *incrementing errors* is shown to be PSPACE-hard but remains solvable in EXPSPACE. This is a notable decrease in complexity over that of insertion-error *channel systems* (with emptiness testing) whose termination problem is known to be non-elementary. Furthermore, by fixing the number of available counters, we obtain a tight NLOGSPACE-complete bound for the termination problem.

Keywords: Termination · Halting problem ·
Unreliable counter machines · Incrementing error · Lossy error

1 Introduction

Reliable (Minsky) counter machines are well-known to be Turing-complete [12] and their reachability and termination problems have served as invaluable 'master' problems in establishing undecidable lower-bounds for a range of diverse decision problems. Furthermore, two counters are sufficient to establish Turing-completeness [12]. *Lossy counter machines (LCMs)*, by contrast, were introduced by Mayr [11] as a weakened version of Minsky's counter machines whose counters are permitted to spontaneously 'leak' their contents, analogous to that of the much-studied lossy FIFO-channel systems [1,2,4,5]. Indeed, LCMs can be seen as a degenerate case of lossy channel systems (with emptiness testing) in which the channel alphabet comprises a single symbol. Mayr showed that the reachability and termination problems for LCMs are both decidable [11], with the exact complexity pinned at being ACKERMANN-complete by Schnoebelen [18] (see [17] for a comprehensive survey of non-elementary complexity classes). Indeed, just five counters are sufficient to establish non-elementary complexity, with each additional counter moving the problem further up the Fast Growing Hierarchy [18].

Less well-studied than LCMs are *incrementing counter machines (ICMs)* which are permitted to spontaneously increase the value of their counters. Incrementing errors have been considered in the context of both counter machines and their more expressive channel systems (both with and without emptiness

© Springer Nature Switzerland AG 2019
E. Filiot et al. (Eds.): RP 2019, LNCS 11674, pp. 137–148, 2019.
https://doi.org/10.1007/978-3-030-30806-3_11

testing) [4,6,13,15], but appear to have received far less attention than their lossy counterparts.

Insertion channel systems (without emptiness testing) were first introduced in [4], where the authors show that the termination problem (among others) is trivially decidable as every transition can be traversed with the aid of timely insertion errors. The problem thus reduces to that of cycle-finding in the underlying control-state diagram. In the presence of emptiness testing—more akin to the operational semantics of incrementing counter machines—the termination problem was shown to be TOWER-complete [3], being among the hardest problems that are primitive recursive but not solvable in elementary time [17].

With regards to the control-state reachability problem, there is no difference between lossy errors and incrementing/insertion errors for counter machines or channel systems, owing to a dualisation that reverses the 'arrow of time' [6,14]. Consequently, the reachability problem for both LCMs and ICMs is ACKERMANN-complete [6,18], while that of both lossy channel systems and incrementing channel systems (with emptiness testing) is HYPERACKERMANN-complete [5,14].

It appears, however, that the problem of termination for incrementing counter machines has remained unaddressed. In what follows we shall establish that the termination problem is, in general, PSPACE-hard but remains decidable in EXPSPACE. Furthermore, we show that the problem is even NLOGSPACE-complete when restricted to a fixed (finite) number of counters. Table 1 summarizes the known results relating to the termination problems for counter machines and channel systems (with emptiness testing) in the presence of lossy and incrementing errors.

Table 1. Summary of termination known results for lossy and incrementing counter machines and channel systems (with emptiness testing).

	Lossy	*Incrementing*
Channel Systems (emptiness testing)	HYPERACKERMANN-complete [5]	TOWER-complete [3]
Counter Machines	ACKERMANN-complete [18]	PSPACE-hard in EXPSPACE Theorems 3.2 & 3.3
Counter Machines with k counters	non-ELEMENTARY for $k > 5$ [18]	NLOGSPACE-complete Theorems 3.4 & 3.4

2   Preliminaries

Definition 2.1. A *counter machine* is a tuple $\mathcal{M} = \langle Q, C, q_{\text{init}}, \Delta \rangle$ where Q is a finite set of control-states with a designated initial state $q_{\text{init}} \in Q$, $C = \{c_1, \ldots, c_n\}$ is a finite set of counters and $\Delta \subseteq Q \times Op_C \times Q$ is a finite set of state transitions labelled with one of the following operations

$Op_C = \{(c_i)^{++}, (c_i)^{--}, (c_i)^{??} : c_i \in C\}$ to *increment, decrement,* or *test* whether a given counter is empty.

A *configuration* of $\mathcal{M}$ is a tuple $(q, \boldsymbol{v}) \in Q \times \mathbb{N}^C$, where $q \in Q$ dictates the state of the machine and $\boldsymbol{v} : C \to \mathbb{N}$ is C-vector describing the contents of each counter. We denote by $\mathrm{Conf}_{\mathcal{M}}$ the set of all possible configurations of $\mathcal{M}$, and define a well-quasiordering (wqo) $\leq$ on $\mathrm{Conf}_{\mathcal{M}}$ by taking

$$(q, \boldsymbol{v}) \leq (q', \boldsymbol{v}') \quad \Longleftrightarrow \quad q = q' \text{ and } \boldsymbol{v}(c_i) \leq \boldsymbol{v}'(c_i) \text{ for all } c_i \in C,$$

for $(q, \boldsymbol{v}), (q', \boldsymbol{v}') \in \mathrm{Conf}_{\mathcal{M}}$. For each $\alpha \in Op_C$, we define a binary consecution relation on the configurations of $\mathcal{M}$ by taking:

- $(q, \boldsymbol{v}) \xrightarrow{(c_i)^{++}} (q', \boldsymbol{v}')$ iff $(q, (c_i)^{++}, q') \in \Delta$ and $\boldsymbol{v}' = \boldsymbol{v} + \boldsymbol{e}_i$,
- $(q, \boldsymbol{v}) \xrightarrow{(c_i)^{--}} (q', \boldsymbol{v}')$ iff $(q, (c_i)^{--}, q') \in \Delta$ and $\boldsymbol{v}' = \boldsymbol{v} - \boldsymbol{e}_i$,
- $(q, \boldsymbol{v}) \xrightarrow{(c_i)^{??}} (q', \boldsymbol{v}')$ iff $(q, (c_i)^{??}, q') \in \Delta$ and $\boldsymbol{v} = \boldsymbol{v}'$ with $\boldsymbol{v}(c_i) = 0$,

for all $(q, \boldsymbol{v}), (q, \boldsymbol{v}') \in \mathrm{Conf}_{\mathcal{M}}$, where $\boldsymbol{e}_i$ is the unit vector with $\boldsymbol{e}_i(c_i) = 1$ and $\boldsymbol{e}_i(c_j) = 0$ for $j \neq i$. Note that transitions of the from $(q, (c_i)^{--}, q')$ are only enabled when $\boldsymbol{v}(c_i)$ is non-zero. We write $(q, \boldsymbol{v}) \xrightarrow{\mathcal{M}} (q', \boldsymbol{v})$ if $(q, \boldsymbol{v}) \xrightarrow{\alpha} (q', \boldsymbol{v}')$ for some $\alpha \in Op_C$. A *(reliable) computation / run* of $\mathcal{M}$ is a sequence $r = \langle(\sigma_{k-1}, \alpha_k, \sigma_k) \in \mathrm{Conf}_{\mathcal{M}} \times Op_C \times \mathrm{Conf}_{\mathcal{M}} : 0 < k < L\rangle$, for some $1 < L \leq \omega$, such that $\sigma_0 = (q_{\mathrm{init}}, \boldsymbol{0})$, where $\boldsymbol{0}$ is the zero vector, and $\sigma_{k-1} \xrightarrow{\alpha_k} \sigma_k$, for all $0 < k < L$. We denote by $\mathrm{Runs}(\mathcal{M})$ the set of all reliable runs of $\mathcal{M}$.

Lossy counter machines (LCMs) and *incrementing counter machines (ICMs)* can be defined by way of a variation on the operational semantics of what it means for two configurations to be consecutive.

Definition 2.2 (Lossy and Incrementing counter machines). Given a counter machine $\mathcal{M}$, we define the relations $\xrightarrow{\alpha\downarrow}$ and $\xrightarrow{\alpha\uparrow}$ on $\mathrm{Conf}_{\mathcal{M}}$ by taking $\sigma_1 \xrightarrow{\alpha\downarrow} \sigma_2$ (resp. $\sigma_1 \xrightarrow{\alpha\uparrow} \sigma_2$) if and only if there are configurations $\sigma_1', \sigma_2' \in \mathrm{Conf}_{\mathcal{M}}$ such that $\sigma_1 \geq \sigma_1' \xrightarrow{\alpha} \sigma_2' \geq \sigma_2$ (resp. $\sigma_1 \leq \sigma_1' \xrightarrow{\alpha} \sigma_2' \leq \sigma_2$), which is to say that we permit the value held in the counters to spontaneously decrease (resp. increase) immediately prior to and subsequent to a reliable transition. We write $(q, \boldsymbol{v}) \xrightarrow{\mathcal{M}\downarrow} (q', \boldsymbol{v})$ (resp. $(q, \boldsymbol{v}) \xrightarrow{\mathcal{M}\uparrow} (q', \boldsymbol{v})$) if $(q, \boldsymbol{v}) \xrightarrow{\alpha\downarrow} (q', \boldsymbol{v}')$ (resp. $(q, \boldsymbol{v}) \xrightarrow{\alpha\uparrow} (q', \boldsymbol{v})$) for some $\alpha \in Op_C$.

However, for the purposes of control-state reachability and termination, it is convenient to work with a more restrictive form of incrementing errors that encroach upon our computations only at the point of decrementing an otherwise empty counter. More precisely, we employ the following alternative definition for $(q, \boldsymbol{v}) \xrightarrow{\alpha\uparrow} (q', \boldsymbol{v})$, for $\alpha \in Op_C$:

- $(q, \boldsymbol{v}) \xrightarrow{(c_i)^{++}\uparrow} (q', \boldsymbol{v}')$ iff $(q, \boldsymbol{v}) \xrightarrow{(c_i)^{++}} (q', \boldsymbol{v}')$,
- $(q, \boldsymbol{v}) \xrightarrow{(c_i)^{--}\uparrow} (q', \boldsymbol{v}')$ iff $(q, \boldsymbol{v}) \xrightarrow{(c_i)^{--}} (q', \boldsymbol{v}')$ or $(q, \boldsymbol{v}) \xrightarrow{(c_i)^{??}} (q', \boldsymbol{v}')$,

$- (q, \boldsymbol{v}) \xrightarrow{(c_i)^{??} \uparrow} (q', \boldsymbol{v}')$ iff $(q, \boldsymbol{v}) \xrightarrow{(c_i)^{??}} (q', \boldsymbol{v}')$.

Such lazy 'just-in-time' incrementing semantics have been introduced for incrementing channel systems in [3] and used implicitly for counter machines in [6].

We define a *lossy (resp. incrementing) computation / run* of $\mathcal{M}$ to be a sequence $r = \langle (\sigma_{k-1}, \alpha_k, \sigma_k) \in \mathrm{Conf}_{\mathcal{M}} \times Op_C \times \mathrm{Conf}_{\mathcal{M}} : 0 < k < L \rangle$, for some $1 < L \leq \omega$, such that $\sigma_0 = (q_{\mathsf{init}}, \boldsymbol{0})$, where $\boldsymbol{0}$ is the zero vector, and $\sigma_{k-1} \xrightarrow{\alpha_k \downarrow} \sigma_k$ (resp. $\sigma_{k-1} \xrightarrow{\alpha_k \uparrow} \sigma_k$), for all $0 < k < L$. We denote by $\mathsf{Runs}^{\downarrow}(\mathcal{M})$ and $\mathsf{Runs}^{\uparrow}(\mathcal{M})$ the set of all lossy and incrementing runs of $\mathcal{M}$, respectively.

In what follows, we will be primarily interested in the following decision problem:

ICM Termination:

Input: Given a counter machine $\mathcal{M} = \langle Q, C, q_{\mathsf{init}}, \Delta \rangle$,
Question: Is every incrementing run $r \in \mathsf{Runs}^{\uparrow}(\mathcal{M})$ finite?

In addition to the termination problem, we will also consider the restricted case where we admit only counter machines with a fixed number of counters.

k-ICM Termination:

Input: Given a counter machine $\mathcal{M} = \langle Q, C, q_{\mathsf{init}}, \Delta \rangle$ such that $|C| = k$,
Question: Is every incrementing run $r \in \mathsf{Runs}^{\uparrow}(\mathcal{M})$ finite?

As noted above, for reliable counter machines the two problems are computationally equivalent for $k \geq 2$, but are known to differ in complexity for lossy counter machines. We will show here that the two problems also differ in complexity for incrementing counter machines.

3   Results

We first show that the termination problem for ICMs is decidable in ExpSpace by establishing a doubly-exponential upper-bound on the length of all finite runs that are possible for a terminating ICM. Any incrementing counter machine exhibiting a finite run exceeding this bound must necessarily possess a non-terminating run. The termination problem can therefore be decided by a non-deterministic search for such a 'long' finite run which, once found, demonstrates non-termination. Such a search can be performed using at most exponential space. It then follows from Savitch's Theorem [16] that the termination problem is decidable in ExpSpace. This is a marked contrast from the Tower-completeness of the termination problem for incrementing channel systems (with emptiness testing) [3].

Lemma 3.1. *Let $\mathcal{M} = \langle Q, C, q_{\mathsf{init}}, \Delta \rangle$ be a counter machine such that every incrementing run $r \in \mathsf{Runs}^{\uparrow}(\mathcal{M})$ is finite. Then the length of each incrementing run is at most $n^{2em!}$, where $n = |Q|$, $m = |C|$, and e is the base of the natural logarithm.*

Proof. Let $\mathcal{M} = \langle Q, C, q_{\text{init}}, \Delta \rangle$ be as described above, with $|Q| = n$ and $|C| = m$, and let $r \in \text{Runs}^\uparrow(\mathcal{M})$ be any incrementing run of $\mathcal{M}$. The case where $n = 1$ is trivial, so we may assume that $n \geq 2$. We shall refer to any sub-sequence of r as an *interval*, with its length being the number of configuration transitions it comprises. For brevity we shall refer to any transition of the form $(\sigma, (c_i)^{??}, \sigma') \in \text{Conf}_{\mathcal{M}} \times Op_C \times \text{Conf}_{\mathcal{M}}$ as a c_i-*gate* and collectively as Σ-*gates* whenever $c_i \in \Sigma$, for $\Sigma \subseteq C$. An interval will be described as *gate-free* whenever it contains no C-gates.

To facilitate the proof, we define a increasing function $T : \mathbb{N} \to \mathbb{N}$ recursively by taking

$$T(0) = 1 \quad \text{and} \quad T(k) = kT(k-1) + 2 \tag{†}$$

for all $k > 0$. It follows from a straightforward induction that

$$T(k) = k! \left(\frac{1}{0!} + \frac{2}{1!} + \cdots + \frac{2}{(k-1)!} + \frac{2}{k!} \right) < 2k! \sum_{t=0}^{\infty} \frac{1}{t!} = 2ek!$$

for all $k \geq 0$, where e is the base of the natural logarithm.

For each subset $\Sigma \subseteq C$, let $\chi^r(\Sigma)$ denote the length of the longest interval in which the only gates traversed belong to Σ. We prove by induction on the size of Σ that

$$\chi^r(\Sigma) < n^{T(|\Sigma|)}.$$

for all subsets $\Sigma \subseteq C$.

- *Base Case.* For the case where $|\Sigma| = 0$, we note that $\chi^r(\emptyset) < n$ since otherwise, by the pigeonhole principle, there would be some gate-free interval I in which the same control-state appears twice. We could then construct a non-terminating run by traversing the resulting loop indefinitely, as every underlying state transition of the form $(q, (c_i)^{++}, q')$ or $(q, (c_i)^{--}, q')$ can always be traversed. This contradicts our assumption that every incrementing computation of $\mathcal{M}$ is terminating.

- *Inductive Case.* Suppose that the claim holds for all subsets of size $\leq k$ and that $|\Sigma| = (k+1)$. Suppose to the contrary that $\chi^r(\Sigma) \geq n^{T(k+1)}$ and let $I = \langle (\sigma_t, \alpha_t, \sigma_{t+1}) : t < \chi \rangle$, be such an interval of length $\chi = \chi^r(\Sigma)$. It follows from that induction hypothesis that I contains at least one c_j-gate, for each $c_j \in \Sigma$.

  Choose $c_i \in \Sigma$ and partition I into subintervals $I_1, \ldots, I_s$ by abscising all c_i-gates, as illustrated in Fig. 1. Note that we can abscise at most n consecutive c_i-gates between each interval, else by the same argument as above, we could construct a non-terminating run by traversing a loop of such gates indefinitely. The resulting subintervals only contain Σ'-gates, where $\Sigma' = \Sigma - \{c_i\}$. Hence, by the induction hypothesis, the length of each subinterval can be at most $\chi^r(\Sigma') < n^{T(k)}$, since $|\Sigma'| = k$. It follows that

  $$|I| \leq s \cdot \chi^r(\Sigma') + (s+1)n$$

which is to say that

$$s \geq \frac{|I| - n}{\chi^r(\Sigma') + n} > \frac{n^{T(k+1)} - n}{n^{T(k)} + n} \geq \frac{1}{2} \cdot \frac{n^{T(k+1)-1}}{n^{T(k)-1}} \geq n^{T(k+1)-T(k)-1}$$

for $n \geq 2$. It then follows from (†) that $s > n^{k\,T(k)+1}$.

For each subinterval I_ℓ, let $\text{start}(\ell) = (q, v \restriction \Sigma')$ denote the configuration at the start of I_ℓ restricted to only those counters occurring in Σ'.

Note that for each $c_j \in \Sigma'$, the first transition of I_ℓ must appear in some interval I' in which no c_j-gate appears, or else must be itself a c_j-gate. In the latter case, we require that the value of $v(c_j)$ at $\text{start}(\ell)$ is zero. In the former case, either the start of I' or the end of I' is contained within I, else I would not contain any c_j-gates.

- If the start of I' is contained in I then the value of $v(c_j)$ at the start of $\text{start}(\ell)$ can be at most $\chi^r(\Sigma')$ since otherwise the counter could not have been incremented (using the lazy semantics) since being emptied to traverse the gate at the start of I'.
- Alternatively, if the end of I' is contained in I then the value of $v(c_j)$ at $\text{start}(\ell)$ can be at most $\chi^r(\Sigma')$ since otherwise the counter could not be depleted in time to traverse the gate at the end of I'.

Hence, for each $\ell = 1, \ldots, s$, there are at most n choices for the state of $\text{start}(\ell)$ and $\chi^r(\Sigma') < n^{T(k)}$ choices for the values of each of the counters $c_j \in \Sigma'$ in $\text{start}(\ell)$. This gives at most $n \cdot (n^{T(k)})^k = n^{k\,T(k)+1}$ possible choices for $\text{start}(\ell)$. However, since $s > n^{k\,T(k)+1}$, by the pigeonhole principle there must be at least two intervals I_ℓ and $I_{\ell'}$ such that $\text{start}(\ell) = \text{start}(\ell')$, where $1 \leq \ell < \ell' \leq s$.

Consequently, it is possible to construct a non-terminating run by traversing the resulting loop indefinitely, as the two partial states agree on all counters from Σ', and the only c_j-gates to be traversed are those from Σ'. Any counters from $(C - \Sigma')$ are free to be incremented or decremented without impeding the computation. Again, this contradicts our assumption that every incrementing run $r' \in \text{Runs}^\dagger(\mathcal{M})$ is terminating. Hence, by contradiction, we must have that $\chi^r(\Sigma) < n^{T(k+1)}$.

Hence, it follows that $\chi^r(\Sigma) < n^{T(|\Sigma|)}$ for all $\Sigma \subseteq C$. In particular, we note that the maximum length of r is given by $\chi^r(C) < n^{T(m)} < n^{2em!}$, as required.   □

With this upper-bound placed on the maximum possible length of runs for terminating ICMs, we are able to secure an ExpSpace upper-bound on the complexity of the termination problem.

Theorem 3.2. *The ICM Termination problem is decidable in* ExpSpace.

Proof. By Lemma 3.1, it is sufficient to identify whether a given counter machine has a finite run whose length exceeds $n^{2em!}$, where $n = |Q|$ and $m = |C|$. This can be achieved via a non-deterministic search using at most exponential space,

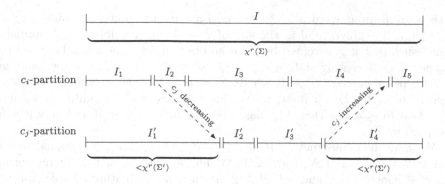

Fig. 1. Illustration of two partitionings of I into subintervals by abscising all c_i-gates and all c_j-gates, respectively.

by storing only the current length and final configuration of the run as the search progresses. Both the length and the final configuration can be encoded as binary strings requiring at most $O(\log_2(n)m!)$ bits of data, which is at most exponential in m and logarithmic in n. Should a run exceeding the aforementioned bound by found, we may conclude that the counter machine has a non-terminating incrementing run.

Hence, the *non*-termination problem for incrementing counter machines is decidable in NEXPSPACE, and so it follows that both the termination and non-termination problems belong to EXPSPACE, as required. □

This result stands in marked contrast to the lofty ACKERMANN-completeness of the termination problem for *lossy* counter machines [18], despite the equivalence of the reachability problem for the two types of unreliable machines [13]. Moreover, this result also highlights a jump in complexity from the relatively modest EXPSPACE for incrementing counter machines to the non-ELEMENTARY complexity for incrementing *channel systems* (with emptiness testing) [3].

Next, we will provide a lower-bound on the complexity of the termination problem for ICMs by showing that an incrementing counter machine with $m = |C|$ counters is capable of simulating a run of a *reliable* counter machine whose counters are bounded by $2^{\lfloor m/2 \rfloor}$. This, in turn, provides us with a mechanism by which we can simulate any Turing machine that operates in space bounded by $\lfloor m/2 \rfloor$, thereby providing us with a PSPACE-hard lower-bound for the termination problem for ICMs.

Theorem 3.3. *The ICM Termination problem is* PSPACE-*hard.*

Proof. Let $X \subseteq \{0,1\}^*$ be an arbitrary problem solvable in PSPACE, which is to say that there is some Turing machine $\mathcal{T}_X$ and polynomial function $p(n)$ such that $\mathcal{T}_X$ terminates on all inputs and accepts $w \in \{0,1\}^*$ if and only if $w \in X$, using at most $p(|w|)$ tape cells. Following Minksy [12], we may translate $\mathcal{T}_X$ together with a given input word $w \in \{0,1\}^*$ into a reliable counter machine $\mathcal{M}_w^X = \langle Q, C, q_{\text{init}}, \Delta \rangle$—polynomial in the size of w and constructible in polynomially time—such that w is accepted by $\mathcal{T}_X$ if and only if $\mathcal{M}_w^X$ has a reliable run that reaches some accepting state $q_{\text{accept}} \in Q$. Moreover, the value of the counters of $\mathcal{M}_w^X$ never exceeds $2^N - 1$, where $N = p(|w|)$ is the maximum length of tape required by $\mathcal{T}_X$ on input w. We may modify $\mathcal{M}_w^X$ by adding a looping transition to q_{accept} so that $\mathcal{M}_w^X$ has a non-terminating run if and only if w is accepted by $\mathcal{T}_X$.

We may then construct an ICM $\mathcal{M}' = \langle Q', C', q'_{\text{init}}, \Delta' \rangle$, polynomial in the size of both $\mathcal{M}_w^X$ and N, such that $\mathcal{M}'$ has a non-terminating incrementing run $r' \in \text{Runs}^\uparrow(\mathcal{M}')$ if and only if $\mathcal{M}_w^X$ has a non-terminating reliable run $r \in \text{Runs}(\mathcal{M}_w^X)$, along which the counters are bounded by 2^N. To achieve this, we first designate counters $c_i^0, \ldots, c_i^{N-1} \in C'$, for each $c_i \in C$, so that the value of counter c_i for a given valuation $v : C \to \mathbb{N}$ can be represented in binary as

$$\theta_v(c_i) = \sum_{j=0}^{N-1} 2^j \min\{1, v(c_i^j)\}$$

In other words, the emptiness (0) or non-emptiness (1) of each of the counters c_i^j collectively represent the value of c_i in binary.

We also require a second copy $\bar{c}_i^0, \ldots, \bar{c}_i^{N-1} \in C'$, for each $c_i \in C$, so that any incrementing errors can be detected and the computation terminated as a result. To achieve this we shall enforce that any increment (resp. decrement) to c_i^j is followed by a decrement (resp. increment) to $\bar{c}_i^j$ so that, over reliable runs, the pair $(c_i^j, \bar{c}_i^j)$ acts like a binary switch with exactly one of the counters being empty at any given time.   For each $(\ell, \alpha, \ell') \in \Delta$, we construct a circuit of

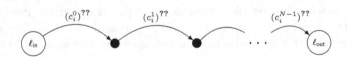

Fig. 2. Circuit emulating the operation $(c_i)^{??}$.

transitions that emulates the effect of α on the corresponding value of θ.

- *Case $\alpha = (c_i)^{??}$*: We can check whether $\theta_v(c_i) = 0$ by a series of emptiness checks to confirm that each of the counters c_i^j are empty, for $j < N$, as illustrated in Fig. 2.

  It is straightforward to check that there is an incrementing path $(\ell_{\text{in}}, v) \xrightarrow{\mathcal{M}\uparrow}^* (\ell_{\text{out}}, v')$ if and only if $\theta_v(c_i) = 0$.
- *Case $\alpha = (c_i)^{++}$*: To increment the value of $\theta_v(c_i)$ by one, we can execute the circuit illustrated in Fig. 3.

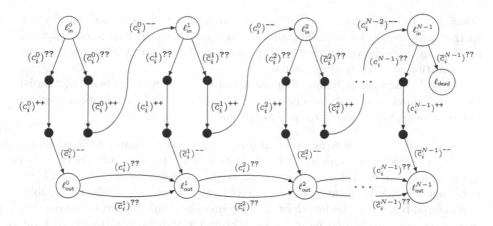

Fig. 3. Circuit emulating the operation $(c_i)^{++}$.

A successful computation through this circuit from ℓ_{in}^0 to ℓ_{out}^{N-1} simulates standard binary addition by one by 'resetting' each c_i counters to zero in turn until the first empty c_i counter is found, which is then set to one, resetting $\bar{c}_i$ to zero in the process. Any remaining counters that have not been inspected are then checked to ensure that exactly one of c_i and $\bar{c}_i$ are non-zero. It follows that there is an incrementing path $(\ell_{in}^0, \boldsymbol{v}) \xrightarrow{\mathcal{M}\uparrow}^* (\ell_{out}^{N-1}, \boldsymbol{v}')$ if and only if:

(i) $\boldsymbol{v}(c_i^j) + \boldsymbol{v}(\bar{c}_i^j) = 1$, for all $j < N$, and

(ii) If $\boldsymbol{v}'(c_i^j) + \boldsymbol{v}'(\bar{c}_i^j) = 1$ then $\theta_{v'}(c_i) = \theta_v(c_i) + 1$.

Note that in the case where $\theta_v(c_i) = 2^N - 1$ (*i.e.* the counter is full) it is not possible to reach ℓ_{out}^{N-1} and instead we terminate in a dead-end state ℓ_{dead}.

– *Case* $\alpha = (c_i)^{--}$: To decrement the value of $\theta_v(c_i)$ by one, we can use an analogous set of a transitions, but with the roles of c_i^j and $\bar{c}_i^j$ exchanged, for $j < N$. The resulting circuit simulates binary subtraction by one and then ensures that exactly one of c_i and $\bar{c}_i$ are non-zero. It is similarly straightforward to check that there is an incrementing path $(\ell_{in}^0, \boldsymbol{v}) \xrightarrow{\mathcal{M}\uparrow}^* (\ell_{out}^{N-1}, \boldsymbol{v}')$ in the resulting circuit if and only if:

(i) $\boldsymbol{v}(c_i^j) + \boldsymbol{v}(\bar{c}_i^j) = 1$, for all $j < N$, and

(ii) If $\boldsymbol{v}'(c_i^j) + \boldsymbol{v}'(\bar{c}_i^j) = 1$ then $\theta_{v'}(c_i) = \theta_v(c_i) - 1$.

Similarly, in the case where $\theta_v(c_i) = 0$ it is not possible to reach ℓ_{out}^{N-1} and we terminate in state ℓ_{dead}.

Note that each of the circuits are cycle-free and so do not allow for non-terminating computations to arise *within* the individual circuit. It follows that we can construct an equivalent ICM $\mathcal{M}'$ by replacing each of the transitions of $\mathcal{M}_w^X$ with a copy of the appropriate circuit described above, each comprising at most $8N$ transitions. The resulting machine is at most polynomial in the size of $\mathcal{M}_w^X$ and N, with $|Q'| \leq 8N \cdot |\Delta| \leq 8Nn^2$ and $|C'| = 2Nm$, and has a non-terminating incrementing run if and only if $\mathcal{M}_w^X$ has a reliable non-terminating run with counters bounded by 2^N. We could, as well, introduce a sequence of transitions to the initial state that first increment each of the $\overline{c}_i$ variables by one so that $v(c_i) + v(\overline{c}_i) = 1$ at the start of the run. However, this is not required as this can be achieved with a timely incrementing error, without which any computation would quickly terminate.

As $\mathcal{M}_w^X$ is at most polynomial in the size of w and constructible in polynomial time, it follows that X is polynomially reducible to ICM Termination, thereby demonstrating the problem to be PSPACE-hard, as required.   □

Note that the above reduction requires an unbounded supply of counters for ever larger values of N. Indeed, such a reduction is not possible using only a fixed number of counters. Taking a closer look at the bound given in Lemma 3.1, we note that it is chiefly the number of counters m that is responsible for the doubly-exponential bound on the length of the incrementing runs. By fixing the number of counters, we obtain a far more tractable bound with little additional overhead.

Theorem 3.4. *The k-Termination problem is* NLOGSPACE-*complete.*

Proof. The lower-bound is trivial and follows from a straightforward logspace reduction from the NLOGSPACE-hard reachability problem for directed graphs [10]. For the upper-bound, the proof is the same as that of Theorem 3.2, noting that for a fixed number of counters the bound given in Lemma 3.1 is logarithmic in the number of states. This, therefore, gives us an NLOGSPACE upper-bound for the *non*-termination problem. However, by the Immerman–Szelepcsényi theorem [9, 19] we have that NLOGSPACE is closed under complements, thereby completing the proof.   □

4   Discussion

1. The main problem left open by this present work is to establish where lies the exact complexity of the termination problem for incrementing counter machines. Using the same principle as in Theorem 3.3, it is not hard to construct a *terminating* ICM with exponentially long runs; for example, by connecting the state ℓ_{out}^{N-1} back to state ℓ_{in}^0 in Fig. 3. The resulting circuit contains no non-terminating computations, but is permitted to cycle through all binary representations from zero to $2^N - 1$ before terminating in state ℓ_{dead}. Unfortunately, it remains unclear whether it is possible to construct terminating ICMs which have *doubly*-exponentially long runs that would be required for the EXPSPACE upper-bound given in Theorem 3.2 to be tight.

2. Reductions from various counter machine reachability problems have been used to establish lower-bounds for several first-order modal and temporal logics endowed with additional counting quantifiers [8]. Their lossy and incrementing counterparts arise naturally in this context when we consider first-order modal logics with decreasing or expanding domains, respectively [7, 8]. In particular, the recurrence problem for ICMs can be reduced to the satisfiability problem for the one-variable fragment of Linear Temporal Logic (LTL)

over expanding domains with both future and next-time operators, thereby providing a Σ_1^0-hard lower-bound. Finite satisfiability, though decidable, can be shown to be ACKERMANN-hard by a reduction from the ICM reachability problem; this remains true even in the absence of the 'next-time' operator. For the fragment having a single future operator, Theorem 3.3 can be utilized to provide a PSPACE-lower bound for the satisfiability problem[1]. However, it is reasonable to suspect that the exact complexity may be much higher and, indeed, the decidability of this fragment still remains open.

Acknowledgements. I would like to thank the anonymous referees for their helpful and invaluable suggestions.

References

1. Abdulla, P.A., Collomb-Annichini, A., Bouajjani, A., Jonsson, B.: Using forward reachability analysis for verification of lossy channel systems. Form. Methods Syst. Des. **25**(1), 39–65 (2004)
2. Abdulla, P.A., Jonsson, B.: Verifying programs with unreliable channels. Inf. Comput. **127**(2), 91–101 (1996)
3. Bouyer, P., Markey, N., Ouaknine, J., Schnoebelen, P., Worrell, J.: On termination and invariance for faulty channel machines. Form. Asp. Comput. **24**(4), 595–607 (2012)
4. Cécé, G., Finkel, A., Iyer, S.P.: Unreliable channels are easier to verify than perfect channels. Inf. Comput. **124**(1), 20–31 (1996)
5. Chambart, P., Schnoebelen, P.: The ordinal recursive complexity of lossy channel systems. In: 2008 23rd Annual IEEE Symposium on Logic in Computer Science, pp. 205–216. IEEE (2008)
6. Demri, S., Lazić, R.: LTL with the freeze quantifier and register automata. ACM Trans. Comput. Log. (TOCL) **10**(3), 16 (2009)
7. Hampson, C.: Two-dimensional modal logics with difference relations. Ph.D. thesis, King's College London (2016)
8. Hampson, C., Kurucz, A.: Undecidable propositional bimodal logics and one-variable first-order linear temporal logics with counting. ACM Trans. Comput. Log. (TOCL) **16**(3), 27:1–27:36 (2015)
9. Immerman, N.: Nondeterministic space is closed under complementation. SIAM J. Comput. **17**(5), 935–938 (1988)
10. Jones, N.D.: Space-bounded reducibility among combinatorial problems. J. Comput. Syst. Sci. **11**(1), 68–85 (1975)
11. Mayr, R.: Undecidable problems in unreliable computations. Theor. Comput. Sci. **297**(1–3), 337–354 (2003)
12. Minsky, M.L.: Computation. Prentice-Hall, Englewood Cliffs (1967)
13. Ouaknine, J., Worrell, J.: On the decidability and complexity of metric temporal logic over finite words. Log. Methods Comput. Sci. **3**(1) (2007)

---

[1] It is incorrectly claimed in [7] that the satisfiability problem for this logic is non-elementary, erroneously stating that the termination problem for ICMs matches that of incrementing channel systems with emptiness testing, as refuted here in Theorem 3.2.

14. Ouaknine, J., Worrell, J.: On the decidability of metric temporal logic. In: 20th Annual IEEE Symposium on Logic in Computer Science (LICS 2005), pp. 188–197. IEEE (2005)
15. Ouaknine, J., Worrell, J.: On metric temporal logic and faulty turing machines. In: Aceto, L., Ingólfsdóttir, A. (eds.) FoSSaCS 2006. LNCS, vol. 3921, pp. 217–230. Springer, Heidelberg (2006). https://doi.org/10.1007/11690634_15
16. Savitch, W.J.: Relationships between nondeterministic and deterministic tape complexities. J. Comput. Syst. Sci. 4(2), 177–192 (1970)
17. Schmitz, S.: Complexity hierarchies beyond elementary. ACM Trans. Comput. Theory (TOCT) 8(1), 3:1–3:36 (2016)
18. Schnoebelen, P.: Revisiting Ackermann-hardness for lossy counter machines and reset petri nets. In: Hliněný, P., Kučera, A. (eds.) MFCS 2010. LNCS, vol. 6281, pp. 616–628. Springer, Heidelberg (2010). https://doi.org/10.1007/978-3-642-15155-2_54
19. Szelepcsényi, R.: The method of forced enumeration for nondeterministic automata. Acta Informatica 26(3), 279–284 (1988)

Reachability Problems on Partially Lossy Queue Automata

Chris Köcher(✉) (iD)

Automata and Logics Group, Technische Universität Ilmenau, Ilmenau, Germany
chris.koecher@tu-ilmenau.de

Abstract. We study the reachability problem for queue automata and lossy queue automata. Concretely, we consider the set of queue contents which are forwards resp. backwards reachable from a given set of queue contents. Here, we prove the preservation of regularity if the queue automaton loops through some special sets of transformations. This is a generalization of the results by Boigelot et al. and Abdulla et al. regarding queue automata looping through a single sequence of transformations. We also prove that our construction is effective and efficient.

Keywords: Partially lossy queue · Reachability · Queue automaton

1 Introduction

Nearly all problems in verification ask whether in a program or automaton one can reach some given configurations from other given configurations. In some computational models this question is decidable, e.g., in finite state machines, pushdown automata [5,8,9] or one-counter automata. In some other, mostly Turing-complete computational models this reachability problem is undecidable.

So, for queue automata reachability is undecidable [6], while this problem is decidable for so-called lossy queue automata [1] which are allowed to forget any parts of their content at any time. In this case, for a regular set of configurations, the set of reachable configurations is regular [10] but it is impossible to compute finite automata accepting these sets [14]. Surprisingly, the set of backwards reachable configurations is effectively regular [1], even though this construction is not primitive recursive [7,15].

Due to the undecidability resp. inefficiency of the reachability problem for reliable and lossy queue automata, one may consider approximations of this problem. One trivial approach is to simulate the automaton's computation step by step until a given configuration (or a given set of configurations) was found. Then, starting from a given set of configurations we simply add or remove a single letter from the queue's contents. An even better and more efficient approach is to consider so-called "meta-transformations" as described in [3,4]. Such a meta-transformation is a combination of multiple transitions of the queue. In particular, given a loop in the queue's control component we combine iterations

© Springer Nature Switzerland AG 2019
E. Filiot et al. (Eds.): RP 2019, LNCS 11674, pp. 149–163, 2019.
https://doi.org/10.1007/978-3-030-30806-3_12

of this loop to one big step of the queue automaton. With this trick it is possible to explore infinitely many contents of the queue in a small amount of time.

Considering reliable queue automata, we know from Boigelot et al. [4] that, starting from a regular language of queue contents, the set of reachable queue contents after application of such meta-transformation is effectively regular. A similar result was proven for lossy queue automata by Abdulla et al. in [2].

In this paper we consider a generalization of this result which regards iterations through certain regular languages. Concretely, we consider so-called *read-write independent* sets where for each two words s, t from this language there is another word from this language consisting of the write actions from s and the read actions from t. For these generalized meta-transformations we prove the preservation of regularity of sets of configurations. We will see that our construction is possible in polynomial time.

Additionally, we consider another type of meta-transformations: sets of transformations which are closed under some special (context-sensitive) commutations of the atomic transformations. For such meta-transformations, the set of reachable configurations is also effectively regular. Moreover, if we start from a context-free set of configurations, the set of reachable configurations is effectively context-free, again. Here, both constructions can be carried out in polynomial time.

In this paper, we first prove the stated results for reliable queue automata. Later we consider so-called partially lossy queue automata which were first introduced in [12,13]. This is a generalization of reliable and lossy queue automata where we can specify which letters can be forgotten at any time. We will see then, that the sets of reachable configurations can be computed from the ones of a reliable queue automaton. Hence, all of our results and the results from [2,4] do also hold for arbitrary partially lossy queue automata.

2 Preliminaries

2.1 Words and Languages

At first, we have to introduce some basic definitions. To this end, let Γ be an alphabet. A word $v \in \Gamma^*$ is a prefix of $w \in \Gamma^*$ iff $w \in v\Gamma^*$. Similarly, v is a *suffix* of w iff $w \in \Gamma^*v$ and v is an *infix* of w iff $w \in \Gamma^*v\Gamma^*$. The *complementary prefix* (resp. *suffix*) of w wrt. v is the word $w/_v \in \Gamma^*$ (resp. $_v\backslash w \in \Gamma^*$) with $w = w/_v \cdot v$ (resp. $w = v \cdot {_v\backslash w}$). The *right quotient* of a language $L \subseteq \Gamma^*$ wrt. $K \subseteq \Gamma^*$ is the language $L/_K = \{u \in \Gamma^* \mid \exists v \in K \colon uv \in L\}$. Similarly, we can define the *left quotient* $_K\backslash L$ of L wrt. K.

For a word $w = a_1a_2 \dots a_n \in \Gamma^*$ we define its *reversal* by $w^{\mathrm{R}} := a_n \dots a_2a_1$. The *reversal* of a language L is $L^{\mathrm{R}} = \{w^{\mathrm{R}} \mid w \in L\}$. The *shuffle* of L and K is the following language:

$$L \sqcup\!\sqcup K := \left\{ v_1w_1v_2w_2 \dots v_nw_n \;\middle|\; \begin{array}{l} n \in \mathbb{N}, v_i, w_i \in \Gamma^*, \\ v_1v_2 \dots v_n \in L, w_1w_2 \dots w_n \in K \end{array} \right\}.$$

The word $v \in \Gamma^*$ is a *subword* of $w \in \Gamma^*$ (denoted by $v \preceq w$) iff $w \in \{v\} \sqcup\!\sqcup \Gamma^*$. Note that the relation $\preceq$ is a partial ordering on Γ^*.

Let $\leq$ be a partial ordering on Γ^* and $L \subseteq \Gamma^*$ be a language. The *upclosure* of L wrt. $\leq$ is $\uparrow_{\leq} L = \{w \in \Gamma^* \mid \exists v \in L \colon v \leq w\}$. Similarly, we can define the *downclosure* $\downarrow_{\leq} L = \{v \in \Gamma^* \mid \exists w \in L \colon v \leq w\}$.

Let $\sim$ be an equivalence relation on Γ^*. The *equivalence class* of $v \in \Gamma^*$ wrt. $\sim$ is $[v]_\sim = \{u \in \Gamma^* \mid u \sim v\}$. A language $L \subseteq \Gamma^*$ is *closed under* $\sim$ if for each $v \in L$ we have $[v]_\sim \subseteq L$.

Let $S \subseteq \Gamma$. Then the *projection* $\pi_S \colon \Gamma^* \to S^*$ to S is the monoid homomorphism induced by $\pi_S(a) = a$ for each $a \in S$ and $\pi_S(a) = \varepsilon$ for each $a \in \Gamma \smallsetminus S$. Additionally, for $w \in \Gamma^*$ we write $|w|_S := |\pi_S(w)|$.

2.2 Automata

A *finite automaton* (*NFA* for short) is a quintuple $\mathcal{A} = (Q, \Gamma, I, \Delta, F)$ where Q is a finite set of states, $I, F \subseteq Q$ are the sets of initial and final states, and $\Delta \subseteq Q \times (\Gamma \cup \{\varepsilon\}) \times Q$ is the transition relation. Then, the *configuration graph* of $\mathcal{A}$ is $\mathfrak{G}_\mathcal{A} := (Q, \Delta)$ which is a finite, edge-labeled, and directed graph. For $p, q \in Q$ and $w \in \Gamma^*$ we write $p \xrightarrow{w}_\mathcal{A} q$ if there is a w-labeled path in $\mathfrak{G}_\mathcal{A}$ from p to q. The *accepted language* of $\mathcal{A}$ is $L(\mathcal{A}) := \{w \in \Gamma^* \mid I \xrightarrow{w}_\mathcal{A} F\}$. A language $L \subseteq \Gamma^*$ is *regular*, if there is an NFA $\mathcal{A}$ accepting L. The class of regular languages is effectively closed under Boolean operations, left and right quotients, shuffle, reversal, up- and downclosures wrt. the subword ordering, and projections.

Let $\mathcal{A} = (Q, \Gamma, I, \Delta, F)$ be an NFA, $Q_i, Q_f \subseteq Q$. Then we set $\mathcal{A}_{Q_i \to Q_f} := (Q, \Gamma, Q_i, \Delta, Q_f)$, i.e., $\mathcal{A}_{Q_i \to Q_f}$ is the NFA constructed from $\mathcal{A}$ with initial states Q_i and final states Q_f. For example, we have $L(\mathcal{A}) = \bigcup_{q \in Q} L(\mathcal{A}_{I \to q}) L(\mathcal{A}_{q \to F})$.

A *pushdown automaton* (*PDA* for short) is a tuple $\mathcal{P} = (Q, \Sigma, \Gamma, \#, I, \Delta, F)$ where Q is a finite set of states, Σ and Γ are alphabets, $\# \in \Gamma$ is the stack bottom, $I, F \subseteq Q$ are the sets of initial and final states, and $\Delta \subseteq Q \times \Gamma \times (\Sigma \cup \{\varepsilon\}) \times Q \times \Gamma^*$ is the *finite* transition relation. A *configuration* of $\mathcal{P}$ is a tuple from $\mathrm{Conf}_\mathcal{P} := Q \times \Gamma^*$. We denote the set of initial configurations by $\mathrm{Init}_\mathcal{P} := I \times \{\#\}$ and the set of accepting configurations by $\mathrm{Final}_\mathcal{P} := F \times \Gamma^*$. For $p, q \in Q$, $x, y \in \Gamma^*$, and $a \in \Sigma \cup \{\varepsilon\}$ we write $(p, x) \xrightarrow{a}_\mathcal{P} (q, y)$ if there are $X \in \Gamma$ and $\gamma, z \in \Gamma^*$ with $(p, X, a, q, \gamma) \in \Delta$, $x = Xz$, and $y = \gamma z$. Then, $\mathfrak{G}_\mathcal{P} := (\mathrm{Conf}_\mathcal{P}, \bigcup_{a \in \Sigma \cup \{\varepsilon\}} \xrightarrow{a}_\mathcal{P})$ is called the *configuration graph* of $\mathcal{P}$. For $(p, x), (q, y) \in \mathrm{Conf}_\mathcal{P}$ and $w \in \Sigma^*$ we write $(p, x) \xrightarrow{w}_\mathcal{P} (q, y)$ if there is a w-labeled path from (p, x) to (q, y) in $\mathfrak{G}_\mathcal{P}$. The accepted language of $\mathcal{P}$ is $L(\mathcal{P}) = \{w \in \Sigma^* \mid \mathrm{Init}_\mathcal{P} \xrightarrow{w}_\mathcal{P} \mathrm{Final}_\mathcal{P}\}$. A language $L \subseteq \Sigma^*$ is *context-free* if there is a PDA $\mathcal{P}$ with $L = L(\mathcal{P})$.

Let $C \subseteq \mathrm{Conf}_\mathcal{P}$ be a set of configurations of $\mathcal{P}$. Then we denote the set of configurations of $\mathcal{P}$ reachable from C by

$$\mathrm{post}^*(C) := \{d \in \mathrm{Conf}_\mathcal{P} \mid \exists w \in \Sigma^* \colon C \xrightarrow{w}_\mathcal{P} d\}.$$

An NFA $\mathcal{A}$ *recognizes* C if $L(\mathcal{A}) = \{q\gamma \,|\, (q,\gamma) \in C\}$ holds. In this case we call C *regular*.

Theorem 2.1 ([8,9]). *Let $\mathcal{P}$ be a PDA and $C \subseteq \mathrm{Conf}_\mathcal{P}$ be a regular set of configurations. Then* $\mathrm{post}^*(C)$ *is effectively regular. An NFA recognizing* $\mathrm{post}^*(C)$ *can be computed from an NFA recognizing C in polynomial time.*   □

3   Queues and Queue Automata

In this section we want to recall basic knowledge on queues and queue automata. A queue can store entries from a given alphabet A. Since A is the alphabet of queue entries, the content of a queue is a word from A^*. For any letter $a \in A$ we have two actions: writing of a at the end of the queue (denoted by a) and reading of a from the head of the queue (denoted by $\bar{a}$). We assume that the alphabet $\overline{A}$ containing each such reading operation $\bar{a}$ is a disjoint copy of A. By $\Sigma := A \cup \overline{A}$ we denote the set of all actions on the queue. For $w = a_1 a_2 \ldots a_n \in A^*$ we also write $\overline{w} := \overline{a_1}\,\overline{a_2}\ldots\overline{a_n}$ and for $L \subseteq A^*$ we write $\overline{L} := \{\overline{w} \,|\, w \in L\}$. Formally, we describe the queue's behavior by a function $\circ$ associating a word $v \in A^*$ and a sequence of atomic transformations $t \in \Sigma^*$ with another word $v \circ t \in A^*$ which is the queue's content after application of t on the content v.

Definition 3.1. *Let A be an alphabet and $\bot \notin A$. Then the map $\circ: (A^* \cup \{\bot\}) \times \Sigma^* \to (A^* \cup \{\bot\})$ is defined for each $v \in A^*$, $a,b \in A$ with $a \neq b$, and $t \in \Sigma^*$ as follows:*

(1) $v \circ \varepsilon = v$

(2) $v \circ at = va \circ t$

(3) $av \circ \bar{a}t = v \circ t$

(4) $bv \circ \bar{a}t = \varepsilon \circ \bar{a}t = \bot \circ t = \bot$

We will say "$v \circ t$ is undefined" if $v \circ t = \bot$.

A *queue automaton* is a finite automaton on Σ equipped with such a queue. Considering the expression "$L \circ T$" then $L \subseteq A^*$ is a set of possible queue inputs, $T \subseteq \Sigma^*$ is the set of transformations, and $(L \circ T) \setminus \{\bot\}$ is the set of outputs of the queue automaton. Since T is represented by a finite automaton, the set T is always a regular language in this paper. All in all, we may define our reachability problems as follows:

Definition 3.2. *Let A be an alphabet, $L \subseteq A^*$ be a set of queue contents, and $T \subseteq \Sigma^*$ be a regular set of transformations. The set of queue contents that are reachable from L via T is*

$$\mathrm{REACH}(L,T) := (L \circ T) \setminus \{\bot\}$$

and the set of queue contents that can reach L via T is

$$\mathrm{BACKREACH}(L,T) := \{v \in A^* \,|\, (v \circ T) \cap L \neq \emptyset\}.$$

In general, for a recursively enumerable language $L \subseteq A^*$ and a regular set $T \subset \Sigma^*$ the language REACH(L,T) is (effectively) recursively enumerable. Since a finite automaton with a queue can simulate a Turing machine [6], the language REACH(L,T) can be any recursively enumerable language. This is true even if $|L| = 1$ and T is the Kleene closure of a finite set of transformations, i.e., if L and T are somewhat "simple" languages:

Remark 3.3. Let $\mathcal{G} = (N, \Gamma, P, S)$ be a (type-0) grammar and $\# \notin N \cup \Gamma$. The set of possible queue entries is $A := N \cup \Gamma \cup \{\#\}$. We construct the set of transformations $T \subseteq \Sigma^*$ as follows:

$$ T := \left(\{ \bar{\ell} r \mid (\ell, r) \in P \} \cup \{ \bar{a} a \mid a \in N \cup \Gamma \cup \{\#\} \} \right)^*, $$

i.e., the queue can apply any rule from $\mathcal{G}$ and move any letter from the head to its end. Then we have REACH$(\{\#S\}, T) \cap \#\Gamma^* = \#L(\mathcal{G})$ which can be any recursively enumerable language.

Due to Remark 3.3 there are sets L and T such that REACH(L,T) is undecidable. Therefore, we need some approximation to decide whether a given regular set of configurations can be reached from the regular language L of queue inputs by application of the transformations from T. A trivial approach is to compute REACH(L, T_n) where T_n is the set of prefixes of T of length at most n for increasing $n \in \mathbb{N}$. Unfortunately, this algorithm is not very efficient: consider $L \subseteq A^*$ be a finite language of queue contents and $T \subseteq \Sigma^*$ be a regular language of transformations. Then T_n is finite for any $n \in \mathbb{N}$ and, hence, REACH(L, T_n) is finite as well.

Boigelot et al. improved this trivial approximation in [3,4] by introduction of so-called *meta-transformations*. This means, that we partition the regular language T into sequences of certain regular languages $S \subseteq \Sigma^*$ such that the mappings $L \mapsto$ REACH(L, S) and $L \mapsto$ BACKREACH(L, S) can be computed efficiently and preserve regularity. For example, such languages can be a regular language of write actions or a regular language of read actions as considered in the following proposition:

Proposition 3.4. *Let A be an alphabet and $L, T \subseteq A^*$. Then the following statements hold:*

(1) REACH$(L, T) = LT$ *(3)* BACKREACH$(L, T) =$ REACH$(L^{\mathrm{R}}, \overline{T^{\mathrm{R}}})^{\mathrm{R}}$
(2) REACH$(L, \overline{T}) = {}_T \backslash L$ *(4)* BACKREACH$(L, \overline{T}) =$ REACH$(L^{\mathrm{R}}, T^{\mathrm{R}})^{\mathrm{R}}$ □

Hence, for regular languages $L \subseteq A^*$ and $T \subseteq A^*$ (or $T \subseteq \overline{A}^*$, resp.) we can compute NFAs accepting REACH(L, T) and BACKREACH(L, T) in polynomial time. In the following two sections we consider two further types of meta-transformations T having efficiently computable mappings $L \mapsto$ REACH(L, T) and $L \mapsto$ BACKREACH(L, T).

4   Behavioral Equivalence

The first type of meta-transformations we want to consider are languages that are closed under the so-called behavioral equivalence. In this connection, we say that two sequences of transformations have the same behavior if for any queue input the application of both transformations lead to the same output of the queue automaton. Formally, this equivalence is defined as follows:

Definition 4.1. *Let A be an alphabet and $s, t \in \Sigma^*$. Then s and t behave equivalently (denoted by $s \equiv t$) if $v \circ s = v \circ t$ for each $v \in A^*$. The relation $\equiv$ is called the behavioral equivalence.*

In other words, we have $s \equiv t$ if the application of s and t have the same effect on any queue's content. For example, for $a \in A$ the sequences $aa\bar{a}$ and $a\bar{a}a$ behave equivalently: let $v \in A^*$ be any queue content. Then we have

$$v \circ aa\bar{a} = vaa \circ \bar{a} = (va \circ \bar{a}) \cdot a = v \circ a\bar{a}a.$$

Nevertheless, we have $a\bar{a} \not\equiv \bar{a}a$ since we have $\varepsilon \circ a\bar{a} = \varepsilon \neq \perp = \varepsilon \circ \bar{a}a$.

This equivalence relation was first introduced by Huschenbett et al. in [11]. They proved in this paper that $\equiv$ is a congruence on Σ^* and is described by a finite set of context-sensitive commutations. We recall these commutations in the following theorem:

Theorem 4.2 ([11]). *Let A be an alphabet. Then $\equiv$ is the least congruence on Σ^* satisfying the following equations for each $a, b \in A$:*

(1) $a\bar{b} \equiv \bar{b}a$ if $a \neq b$,   (2) $a\bar{a}\bar{b} \equiv \bar{a}a\bar{b}$, and   (3) $ba\bar{a} \equiv b\bar{a}a$.   □

The behavioral equivalence was further considered in [12]. Concretely, we regarded the languages which are regular and closed under the behavioral equivalence $\equiv$ and gave some interesting properties of these languages. In that paper, we defined some kind of rational expressions constructing these sets as well as some MSO-logic describing them. In particular, let $T \subseteq \Sigma^*$ be a language that is closed under $\equiv$. Then, we know that T is regular if, and only if, $T \cap \overline{A}^* A^* \overline{A}^*$ is regular.

Example 4.3 ([12]). Let $W, R \subseteq A^*$ be regular languages. Then $[W \sqcup \overline{R}]_{\equiv} = W \sqcup \overline{R}$ is regular and closed under $\equiv$.

Now, let $a \in A$. Then $[(a\bar{a})^*]_{\equiv}$ is not regular since (by Theorem 4.2) we can prove $[(a\bar{a})^*]_{\equiv} \cap \overline{A}^* A^* \overline{A}^* = \{a^n \bar{a}^n \mid n \in \mathbb{N}\}$ which is not regular.

Let $T \subseteq \Sigma^*$ be regular. Using the equations from Theorem 4.2, we can decide whether T is closed under behavioral equivalence:

Remark 4.4. We can understand the equations from Theorem 4.2 as a finite Thue-system. Then for each rule $(\ell \to r)$ we can compute $T_\ell := T \cap \Sigma^* \ell \Sigma^*$ and $T_r := T \cap \Sigma^* r \Sigma^*$. Applying $(\ell \to r)$ on T_ℓ we obtain a regular language T_r'. Finally, we have to check whether $T_r = T_r'$ holds. The language T is closed under behavioral equivalence if, and only if, all of these tests succeed.

However, given a regular language $T \subseteq \Sigma^*$, it is impossible to compute the closure of T under behavioral equivalence. Moreover, it is undecidable whether the closure of T under $\equiv$ is regular, again [12].

Next, we want to prove that, for meta-transformations $T \subseteq \Sigma^*$ that are regular and closed under behavioral equivalence, the mappings $L \mapsto \mathrm{REACH}(L, T)$ and $L \mapsto \mathrm{BACKREACH}(L, T)$ preserve regularity. We do this with the help of some corollary of Theorem 4.2:

Proposition 4.5 ([11,12]). *Let A be an alphabet and $t \in \Sigma^*$. Then there is $s \in \overline{A}^* A^* \overline{A}^*$ with $s \equiv t$. From a given word t we can compute such a word s in polynomial time.* $\qquad\square$

Now, we can prove the main theorem in this section:

Theorem 4.6. *Let A be an alphabet, $L \subseteq A^* \cup \{\bot\}$ be regular, and $T \subseteq \Sigma^*$ be regular and closed under $\equiv$. Then $\mathrm{REACH}(L, T)$ and $\mathrm{BACKREACH}(L, T)$ are effectively regular. In particular, from NFAs accepting L and T we can construct NFAs accepting $\mathrm{REACH}(L, T)$ and $\mathrm{BACKREACH}(L, T)$ in polynomial time.*

Proof. Let $\mathcal{T} = (Q, \Sigma, I, \Delta, F)$ be an NFA with $L(\mathcal{T}) = T$. Since T is closed under $\equiv$ we have, by Proposition 4.5,

$$\mathrm{REACH}(L, T) = \mathrm{REACH}(L, T \cap \overline{A}^* A^* \overline{A}^*).$$

We partition $T \cap \overline{A} * A^* \overline{A}^*$ as follows: let $p, q \in Q$ be any pair of states. Then we can compute the following three regular languages in polynomial time:

$$\overline{K_1^{p,q}} = L(\mathcal{T}_{I \to p}) \cap \overline{A}^*, \quad K_2^{p,q} = L(\mathcal{T}_{p \to q}) \cap A^*, \text{ and } \quad \overline{K_3^{p,q}} = L(\mathcal{T}_{q \to F}) \cap \overline{A}^*.$$

Then it is easy to see that $T \cap \overline{A}^* A^* \overline{A}^* = \bigcup_{p,q \in Q} \overline{K_1^{p,q}} K_2^{p,q} \overline{K_3^{p,q}}$ holds.

Hence, due to Proposition 3.4 and the closure properties of the class of regular languages $\mathrm{REACH}(L, \overline{K_1^{p,q}} K_2^{p,q} \overline{K_3^{p,q}})$ is effectively regular and, hence, $\mathrm{REACH}(L, T)$. $\qquad\square$

Remark 4.7. Since the left or right quotient of a context-free language with a regular language is again context-free, we can compute, from a PDA accepting L and an NFA accepting T, PDAs accepting $\mathrm{REACH}(L, T)$ resp. $\mathrm{BACKREACH}(L, T)$ in polynomial time.

5 Read-Write Independence

Another kind of meta-transformations was first considered in the research of Boigelot et al. [4] (and similarly for lossy queue automata by Abdulla et al. [2]). There, the authors considered queue automata looping through a single sequence of transformations. This means, we consider queue automata having exactly one initial state which is the only final state and there is exactly one labeled path from the initial state back to itself, again.

Concretely, in that paper the authors have proven that beginning with a regular language of queue contents we reach a regular set of queue contents, again. In particular, one can compute infinitely many succeeding queue contents at once in polynomial time. So, a natural question would be, whether this result can be generalized to meta-transformations consisting of multiple such loops starting from a single initial state. Unfortunately, already for queue automata having two loops the set of reachable queue contents is not regular in general:

Example 5.1. Let A be an alphabet and $a, b \in A$ be distinct letters. Then we have $\text{REACH}(\{a\}, \{\overline{a}bb, \overline{b}a\}^*) \cap \{a\}^* = \{a^{2^n} \mid n \in \mathbb{N}\}$ which is not even context-free.

Moreover, in Remark 3.3 we have seen that such queue automata consisting of a finite number of such loops are Turing-complete.

In both cases, there are two words $s, t \in T$ having different sub-sequences of write or read actions. One trivial solution would be considering only words having the same sub-sequences of write and read actions. Another even stronger approach is to choose a set T such that independently of which word from $\pi_{\overline{A}}(T)$ we read from the queue, we can write any word from $\pi_A(T)$. In this case, it is impossible that a special queue content can enforce a unique, complicated, infinite run of the queue automaton since we can now write any word from $\pi_A(T)$ at any time into the queue. This can be understood as lifting of a word having sub-sequences of write and read actions to an object having a set of sub-sequences of write and read actions. Formally, we are considering the following sets of transformations:

Definition 5.2. *Let A be an alphabet. A set $T \subseteq \Sigma^*$ is read-write independent if for each $s, t \in T$ we have $\pi_A(s)\pi_{\overline{A}}(t) \in T$.*

We may see read-write independent sets as some kind of a Cartesian product of a set of write actions $W \subseteq A^*$ with a set of read actions $\overline{R} \subseteq \overline{A}^*$ where for each element $(w, \overline{r}) \in W \times \overline{R}$ we have the transformation $w\overline{r}$. Some simple read-write independent sets are given in the following example:

Example 5.3. Let $W, R \subseteq A^*$. Then $W\overline{R}$ and $W \sqcup\!\sqcup \overline{R}$ are read-write independent.

Obviously, each language $T \subseteq \Sigma^*$ with $\pi_A(T)\pi_{\overline{A}}(T) \subseteq T$ is read-write independent. Hence, for a given regular language it is clear how to check read-write independency.

In the following we will prove that the mapping $L \mapsto \text{REACH}(L, T^*)$, for any regular, read-write independent set $T \subseteq \Sigma^*$, preserves regularity and is computable in polynomial time. But first we focus on a special case where the read-write independent set is the product of a language of write actions W with a language of read actions . Here, we consider regular subsets $W\overline{R} \subseteq A^*\overline{A}^*$ where A is some alphabet having a special letter \$ which marks the begin of a word from W and is used for synchronization between writing and reading actions.

Theorem 5.4. *Let A be an alphabet and $\$ \in A$ be some letter. Additionally, let $L \subseteq (A \setminus \{\$\})^*$, $W \subseteq \$(A \setminus \{\$\})^*$, and $R \subseteq A^*$ be regular languages such* *that* $R = \$^* \sqcup \pi_{A \smallsetminus \{\$\}}(R)$ *holds. Then* $\text{REACH}(L, (W\overline{R})^*)$ *is effectively regular. In particular, from NFAs accepting* L, W, *and* R *we can construct an NFA accepting* $\text{REACH}(L, (W\overline{R})^*)$ *in polynomial time.*

We prove Theorem 5.4 by reduction to the reachability problem in pushdown automata. A first, trivial idea would be a simple replacement of the queue by a stack, i.e., from the queue's content v we reach w if, and only if, the PDA reaches $w\#$ from $v\#$. Unfortunately, this construction is not possible since our queue automaton modifies its content at both ends which cannot be simulated with a single stack. Hence, we need a more abstract presentation of the queue's contents. To this end, we consider some computation of the queue on a word $v \in L$. So, let $v_0, \ldots, v_k \in A^*$ and $\alpha_0, \ldots, \alpha_{k-1} \in \Sigma$ with $v_0 = v$, $v_{i+1} = v_i \circ \alpha_i \neq \bot$ for each $0 \leq i < k$, and $\alpha_0 \ldots \alpha_{k-1}$ be some prefix of $(W\overline{R})^*$. Consider an NFA $\mathcal{C}$ accepting LW^*. Then, there is some path from an initial state p_0 to a final state q_0 of $\mathcal{C}$ with label v_0. When applying α_0 to v_0 this corresponds either to moving q_0 by one edge labeled with $\alpha_0 \in A$ to state q_1 or to moving p_0 by one edge labeled with a (where $\alpha_0 = \overline{a} \in \overline{A}$) to state p_1. Application of the following actions α_i similarly moves one of the states p_i and q_i by one edge to p_{i+1} resp. q_{i+1}. The result is that v_k is the labeling of some path from p_k to q_k in $\mathcal{C}$. In this sense, we can abstract v_k and its corresponding path in $\mathcal{C}$ by these two states p_k and q_k and a number $n \in \mathbb{N}$ representing the number of W-loops in this path (and hence, the number of words from W to be contained in v_k or the number of $\$$ on this path). Additionally, since $\alpha_0 \ldots \alpha_{n-1}$ is some prefix of $(W\overline{R})^*$ there is some path in an NFA $\mathcal{T}$ accepting $(W\overline{R})^*$ from an initial state to some state s labeled with $\alpha_0 \ldots \alpha_{n-1}$.

   Alternatively, we can understand the components p_k, q_k, and n as follows: since the queue automaton starts with some word from L, adds a prefix of W^* at the end, and removes some prefix of R^* from the head, the word v_k is some infix of LW^*. Hence, there is a suffix w_0 of $L \cup W$, some words $w_1, \ldots, w_{n-1} \in W$, and a prefix w_n of W with $v_k = w_0 w_1 \ldots w_{n-1} w_n$. In this case, w_0 is the labeling

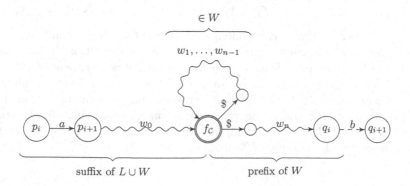

Fig. 1. A path labeled with v_i from p_i to q_i in $\mathcal{C}$ and its three components.

of some path from p_k to the final states of C and w_n is the labeling of some path from C's final states to q_k (cf. Fig. 1).

Now, we want to construct a PDA P which handles exactly the four components named above. In this sense, P's states contain the three states p_k, q_k, and s of C and T and the number n is stored in the stack of P. To this end, let $C = (Q_C, A, I_C, \Delta_C, F_C)$ be an NFA accepting LW^* (i.e., the possible queue Contents) and $T = (Q_T, \Sigma, I_T, \Delta_T, F_T)$ be an NFA accepting $(W\overline{R})^*$ (i.e., the possible Transformations). W.l.o.g., we can assume that both, C and T, are reduced in the sense that each state is reachable from the initial state and can reach some final state. Additionally, we assume that C and T have exactly one final state called f_C resp. f_T. Note that we can compute these two automata in polynomial time from NFAs accepting L, W, and R.

Recall that the queue's content is abstracted by three states from C and T and by some natural number. Then the PDA $P = (Q_P, \Sigma, \Gamma, \#, I_P, \Delta_P, F_P)$ is defined as follows:

- $\Gamma := \{\$, \#\}$
- $Q_P := Q_C \times Q_C \times Q_T$. Here, the first and second component represent the two states characterizing the queue's content as described above. The third component represents the actions we have already executed on the queue.
- $I_P := I_C \times Q_L \times I_T$ where $Q_L := \{q \in Q_C \mid \exists v \in L \colon I_C \xrightarrow{v}_C q\}$ is the set of states being reachable via L (i.e., the final states of the NFA accepting L)
- $F_P := Q_C \times F_C \times F_T$
- Δ_P contains exactly the following transitions for $a \in A \smallsetminus \{\$\}$, $X \in \Gamma$, $p, p, q, q' \in Q_C$, and $s, s' \in Q_T$:
  (I) *Simulate writing of the letter a into the queue*:
      $((p, q, s), X, a, (p, q', s'), X) \in \Delta_P$ if $(q, a, q') \in \Delta_C$ and $(s, a, s') \in \Delta_T$.
  (II) *Simulate writing of the letter $\$$ into the queue*:
      $((p, q, s), X, \$, (p, q', s'), \$X) \in \Delta_P$ if $(q, \$, q') \in \Delta_C$ and $(s, \$, s') \in \Delta_T$.
  (III) *Simulate reading of the letter a from the queue*:
      $((p, q, s), X, \overline{a}, (p', q, s'), X) \in \Delta_P$ if $(p, a, p') \in \Delta_C$ and $(s, \overline{a}, s') \in \Delta_T$.
  (IV) *Simulate reading of the letter $\$$ from the queue*:
      $((p, q, s), \$, \overline{\$}, (p', q, s'), \varepsilon) \in \Delta_P$ if $(p, \$, p') \in \Delta_C$ and $(s, \overline{\$}, s') \in \Delta_T$.

It is easy to see that the stack's contents are words from $\$^*\#$ at any time.

Now, we assign the configuration $((p, q, s), \$^n\#)$ to the set of all words being the labeling of some path from p to q in C, containing n appearances of the letter $\$$ (which marks the beginning of a word from W), and are reachable by application of some infix of $(W\overline{R})^*$ that is the labeling of some path from I_T to s in T. Since we do not care about the queue's control component and its states, we only focus on the path from p to q in C and the n appearances of $\$$. Formally, our assignment is the mapping $[\![.]\!] \colon \mathrm{Conf}_P \to 2^{A^*}$ with

$$[\![(p, q, s), \$^n\#]\!] = L(C_{p \to q}) \cap (\$^n \sqcup\!\sqcup (A \smallsetminus \{\$\})^*)$$

for each $p, q \in Q_C$, $s \in Q_T$, and $n \in \mathbb{N}$.

Next, we can prove that the set of reachable queue contents coincides with this semantics of the reachable, accepting configurations of the PDA P.

Proposition 5.5. *We have* $\text{REACH}(L,(W\overline{R})^*) = \bigcup_{\sigma \in \text{post}^*(\text{Init}_\mathcal{P}) \cap \text{Final}_\mathcal{P}} [\![\sigma]\!]$. $\qquad\square$

With Proposition 5.5 in mind, we are ready to prove the effective regularity of the set of reachable configurations of our special queue automata:

Proof (of Theorem 5.4). From Theorem 2.1 we know that $\text{post}^*(\text{Init}_\mathcal{P})$ is effectively regular. Let $\mathcal{A}$ be the NFA recognizing $\text{post}^*(\text{Init}_\mathcal{P})$ which can be computed in polynomial time. Then the following language is effectively regular as well:

$$K := \bigcup_{(p,q,s)\in F_\mathcal{P}} (L(\mathcal{C}_{p\to q}) \cap (\pi_\$(L(\mathcal{A}) \cap (p,q,s)\$^*\#) \sqcup (A \smallsetminus \{\$\})^*)).$$

Hence, using Proposition 5.5, we can prove

$$K = \bigcup_{\sigma \in \text{post}^*(\text{Init}_\mathcal{P}) \cap \text{Final}_\mathcal{P}} [\![\sigma]\!] = \text{REACH}(L,T^*).$$

$$\square$$

Until now we have seen the effective preservation of regularity if our read-write independent set $T \subseteq \Sigma^*$ satisfies a special condition. From this special case we infer now the effective preservation of regularity for arbitrary read-write independent sets.

Theorem 5.6. *Let A be an alphabet, $L \subseteq A^*$ be regular, and $T \subseteq \Sigma^*$ be read-write independent and regular. Then $\text{REACH}(L,T^*)$ is effectively regular. In particular, from NFAs accepting L and T we can compute an NFA accepting $\text{REACH}(L,T^*)$ in polynomial time.*

Proof. First, we can prove that for each $t \in T$ and $v \in L$ with $v \circ t \neq \bot$ there is $t' \in \pi_A(T)\pi_{\overline{A}}(T)$ with $v \circ t' = v \circ t$. Hence, we have $\text{REACH}(L,T^*) = \text{REACH}(L,(\pi_A(T)\pi_{\overline{A}}(T))^*)$. Now, let $\$ \notin A$ be a new letter. Then we set $W := \$\pi_A(T)$ and $R := \pi_{\overline{A}}(T) \sqcup \overline{\$}^*$ which are effectively regular. By Theorem 5.4 the set $\text{REACH}(L,(W\overline{R})^*)$ is effectively regular as well. Finally, we can prove $\text{REACH}(L,(W\overline{R})^*) = \text{REACH}(L,(\pi_A(T)\pi_{\overline{A}}(T))^*) = \pi_A(\text{REACH}(L,T^*))$. $\qquad\square$

From Theorem 5.6 and Proposition 3.4 we can infer that also the set of backwards reachable queue contents is effectively regular.

Corollary 5.7. *Let A be an alphabet, $L \subseteq A^*$ be regular, and $T \subseteq \Sigma^*$ be read-write independent and regular. Then $\text{BACKREACH}(L,T^*)$ is effectively regular. In particular, from NFAs accepting L and T we can construct an NFA accepting $\text{BACKREACH}(L,T^*)$ in polynomial time.* $\qquad\square$

Theorem 5.6 can also be used to prove the effective regularity of other language classes. First, with the help of the behavioral equivalence $\equiv$ we can see that the result of [4] is a direct corollary of the result above.

Corollary 5.8. *Let A be an alphabet, $L \subseteq A^*$ be regular, and $T \subseteq \Sigma^*$ be regular. Then* REACH(L, T^*) *and* BACKREACH(L, T^*) *are effectively regular, if*

(1) $T = \{t\}$ for some $t \in \Sigma^*$ (cf. [4]),
(2) $T = \overline{R_1} W \overline{R_2}$ for some regular sets $W, R_1, R_2 \subseteq A^*$, or
(3) $T \subseteq A^* \cup \overline{A}^*$.

In all of these cases the computation of NFAs accepting REACH(L, T^*) *and* BACKREACH(L, T^*), *respectively, is possible in polynomial time.*

Proof. First, we prove (1). To this end, let $s = \overline{u}v\overline{w} \in \overline{A}^* A^* \overline{A}^*$ with $s \equiv t$ as in Proposition 4.5. Then we have $t^* \equiv s^* = \overline{u}(v\overline{w}\overline{u})^* v\overline{w} \cup \{\varepsilon\}$. Since $\{v\overline{w}\overline{u}\}$ is read-write independent, REACH$(L, t^*) = $ REACH(L, s^*) is effectively regular by Proposition 3.4 and Theorem 5.6.

The proof of (2) is similar to (1).

Finally, we prove (3). Set $S := (T \cap A^*)^* (T \cap \overline{A}^*)^*$. Then S is effectively regular and read-write independent. Additionally, we have $S^* = T^*$ and, hence, REACH$(L, T^*) = $ REACH(L, S^*). □

Note that Corollary 5.8(2) also implies that REACH$(L, (\overline{R}W)^*)$ is effectively regular for some regular languages $W, R \subseteq A^*$.

Though, it is still open whether REACH(L, T^*) is regular for each regular $T \subseteq \Sigma^*$ with $\pi_{\overline{A}}(T)\pi_A(T) \subseteq T$. At least the reduction in Theorem 5.6, where we have de-shuffled the words from T, does not hold in this case. E.g., we have REACH$(\{\varepsilon\}, \{a\overline{a}a, \overline{a}aa\}^*) = a^* \neq \{\varepsilon\} = $ REACH$(\{\varepsilon\}, \{\overline{a}aa\}^*)$. However, we believe that REACH(L, T^*) is effectively (and efficiently) regular for each $T \subseteq \Sigma^*$ such that for each $s, t \in T$ there is $r \in T$ with $\pi_A(r) = \pi_A(s)$ and $\pi_{\overline{A}}(r) = \pi_{\overline{A}}(t)$. Possibly, the construction of our PDA $\mathcal{P}$ can be generalized to this case.

6   Partially Lossy Queues

Until now we have only considered queue automata which are reliable. We can also prove the results from the previous sections for (partially) lossy queue automata. These partially lossy queue automata are queue automata with an additional uncontrollable action which is forgetting parts of its contents that are specified by a so-called lossiness alphabet.

Definition 6.1. *A lossiness alphabet is a tuple $\mathcal{L} = (A, U)$ where A is an alphabet (with $|A| \geq 2$) and $U \subseteq A$.*

In this connection, U contains the *unforgettable* letters of the partially lossy queue and $A \smallsetminus U$ contains the *forgettable* letters.

In fact, a partially lossy queue automaton is allowed to forget any letter from $A \smallsetminus U$ in its content at any time. Here, we first consider partially lossy queues with restricted lossiness. Concretely, we consider only the computations of the automata where the queue forgets letters when necessary. That is, if the queue tries to read some letter which is preceded by some forgettable letters.

Formally, the transformations of a restricted partially lossy queue are defined as follows:

Definition 6.2. *Let $\mathcal{L} = (A, U)$ be a lossiness alphabet and $\bot \notin A$. Then the map $\circ_{\mathcal{L}} \colon (A^* \cup \{\bot\}) \times \Sigma^* \to (A^* \cup \{\bot\})$ is defined for each $v \in A^*$, $a, b \in A$, and $t \in \Sigma^*$ as follows:*

(1) $v \circ_{\mathcal{L}} \varepsilon = v$

(2) $v \circ_{\mathcal{L}} at = va \circ t$

(3) $av \circ_{\mathcal{L}} \bar{a}t = v \circ_{\mathcal{L}} t$

(4) $bv \circ_{\mathcal{L}} \bar{a}t = v \circ_{\mathcal{L}} \bar{a}t$ *if* $b \in A \setminus (U \cup \{a\})$

(5) $bv \circ_{\mathcal{L}} \bar{a}t = \bot$ *if* $b \in U \setminus \{a\}$

(6) $\varepsilon \circ_{\mathcal{L}} \bar{a}t = \bot \circ_{\mathcal{L}} t = \bot$

Let $\mathcal{L} = (A, U)$ be a lossiness alphabet and $u, v \in A^*$. We say that v is an $\mathcal{L}$-*subword* of w (denoted by $v \preceq_{\mathcal{L}} w$) if $\pi_U(w) \preceq v \preceq w$ holds. It is easy to see, that $\preceq_{(A,A)}$ is the equality relation and $\preceq_{(A,\emptyset)}$ is the subword relation on A as defined in the preliminaries.

Then a (non-restricted) partially lossy queue with some content $w \in A^*$ may contain any $\mathcal{L}$-subword of w after a single forgetting action. Moreover, for $v \in A^*$ and $t \in \Sigma^*$ with $v \circ_{\mathcal{L}} t \neq \bot$ the set $\downarrow_{\preceq_{\mathcal{L}}}(v \circ_{\mathcal{L}} t)$ is the set of all reachable queue contents after application of the transformation t on v (cf. [13]). Hence, we define our reachability problems as follows:

Definition 6.3. *Let $\mathcal{L} = (A, U)$ be a lossiness alphabet, $L \subseteq A^*$ be a set of queue contents, and $T \subseteq \Sigma^*$ be a regular set of transformations. The set of queue contents that are reachable from L via T is*

$$\text{REACH}_{\mathcal{L}}(L, T) := \downarrow_{\preceq_{\mathcal{L}}}((L \circ_{\mathcal{L}} T) \setminus \{\bot\})$$

and the set of queue contents that can reach L via T is

$$\text{BACKREACH}_{\mathcal{L}}(L, T) := \uparrow_{\preceq_{\mathcal{L}}}\{v \in A^* \mid (v \circ_{\mathcal{L}} T) \cap L \neq \emptyset\}.$$

Now, we consider fully lossy queues: let $\mathcal{L} = (A, \emptyset)$ be a lossiness alphabet. Then, for regular languages $L \subseteq A^*$ and $T \subseteq \Sigma^*$, the set $\text{REACH}_{\mathcal{L}}(L, T)$ has a decidable membership problem [1] and, since it is downwards closed under the subword ordering $\preceq$ [10], it is regular. Though, we cannot compute an NFA accepting this set - even if $L = \{w\}$ [14]. Surprisingly, the set $\text{BACKREACH}_{\mathcal{L}}(L, T)$ is effectively regular [1], but the computation of an NFA accepting this set is not primitive recursive [7,15].

Hence, again we try to approximate the reachability problem with the help of meta-transformations. To this end, we need the following partial ordering: we say v is a *reduced* $\mathcal{L}$-*subword* of w (denoted by $v \sqsubseteq_{\mathcal{L}} w$) if, and only if, there are $a_1, \ldots, a_n \in A$ and $w_i \in (A \setminus (U \cup \{a_i\}))^*$ with $v = a_1 \ldots a_n$ and $w = w_1 a_1 \ldots w_n a_n$. Note that $v \sqsubseteq_{\mathcal{L}} w$ implies $v \preceq_{\mathcal{L}} w$ but not vice versa, since for $v \preceq_{\mathcal{L}} w$ it is allowed to add some forgettable letters at the end of v. It is very easy to verify that in the reliable case (i.e., $A = U$) this ordering is the equality relation on A^*. With the help of $\sqsubseteq_{\mathcal{L}}$ we can prove the following statement:

Lemma 6.4. *Let $\mathcal{L} = (A, U)$ be a lossiness alphabet and $v, w, t \in A^*$. Then we have $v \circ_{\mathcal{L}} \bar{t} = w$ if, and only if, there is $s \in A^*$ with $t \sqsubseteq_{\mathcal{L}} s$ and $v = sw$.* □

With the help of Lemma 6.4 we can finally prove the following reductions from reachability in partially lossy queues to reachability in reliable queues:

Proposition 6.5. *Let $\mathcal{L} = (A, U)$ and $\mathcal{K} = (A, A)$ be lossiness alphabets and $L, T \subseteq A^*$. Then the following statements hold:*

(1) $L \circ_{\mathcal{L}} T = L \circ_{\mathcal{K}} T$
(2) $L \circ_{\mathcal{L}} \overline{T} = L \circ_{\mathcal{K}} \uparrow_{\sqsubseteq_{\mathcal{L}}} T$
(3) $\text{REACH}_{\mathcal{L}}(L, T) = \downarrow_{\preceq_{\mathcal{L}}} \text{REACH}_{\mathcal{K}}(L, T)$
(4) $\text{REACH}_{\mathcal{L}}(L, \overline{T}) = \downarrow_{\preceq_{\mathcal{L}}} \text{REACH}_{\mathcal{K}}(L, \uparrow_{\sqsubseteq_{\mathcal{L}}} T)$
(5) $\text{BACKREACH}_{\mathcal{L}}(L, T) = \uparrow_{\preceq_{\mathcal{L}}} \text{BACKREACH}_{\mathcal{K}}(L, T)$
(6) $\text{BACKREACH}_{\mathcal{L}}(L, \overline{T}) = \uparrow_{\preceq_{\mathcal{L}}} \text{BACKREACH}_{\mathcal{K}}(L, \uparrow_{\sqsubseteq_{\mathcal{L}}} T)$ $\qquad\qquad$ □

Finally, we can prove that our results from the previous sections also hold for arbitrary partially lossy queues:

Theorem 6.6. *Let $\mathcal{L} = (A, U)$ be a lossiness alphabet, $L \subseteq A^*$ be regular, and $T \subseteq \Sigma^*$ be regular. Then $\text{REACH}_{\mathcal{L}}(L, T)$ and $\text{BACKREACH}_{\mathcal{L}}(L, T)$ are effectively regular, if*

(1) T *is closed under* $\equiv_{\mathcal{L}}$ *(where* $s \equiv_{\mathcal{L}} t$ *if* $v \circ_{\mathcal{L}} s = v \circ_{\mathcal{L}} t$ *for each* $v \in A^*$*),*
(2) $T = S^*$ *for some regular, read-write independent* $S \subseteq \Sigma^*$*,*
(3) $T = t^*$ *for some* $t \in \Sigma^*$ *(cf. [2,4]),*
(4) $T = (\overline{R_1} W \overline{R_2})^*$ *for some regular sets* $W, R_1, R_2 \subseteq A^*$*, or*
(5) $T = S^*$ *where* $S \subseteq A^* \cup \overline{A}^*$ *is regular.*

In all of these cases the computation of NFAs accepting $\text{REACH}_{\mathcal{L}}(L, T)$ *and* $\text{BACKREACH}_{\mathcal{L}}(L, T)$*, respectively, is possible in polynomial time.* $\qquad$ □

7   Conclusion

In this paper we introduced so-called partially lossy queue automata (plq automata for short) which are queue automata that are allowed to forget specified parts of their contents at any time. Here, we considered the forwards and backwards reachability problem of such plq automata. Since those automata are Turing-complete (except of the ones allowed to forget everything) Boigelot et al. [4] and Abdulla et al. [1] tried to approximate the reachability problem with the help of so-called meta-transformations. These are regular languages of transformations such that we can easily compute the set of reachable queue contents. Here, we considered two special kinds of meta-transformations:

1. the set of possible sequences of queue transformations is closed under certain (context-sensitive) commutations of the atomic transformations.
2. the plq automaton alternates between writing of words from a regular language and reading of words from another regular language. This is a generalization of the results [2,4] where the authors considered queue automata looping through a single sequence of transformations.

In both cases we could prove that, starting with a regular language of queue contents the queue reaches a regular set of new contents.

Acknowledgment. The author would like to thank Dietrich Kuske and the anonymous reviewers of this paper for their helpful suggestions to improve this paper.

References

1. Abdulla, P.A., Jonsson, B.: Verifying programs with unreliable channels. Inf. Comput. **127**(2), 91–101 (1996). https://doi.org/10.1006/inco.1996.0053
2. Abdulla, P.A., Collomb-Annichini, A., Bouajjani, A., Jonsson, B.: Using forward reachability analysis for verification of lossy channel systems. Formal Methods Syst. Des. **25**(1), 39–65 (2004). https://doi.org/10.1023/B:FORM.0000033962.51898.1a
3. Boigelot, B., Godefroid, P.: Symbolic verification of communication protocols with infinite state spaces using QDDs. Formal Methods Syst. Des. **14**(3), 237–255 (1999). https://doi.org/10.1023/A:1008719024240
4. Boigelot, B., Godefroid, P., Willems, B., Wolper, P.: The power of QDDs (extended abstract). In: Van Hentenryck, P. (ed.) SAS 1997. LNCS, vol. 1302, pp. 172–186. Springer, Heidelberg (1997). https://doi.org/10.1007/BFb0032741
5. Bouajjani, A., Esparza, J., Maler, O.: Reachability analysis of pushdown automata: application to model-checking. In: Mazurkiewicz, A., Winkowski, J. (eds.) CONCUR 1997. LNCS, vol. 1243, pp. 135–150. Springer, Heidelberg (1997). https://doi.org/10.1007/3-540-63141-0_10
6. Brand, D., Zafiropulo, P.: On communicating finite-state machines. J. ACM **30**(2), 323–342 (1983). https://doi.org/10.1145/322374.322380
7. Chambart, P., Schnoebelen, P.: The ordinal recursive complexity of lossy channel systems. In: LICS 2008, pp. 205–216. IEEE Computer Society Press (2008). https://doi.org/10.1109/LICS.2008.47
8. Esparza, J., Hansel, D., Rossmanith, P., Schwoon, S.: Efficient algorithms for model checking pushdown systems. In: Emerson, E.A., Sistla, A.P. (eds.) CAV 2000. LNCS, vol. 1855, pp. 232–247. Springer, Heidelberg (2000). https://doi.org/10.1007/10722167_20
9. Finkel, A., Willems, B., Wolper, P.: A direct symbolic approach to model checking pushdown systems. Electron. Notes Theor. Comput. Sci. **9**, 27–37 (1997). https://doi.org/10.1016/S1571-0661(05)80426-8
10. Haines, L.H.: On free monoids partially ordered by embedding. J. Comb. Theory **6**(1), 94–98 (1969). https://doi.org/10.1016/S0021-9800(69)80111-0
11. Huschenbett, M., Kuske, D., Zetzsche, G.: The monoid of queue actions. Semigroup Forum **95**(3), 475–508 (2017). https://doi.org/10.1007/s00233-016-9835-4
12. Köcher, C.: Rational, recognizable, and aperiodic sets in the partially lossy queue monoid. In: STACS 2018. LIPIcs, vol. 96, pp. 45:1–45:14. Dagstuhl Publishing (2018). https://doi.org/10.4230/LIPIcs.STACS.2018.45
13. Köcher, C., Kuske, D., Prianychnykova, O.: The inclusion structure of partially lossy queue monoids and their trace submonoids. RAIRO - Theor. Inf. Appl. **52**(1), 55–86 (2018). https://doi.org/10.1051/ita/2018003
14. Mayr, R.: Undecidable problems in unreliable computations. Theoret. Comput. Sci. **297**(1), 337–354 (2003). https://doi.org/10.1016/S0304-3975(02)00646-1
15. Schnoebelen, P.: Verifying lossy channel systems has nonprimitive recursive complexity. Inf. Process. Lett. **83**(5), 251–261 (2002). https://doi.org/10.1016/S0020-0190(01)00337-4

On the Computation of the Minimal Coverability Set of Petri Nets

Pierre-Alain Reynier[1]([✉]) and Frédéric Servais[2]

[1] Aix-Marseille Univ, Université de Toulon, CNRS, LIS, Marseille, France
pierre-alain.reynier@univ-amu.fr
[2] École Supérieure d'Informatique de Bruxelles, Bruxelles, Belgium
frederic.servais@gmail.com,
https://pageperso.lis-lab.fr/~pierre-alain.reynier/

Abstract. The verification of infinite-state systems is a challenging task. A prominent instance is reachability analysis of Petri nets, for which no efficient algorithm is known. The *minimal coverability set* of a Petri net can be understood as an approximation of its reachability set described by means of ω-markings (*i.e.* markings in which some entries may be set to infinity). It allows to solve numerous decision problems on Petri nets, such as any coverability problem. In this paper, we study the computation of the minimal coverability set.

This set can be computed using the Karp and Miller trees, which perform accelerations of cycles along branches [10]. The resulting algorithm may however perform redundant computations. In a previous work [17], we proposed an improved algorithm allowing pruning between branches of the Karp and Miller tree, and proved its correctness. The proof of its correctness was complicated, as the introduction of pruning between branches may yield to incompleteness issues [5,9].

In this paper, we propose a new proof of the correctness of our algorithm. This new proof relies on an original invariant of the algorithm, leading to the following assets:
1. it is considerably shorter and simpler,
2. it allows to prove the correctness of a more generic algorithm, as the acceleration used is let as a parameter. Indeed, we identify the property that the acceleration should satisfy to ensure completeness.
3. it opens the way to a generalization of our algorithm to extensions of Petri nets.

Keywords: Petri nets · Coverability · Acceleration

1 Introduction

Verification of Infinite-State Systems. Petri nets [14] constitute one of the most popular formalism for the description and analysis of concurrent systems. While their state space may be infinite, many verification problems are decidable. Dealing with infinite-state systems is useful in numerous situations, such as considering an unbounded number of agents or modelling ressources.

© Springer Nature Switzerland AG 2019
E. Filiot et al. (Eds.): RP 2019, LNCS 11674, pp. 164–177, 2019.
https://doi.org/10.1007/978-3-030-30806-3_13

When considering the verification of safety properties for Petri nets, an important problem is the *coverability problem*, which can be understood as a weakening of the reachability problem. It asks whether it is possible to reach a marking larger than or equal to a given target marking, and thus exactly corresponds to fireability of a transition. This problem is ExpSpace-complete [3,10,16] and has attracted a lot of interest (see for instance [2,8,11]).

The Minimal Coverability Set. In this work, we are interested in a related problem, which consists in computing the so-called *minimal coverability set* of a Petri net (MCS for short) [5]. This set can be understood as an approximation of its reachability set described by means of ω-markings (*i.e.* markings in which some entries may be set to infinity). Once it is computed, this set allows to solve any coverability problem, and several other problems such as the *(place) boundedness* and *regularity* problems (see [18]).

The MCS can be derived from the classical Karp and Miller algorithm [10]. This algorithm builds a finite tree representation of the (potentially infinite) unfolding of the reachability graph of the given Petri net. It uses acceleration techniques to collapse branches of the tree and ensure termination. By taking advantage of the fact that Petri nets are strictly monotonic transition systems, the acceleration essentially computes the limit of repeatedly firing a sequence of transitions. However, this algorithm is not efficient as several branches may perform similar computations. This observation led to the Minimal Coverability Tree (MCT) algorithm [5], which introduces comparisons (and pruning) between branches of the tree. However, it was shown that the MCT algorithm is incomplete [9,13]. The flaw is intricate and, according to [9], difficult to patch, with wrong previous attempts [13].

The Monotone-Pruning Algorithm. As a solution to this problem, we introduced in [17] the Monotone-Pruning algorithm (MP), an improved Karp and Miller algorithm with pruning. This algorithm can be viewed as the MCT Algorithm with a slightly more aggressive pruning strategy which ensures completeness. The MP algorithm constitutes a simple modification of the Karp and Miller algorithm and thus enjoys the following assets: it is easily amenable to implementation, any strategy of exploration of the Petri net is correct: depth first, breadth first, random ..., and experimental results based on a prototype implementation in Python show promising results [17]. Recently, the MP algorithm has been used successfully in the context of the verification of data-driven workflows [12].

While MP algorithm is simple and includes the elegant ideas of the original MCT Algorithm, the proof of its correctness presented in [17] is long and technical. The main difficulty is to prove the completeness of the algorithm, i.e. to show that the set returned by the algorithm covers every reachable marking (recall that the flaw of MCT algorithm identified in [9] is precisely its incompleteness). In [17], to overcome this difficulty, we reduce the problem to the completeness of the algorithm for a particular class of finite state systems, which we call widened Petri nets (WPN). Yet, the proof of the completeness of MP algorithm for WPN provided in [17] is approximately ten pages long, and goes through several technical lemmas, making it hard to understand and to generalise.

Contributions of the Paper. In this paper, we present a new proof of the completeness of MP Algorithm for WPN. More precisely, we consider a more general version of MP Algorithm, in which the acceleration used is considered as a parameter. In the context of WPN, a concretisation function can be associated with an acceleration: it gives a concrete sequence of transitions allowing to reach the ω-marking resulting from the acceleration. We identify a property of the acceleration by means of its concretisation function, which we call *coherence* and prove the completeness of MP Algorithm for WPN provided the acceleration used is coherent. This new proof relies on a simple invariant of the property, whose proof is less than two pages long.

We argue that this new proof has the following assets:

1. it is much more readable, increasing its confidence,
2. it is more general, as the acceleration is now a parameter of the algorithm,
3. it opens the way to a generalisation of MP algorithm to other classes of well-structured transition systems [1, 4, 6, 7].

Related Work. Other algorithms have been proposed to compute the MCS. First, the CoverProc algorithm has been introduced in [9]. This algorithm follows a different approach and is not based on the Karp and Miller Algorithm. Instead, it relies on pairs of markings, yielding an important overhead in terms of complexity. Another algorithm has been proposed in [15]. This algorithm is however very tailored to Petri nets and relies on ad-hoc tricks to improve its efficiency. In addition, it does not offer the possibility to modify the exploration strategy: it should be depth-first search.

Organisation of the Paper. Definitions of Petri nets are given in Sect. 2, together with the notion of minimal coverability set. The Monotone-Pruning algorithm is presented in Sect. 3, and the overall proof structure of its correctness is given in Sect. 4. In Sect. 5, we present our new arguments to prove its completeness. In Sect. 6, we show that a simple acceleration function satisfies the expected property to ensure completeness of MP Algorithm.

2   Preliminaries

$\mathbb{N}$ denotes the set of natural numbers. A quasi order $\leq$ on a set S is a reflexive and transitive relation on S. Given a quasi order $\leq$ on S, a state $s \in S$ and a subset X of S, we write $s \leq X$ iff there exists an element $s' \in X$ s.t. $s \leq s'$.

Given a finite alphabet Σ, we denote by Σ^* the set of words on Σ, and by ε the empty word. We denote by $\prec$ the (strict) prefix relation on Σ^*: given $u, v \in \Sigma^*$ we have $u \prec v$ iff there exists $w \in \Sigma^*$ such that $uw = v$ and $w \neq \varepsilon$. We denote by $\preceq$ the relation obtained as $\prec \cup =$.

2.1   Markings, ω-markings and Labelled Trees

Given a finite set P, a *marking on P* is an element of the set $\mathsf{Mark}(P) = \mathbb{N}^P$. The set $\mathsf{Mark}(P)$ is naturally equipped with a partial order denoted $\leq$.

Given a marking $m \in \mathsf{Mark}(P)$, we represent it by giving only the positive components. For instance, $(1, 0, 0, 2)$ on $P = (p_1, p_2, p_3, p_4)$ is represented by the multiset $\{p_1, 2p_4\}$. An *ω-marking on P* is an element of the set $\mathsf{Mark}^\omega(P) = (\mathbb{N} \cup \{\omega\})^P$. The order $\leq$ on $\mathsf{Mark}(P)$ is naturally extended to this set by letting $n < \omega$ for any $n \in \mathbb{N}$, and $\omega \leq \omega$. Addition and subtraction on $\mathsf{Mark}^\omega(P)$ are obtained using the rules $\omega + n = \omega - n = \omega$ for any $n \in \mathbb{N}$. The ω-marking $(\omega, 0, 0, 2)$ on $P = (p_1, p_2, p_3, p_4)$ is represented by the multiset $\{\omega p_1, 2p_4\}$.

Given two sets Σ_1 and Σ_2, a labelled tree is a tuple $\mathcal{T} = (N, n_0, E, \Lambda)$ where N is the set of nodes, $n_0 \in N$ is the root, $E \subseteq N \times \Sigma_2 \times N$ is the set of edges labelled with elements of Σ_2, and $\Lambda : N \to \Sigma_1$ labels nodes with elements of Σ_1. We extend the mapping Λ to sets of nodes: for $S \subseteq N$, $\Lambda(S) = \{\Lambda(n) \mid n \in S\}$. Given a node $n \in N$, we denote by $\mathsf{Ancestor}_{\mathcal{T}}(n)$ the set of ancestors of n in $\mathcal{T}$ (n included). If n is not the root of $\mathcal{T}$, we denote by $\mathsf{parent}_{\mathcal{T}}(n)$ its first ancestor in $\mathcal{T}$. Finally, given two nodes x and y such that $x \in \mathsf{Ancestor}_{\mathcal{T}}(y)$, we denote by $\mathsf{path}_{\mathcal{T}}(x, y) \in E^*$ the sequence of edges leading from x to y in $\mathcal{T}$. We also denote by $\mathsf{pathlabel}_{\mathcal{T}}(x, y) \in \Sigma_2^*$ the label of this path.

2.2   Petri Nets

Definition 1 (Petri net (PN)). *A Petri net $\mathcal{N}$ is a tuple (P, T, I, O, m_0) where P is a finite set of places, T is a finite set of transitions with $P \cap T = \emptyset$, $I : T \to \mathsf{Mark}(P)$ is the backward incidence mapping, representing the input tokens, $O : T \to \mathsf{Mark}(P)$ is the forward incidence mapping, representing output tokens, and $m_0 \in \mathsf{Mark}(P)$ is the initial marking.*

The semantics of a PN is usually defined on markings, but can easily be extended to ω-markings. We define the semantics of $\mathcal{N} = (P, T, I, O, m_0)$ by its associated labelled transition system $(\mathsf{Mark}^\omega(P), m_0, \Rightarrow)$ where $\Rightarrow \subseteq \mathsf{Mark}^\omega(P) \times \mathsf{Mark}^\omega(P)$ is the transition relation defined by $m \Rightarrow m'$ iff $\exists t \in T$ s.t. $m \geq I(t) \wedge m' = m - I(t) + O(t)$. For convenience we will write, for $t \in T$, $m \overset{t}{\Rightarrow} m'$ if $m \geq I(t)$ and $m' = m - I(t) + O(t)$. In addition, we also write $m' = \mathsf{Post}(m, t)$, this defines the operator Post which computes the successor of an ω-marking by a transition. We naturally extend this operator to sequences of transitions. Given an ω-marking m and a transition t, we write $m \overset{t}{\Rightarrow} \cdot$ iff there exists $m' \in \mathsf{Mark}^\omega(P)$ such that $m \overset{t}{\Rightarrow} m'$. The relation $\Rightarrow^*$ represents the reflexive and transitive closure of $\Rightarrow$. We say that a marking m is *reachable in $\mathcal{N}$* iff $m_0 \Rightarrow^* m$. We say that a Petri net is bounded if the set of reachable markings is finite.

Example 1. We consider the Petri net $\mathcal{N}$ depicted on Fig. 1, which is the example used in [17]. The initial marking is $\{p_1\}$, depicted by the token in the place p_1. For any integer n, we have $\mathsf{Post}(\{p_1\}, t_1(t_3t_4)^n) = \{p_3, np_5\}$. In particular, this net is not bounded as place p_5 is not.   ⌐

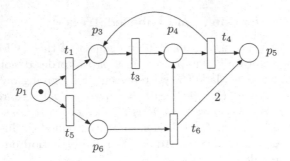

Fig. 1. A Petri net $\mathcal{N}$.

2.3   Minimal Coverability Set of Petri Nets

We recall the definition of minimal coverability set introduced in [5].

Definition 2. *A coverability set of a Petri net* $\mathcal{N} = (P, T, I, O, m_0)$ *is a finite subset* C *of* $\mathsf{Mark}^\omega(P)$ *such that the two following conditions hold:*

(1) for every reachable marking m *of* $\mathcal{N}$, *there exists* $m' \in C$ *such that* $m \leq m'$,
(2) for every $m' \in C$, *either* m' *is reachable in* $\mathcal{N}$ *or there exists an infinite strictly increasing sequence of reachable markings* $(m_n)_{n \in \mathbb{N}}$ *converging to* m'.

A coverability set is minimal iff no proper subset is a coverability set.

One can prove (see [5]) that a PN $\mathcal{N}$ admits a unique minimal coverability set, which we denote by $\mathrm{MCS}(\mathcal{N})$.

Note that every two elements of a minimal coverability set are incomparable. Computing the minimal coverability set from a coverability set is easy. Note also that if the PN is bounded, then the set of reachable markings is finite, and thus the notion of reachable maximal marking is well-defined. In this case, a set of markings is a coverability set iff it contains all maximal reachable markings.

Example 2. (Example 1 continued). The MCS of the Petri net $\mathcal{N}$ is composed of the following ω-markings: $\{p_1\}$, $\{p_6\}$, $\{p_3, \omega p_5\}$, and $\{p_4, \omega p_5\}$.   ⌐

3   Presentation of the Monotone-Pruning Algorithm

3.1   Acceleration(s)

Following previous works, as Karp and Miller algorithm, MP algorithm involves an acceleration function. Such a function takes as input a set of ω-markings M and an ω-marking m, and returns an ω-marking m', which can be used to replace m. Several such functions have been considered in the literature. A classical one is the mapping $\mathsf{Acc}_{\mathsf{all}} : 2^{\mathsf{Mark}^\omega(P)} \times \mathsf{Mark}^\omega(P) \to \mathsf{Mark}^\omega(P)$ which is defined as follows:

$$\forall p \in P, \mathsf{Acc}_{\mathsf{all}}(M, m)(p) = \begin{cases} \omega & \text{if } \exists m' \in M \mid m' < m \wedge m'(p) < m(p) < \omega \\ m(p) & \text{otherwise.} \end{cases}$$

A weaker acceleration computes the acceleration w.r.t. a single ω-marking chosen in the set M. The mapping $\mathsf{Acc_{one}} : 2^{\mathsf{Mark}^\omega(P)} \times \mathsf{Mark}^\omega(P) \to \mathsf{Mark}^\omega(P)$ is defined as follows:

– if there exists $m' \in M$ such that $m' < m$, then we fix one such ω-marking m', and define $\mathsf{Acc_{one}}(M, m)$ as follows:

$$\forall p \in P, \mathsf{Acc_{one}}(M, m)(p) = \begin{cases} \omega & \text{if } m'(p) < m(p) < \omega \\ m(p) & \text{otherwise.} \end{cases}$$

– otherwise, we define $\mathsf{Acc_{one}}(M, m) = m$.

In both functions, the acceleration *uses* one (or several) of the ω-markings in M to build the new ω-marking. Note that these accelerations will always be used along a branch of the tree constructed by the algorithm.

3.2 Definition of the Algorithm

The K&M Algorithm uses comparisons along the same branch to compute the acceleration and stop the exploration. We present in this section the Monotone-Pruning Algorithm which includes a comparison (and a pruning) between branches. We denote this algorithm by MP. It has as a parameter an acceleration function Acc as defined in the previous section.

Algorithm 1. Monotone Pruning Algorithm for Petri Nets.

Require: A Petri net $\mathcal{N} = (P, T, I, O, m_0)$ and an acceleration function Acc.
Ensure: A labelled tree $\mathcal{C} = (X, x_0, B, \Lambda)$ with nodes (resp. edges) labelled with elements in $\mathsf{Mark}^\omega(P)$ (resp. T), and a set $\mathsf{Act} \subseteq X$ such that $\Lambda(\mathsf{Act}) = \mathrm{MCS}(\mathcal{N})$.
1: Let x_0 be a new node such that $\Lambda(x_0) = m_0$;
2: $X := \{x_0\}$; $\mathsf{Act} := X$; $\mathsf{Wait} := \{(x_0, t) \mid \Lambda(x_0) \overset{t}{\Rightarrow} \cdot\}$; $B := \emptyset$;
3: **while** $\mathsf{Wait} \neq \emptyset$ **do**
4:    Pop (n, t) from Wait. $m := \mathsf{Post}(\Lambda(n), t)$;
5:    **if** $n \in \mathsf{Act}$ and $m \not\leq \Lambda(\mathsf{Act})$ **then**
6:       Let n' be a new node such that $\Lambda(n') = \mathsf{Acc}(\Lambda(\mathsf{Ancestor}_\mathcal{C}(n) \cap \mathsf{Act}), m)$;
7:       $X := X \cup \{n'\}$; $B := B \cup \{(n, t, n')\}$;
8:       $\mathsf{Act} := \mathsf{Act} \setminus \{x \mid \exists y \in \mathsf{Ancestor}_\mathcal{C}(x) \text{ s.t. } \Lambda(y) \leq \Lambda(n') \wedge (y \in \mathsf{Act} \vee y \notin \mathsf{Ancestor}_\mathcal{C}(n'))\}$;
9:       $\mathsf{Act} := \mathsf{Act} \cup \{n'\}$; $\mathsf{Wait} := \mathsf{Wait} \cup \{(n', t') \mid \Lambda(n') \overset{t'}{\Rightarrow} \cdot\}$;
10:   **end if**
11: **end while**
12: Return $\mathcal{C} = (X, x_0, B, \Lambda)$ and Act.

As Karp and Miller Algorithm, the MP Algorithm builds a tree $\mathcal{C}$ in which nodes are labelled by ω-markings and edges by transitions of the Petri net. Therefore it proceeds in an exploration of the reachability tree of the Petri net, and uses acceleration along branches to reach the "limit" markings. In addition, it can prune branches that are covered by nodes on other branches. This additional pruning is the source of efficiency, as it avoids to perform redundant computations. It is also the source of difficulty, as a previous attempt of introduction of such pruning led to an incomplete algorithm (MCT Algorithm [5]). In order to obtain a complete algorithm, nodes of the tree are partitioned into two subsets: active nodes, and inactive ones. Intuitively, active nodes will form the minimal coverability set of the Petri net, while inactive ones are kept to ensure completeness of the algorithm.

Given a pair (n, t) popped from Wait, the introduction in C of the new node obtained from (n, t) proceeds in the following steps:

1. the "regular" successor marking is computed: $m = \mathsf{Post}(\Lambda(n), t)$ (Line 4) ;
2. node n should be active and marking m should not be covered by some active node (test of Line 5) ;
3. the marking resulting from the acceleration of m w.r.t. the *active ancestors* of node n is computed and associated with a new node n': $\Lambda(n') = \mathsf{Acc}(\Lambda(\mathsf{Ancestor}_C(n) \cap \mathsf{Act}), m)$ (Line 6) ;
4. update of Act: some nodes are "deactivated", i.e. removed from Act (Line 8).
5. the new node n' is declared as active and Wait is updated (Line 9) ;

We detail the update of the set Act. Intuitively, one wants to deactivate nodes (and their descendants) that are covered by the new node n'. In MP Algorithm (see Line 8), node x is deactivated iff its ancestor y is either active ($y \in \mathsf{Act}$), or is not itself an ancestor of n' ($y \notin \mathsf{Ancestor}_C(n')$). In this case, we say that x *is deactivated by* n'. This subtle condition constitutes the main difference between MP and MCT Algorithms.

To illustrate the behaviour of MP Algorithm, consider the introduction of a new node n' obtained from $(n, t) \in \mathsf{Wait}$, and a node y such that $\Lambda(y) \leq \Lambda(n')$, y can be used to deactivate nodes in two ways:

- if $y \notin \mathsf{Ancestor}_C(n')$, then no matter whether y is active or not, all its descendants are deactivated (represented in gray on Fig. 2(a)),
- if $y \in \mathsf{Ancestor}_C(n')$, then y must be active ($y \in \mathsf{Act}$), and in that case all its descendants are deactivated, except node n' itself as it is added to Act at Line 9 (see Fig. 2(b)).

4   Structure of the Proof of Correction of MP Algorithm

In this section, we describe the overall structure of the proof of [17]. Given an input Petri net $\mathcal{N}$, MP Algorithm returns a set Act of ω-markings. We say that MP Algorithm is:

- *sound* if for every $m \in \mathsf{Act}$, there exists $n \in \mathrm{MCS}(\mathcal{N})$ such that $m \leq n$,
- *complete* if for every $n \in \mathrm{MCS}(\mathcal{N})$, there exists $m \in \mathsf{Act}$ such that $n \leq m$.

It is easy to show that Act is composed of pairwise incomparable ω-markings. Hence, if MP Algorithm is both sound and complete, then it returns exactly the set $\mathrm{MCS}(\mathcal{N})$.

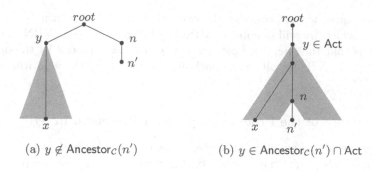

(a) $y \notin \mathsf{Ancestor}_C(n')$ (b) $y \in \mathsf{Ancestor}_C(n') \cap \mathsf{Act}$

Fig. 2. Deactivations of MP Algorithm.

4.1 Widened Petri Nets

Our proof involves a widening operation which turns a Petri net into a finite state system. Let P be a finite set, and $\varphi \in \mathsf{Mark}(P)$ be a marking. We consider the *finite* set of ω-markings whose finite components (*i.e.* values different from ω) are less or equal than φ. Formally, we define:

$$\mathsf{Mark}^\omega_\varphi(P) = \{m \in \mathsf{Mark}^\omega(P) \mid \forall p \in P, m(p) \le \varphi(p) \vee m(p) = \omega\}.$$

The widening operator Widen_φ maps an ω-marking to an element of $\mathsf{Mark}^\omega_\varphi(P)$:

$$\forall m \in \mathsf{Mark}^\omega(P), \forall p \in P, \mathsf{Widen}_\varphi(m)(p) = \begin{cases} m(p) & \text{if } m(p) \le \varphi(p) \\ \omega & \text{otherwise.} \end{cases}$$

Note that this operator trivially satisfies $m \le \mathsf{Widen}_\varphi(m)$.

Definition 3 (Widened Petri net). *A widened Petri net (WPN for short) is a pair $(\mathcal{N}, \varphi)$ composed of a PN $\mathcal{N} = (P, T, I, O, m_0)$ and of a marking $\varphi \in \mathsf{Mark}(P)$ such that $m_0 \le \varphi$.*

The semantics of $(\mathcal{N}, \varphi)$ is given by its associated labelled transition system $(\mathsf{Mark}^\omega_\varphi(P), m_0, \Rightarrow_\varphi)$ where for $m, m' \in \mathsf{Mark}^\omega_\varphi(P)$, and $t \in T$, we have $m \xrightarrow{t}_\varphi m'$ iff $m' = \mathsf{Widen}_\varphi(\mathsf{Post}(m, t))$. We carry over from PN to WPN the relevant notions, such as reachable marking. We define the operator Post_φ by $\mathsf{Post}_\varphi(m, t) = \mathsf{Widen}_\varphi(\mathsf{Post}(m, t))$. Subscript φ may be omitted when it is clear from the context. Finally, the minimal coverability set of a widened Petri net $(\mathcal{N}, \varphi)$ is simply the set of its maximal reachable states as its reachability set is finite. It is denoted $\mathrm{MCS}(\mathcal{N}, \varphi)$.

Example 3 (Example 1 continued). Consider the mapping φ associating 1 to places p_1, p_3, p_4 and p_6, and 3 to place p_5, and the widened Petri net $(\mathcal{N}, \varphi)$. Then from marking $\{p_4, 3p_5\}$, the firing of t_4 results in the marking $\{p_3, \omega p_5\}$, instead of the marking $\{p_3, 4p_5\}$ in the standard semantics. One can compute the MCS of this WPN. Due to the choice of φ, it coincides with $\mathrm{MCS}(\mathcal{N})$. ⌐

In the sequel, we will consider the execution of MP Algorithm on widened Petri nets, which we will denote by MP_{WPN}. Let $(\mathcal{N}, \varphi)$ be a WPN. The only difference is that the operator Post (resp. $\Rightarrow$) must be replaced by the operator $Post_\varphi$ (resp. $\Rightarrow_\varphi$). Thus, all the ω-markings computed by the algorithm belong to $Mark_\varphi^\omega(P)$.

4.2 Structure of the Proof of Correction Presented in [17]

The structure of the proof of correction presented in [17] is depicted in Fig. 3. In this proof, all results have rather simple proofs, except the completeness of MP_{WPN}. The proof of this property presented in [17] is approximately ten pages long. The main contribution of this paper is a very short proof of this property. It is presented in the next section, and stated as Theorem 1.

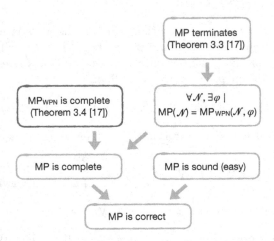

Fig. 3. Structure of the proof of [17]. The main difficulty lies in the completeness of MP_{WPN}, depicted in red. (Color figure online)

5   Completeness of MP Algorithm for WPN

In this section, we present the main contribution of this article, which is a new and simple proof of the completeness of MP Algorithm for WPN.

5.1   Coherence of an Acceleration

MP Algorithm builds a labelled tree $\mathcal{C}$. In this context, the acceleration is applied along a branch β starting from the root, and leading to a node n whose marking is m. More precisely, there exists a set N of nodes on β, which are active ancestors of node n, and such that M is the set of markings of N. We recall the notion of concretization that "explains" how the accelerated marking is computed, by giving an explicit sequence of transitions leading to the accelerated marking.

Definition 4. *We consider an acceleration* Acc *and a labelled tree* $\mathcal{C} = (X, x_0, B, \Lambda)$ *obtained from a WPN* $(\mathcal{N}, \varphi)$ *using* Acc. *A concretization function is a mapping* $\gamma : B^* \to T^*$ *associating to every path in* $\mathcal{C}$ *a sequence of transitions of* $\mathcal{N}$. *In addition, given* $x, y \in X$ *such that* $y \in$ Ancestor$_\mathcal{C}(x)$, *we have* $\Lambda(x) = \mathsf{Post}_\varphi(\Lambda(y), \gamma(\mathsf{path}_\mathcal{C}(y, x)))$.

In order to have a generic proof, independent of the acceleration considered, we identify a property of the acceleration together with its concretization function which ensures that the algorithm is correct.

Definition 5. *We consider an acceleration function* Acc. *We say that* Acc *is* coherent *if it admits a concretization function* γ *such that the following property holds:* (†) *Let* x, y *in* $\mathcal{C}$ *and* $w = \mathsf{path}_\mathcal{C}(y, x) \in B^*$. *Then for every* $\rho_p \preceq \gamma(w)$, *there exist two nodes* x' *and* y' *such that:*

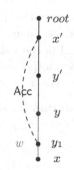

- $\mathsf{Post}(\Lambda(y), \rho_p) \geq \Lambda(y')$,
- x' *is an ancestor of* x, *used by some acceleration for node* y_1 *on the path from* y *to* x,
- y' *lies between* x' *and* y_1.

We prove now that when an acceleration Acc is coherent, then MP Algorithm satisfies a property that we call the coherence of this algorithm.

Lemma 1 (MP Algorithm is coherent). *We consider MP Algorithm with a coherent acceleration* Acc, *with concretization* γ. *Then the following property holds: consider three nodes* x, y, z *such that* $x, z \in$ Act *and* $y \in$ Ancestor$_\mathcal{C}(x)$, *and define* $\rho = \gamma(\mathsf{path}_\mathcal{C}(y, x))$. *If* $\Lambda(z) \geq \mathsf{Post}(\Lambda(y), \rho_p)$ *for some* $\rho_p \preceq \rho$, *then* $y \in$ Ancestor$_\mathcal{C}(z)$.

Proof. As the acceleration is coherent, we fix an adequate concretization function γ. We consider nodes x, y, z and some $\rho_p \preceq \rho = \gamma(\mathsf{path}_\mathcal{C}(y, x))$ as in the premises of the statement. Thanks to property (†), there exist two nodes x', y' such that:

- $\mathsf{Post}(\Lambda(y), \rho_p) \geq \Lambda(y')$,
- x' is an ancestor of x, used by some acceleration for node y_1 on the path from y to x,
- y' lies between x' and y_1.

By contradiction, assume that $y' \notin$ Ancestor$_\mathcal{C}(z)$. We have $\Lambda(z) \geq \mathsf{Post}(\Lambda(y), \rho_p)$ and $\mathsf{Post}(\Lambda(y), \rho_p) \geq \Lambda(y')$, hence $\Lambda(z) \geq \Lambda(y')$. Then, by definition of MP Algorithm, x is deactivated by z. This is in contradiction with our assumption that x and z are active. Thus, we have $y' \in$ Ancestor$_\mathcal{C}(z)$.

Assume now that $y_1 \notin$ Ancestor$_\mathcal{C}(z)$. Recall that y_1 used the node x' when the acceleration has been applied. By definition of MP Algorithm, it deactivated everything below x', except itself. As we have that y' is between x' and y_1, y' is an ancestor of z, and y_1 is not an ancestor of z, this entails that y_1 deactivated z, which is a contradiction. Thus, we have $y_1 \in$ Ancestor$_\mathcal{C}(z)$.

In particular, this implies $y \in$ Ancestor$_\mathcal{C}(z)$, as expected. $\qquad\square$

5.2 New Proof

Our new proof relies on a simple invariant of the algorithm, from which completeness easily follows. This invariant is defined as the following property ($\mathcal{P}$):

$$\forall m \in \mathsf{Reach}(\mathcal{N}), \exists (x, \rho) \in \mathsf{Act} \times T^* \text{ such that:}$$

$$\begin{cases} (1) & \mathsf{Post}(\Lambda(x), \rho) \geq m \\ (2) & \rho \neq \epsilon \Rightarrow (x, \mathsf{first}(\rho)) \in \mathsf{Wait} \\ (3) & \forall \epsilon \neq \rho_p \preceq \rho, \neg \exists z \in \mathsf{Act}. \Lambda(z) \geq \mathsf{Post}(\Lambda(x), \rho_p) \end{cases}$$

Intuitively, the invariant states that for every reachable marking m, there exists a pair (x, ρ) which allows to cover m (property (1)), whose exploration is still in the waiting list (property (2)), and whose exploration will not be stopped by another active node (property (3)).

We now prove that the MP Algorithm satisfies the invariant ($\mathcal{P}$):

Lemma 2. *When used with a coherent acceleration, the MP Algorithm satisfies the property ($\mathcal{P}$) at every step of its execution.*

Proof. We proceed by induction on the number of steps of the algorithm.
Base case. Initially, the invariant is trivially satisfied as there is a single active node corresponding to the initial marking.
Induction. $\mathcal{P}(k) \Rightarrow \mathcal{P}(k+1)$

Let $m \in \mathsf{Reach}(\mathcal{N})$. By $\mathcal{P}(k)$, there exists (x, ρ) as given by $\mathcal{P}$.

We consider different cases depending on what happens in the While loop of the algorithm. At Line 4, a pair (n, t) is popped from the waiting list. If this pair does not pass the test of Line 5, then nothing changes and the pair (x, ρ) still satisfies the properties. The interesting case is when this pair passes the test of Line 5. We distinguish three cases:

1. if x is not deactivated and no successor of x by prefixes of ρ is covered by n', then we can simply choose the pair (x, ρ).
2. otherwise, assume that some successor of x by a (possibly empty) prefix of ρ is covered by n'. Then, let ρ_1 be the longest prefix of ρ such that $\Lambda(n') \geq \mathsf{Post}(\Lambda(x), \rho_1)$. We claim that we can choose the pair (n', ρ') where $\rho' = \rho_1^{-1}.\rho$. Indeed:
   - $n' \in \mathsf{Act}$ (Line 9),
   - Property (1) follows from monotonicity of Petri nets and from $\Lambda(n') \geq \mathsf{Post}(\Lambda(x), \rho_1)$,
   - Property (2) follows from Line 9,
   - In order to show that Property (3) holds, we proceed by contradiction. Assume that there exists ρ'_p a non-empty prefix of ρ' and an active node z such that $\Lambda(z) \geq \mathsf{Post}(\Lambda(n'), \rho'_p)$. Then we have:

$$\mathsf{Post}(\Lambda(x), \rho_1\rho'_p) \leq \mathsf{Post}(\Lambda(n'), \rho'_p) \leq \Lambda(z)$$

As $\rho_1\rho'_p \neq \epsilon$ and $\rho_1\rho'_p \preceq \rho$, Property (3) of our invariant for (x, ρ) implies that z is a new active node, *i.e.* $z = n'$. This is a contradiction with our choice of ρ_1 of maximal length.

3. otherwise, x is deactivated by n': n' dominates a strict ancestor y of x such that $y \in$ Act or $y \notin$ Ancestor$_C(n')$ (see Line 8 of the algorithm). We fix such a node y and let $w =$ path$_C(y, x)$. We define $\rho_0 = \gamma(w)$ and ρ_1 as the longest prefix of ρ_0 such that $\Lambda(n') \geq$ Post$(\Lambda(n'), \rho_1)$. We write $\rho_0 = \rho_1\rho_2$ and claim that the pair $(n', \rho_2\rho)$ satisfies the properties of the invariant. Property (1) follows directly from monotonicity of Petri nets. Property (2) follows from the fact that n' has just been added to C.

We prove now Property (3). By contradiction, assume that there exists ρ_p a non-empty prefix of $\rho_2\rho$ and an active node z such that $\Lambda(z) \geq$ Post$(\Lambda(n'), \rho_p)$. Then we also have, by monotonicity, $\Lambda(z) \geq$ Post$(\Lambda(y), \rho_p)$. First case: $z = n'$. As we are not in Case 2, ρ_p should be a prefix of ρ_2. But this in contradiction with the definition of ρ_2.

Second case: $z \neq n'$. In particular, z is already active at the previous iteration of the algorithm. By Property (3) of the invariant for (x, ρ), ρ_p is a prefix of ρ_2.

We consider the prefix $\rho_1\rho_p$ of $\rho_0 = \gamma(w)$. We can apply Lemma 1 and deduce that $y \in$ Ancestor$_C(z)$. Thus z is deactivated by the construction of n' (see Line 8 of the algorithm), yielding the contradiction as we supposed z is active.

$\square$

Theorem 1. *If* Acc *is coherent, then* MP *Algorithm for* WPN *is complete.*

Proof. The result directly follows from Lemma 2 and from the termination of the algorithm. Consider some reachable marking m and the set Act returned by MP Algorithm after its termination. Thanks to Lemma 2, there exists some pair (x, ρ) as given by property $(\mathcal{P})$. As the waiting list is empty, we have $\rho = \epsilon$, hence property (1) directly gives the completeness of Act. $\square$

6   Coherence of the Acceleration Acc$_{one}$

In this section, we exhibit a concretization function for the acceleration Acc$_{one}$ which allows to show that this acceleration is coherent.

Definition 6 (Concretization function for Acc$_{one}$). *The concretization function is a morphism γ_{one} from B^* to T^*. We let $M = \max\{\varphi(p) \mid p \in P\}+1$.*

Let $b = (n, t, n') \in B$. We assume γ_{one} is defined on all edges $(x, u, y) \in B$ such that $y \in$ Ancestor$_C(n)$.

Let $m =$ Post$_\varphi(\Lambda(n), t)$, then there are two cases, either :

1. *$\Lambda(n') = m$ (t is not accelerated), then we define $\gamma(b) = t$, or*
2. *$\Lambda(n') > m$. Let x be the ancestor of n used for this acceleration, and $w =$ path$_C(x, n) \in B^*$. Then we define:*
   $\gamma_{one}(b) = t.(\gamma_{one}(w).t)^M$

The following property can easily be proved by induction:

Lemma 3. *The mapping γ_{one} is a concretization of the acceleration* Acc$_{one}$.

By reasoning on γ_{one} and using again an induction, we can show the existence of adequate ancestors, to prove the following property:

Lemma 4. *The acceleration* Acc_{one} *is coherent.*

7   Conclusion

In this paper, we have provided a new proof of the completeness of MP Algorithm, an algorithm introduced in [17] to compute the minimal coverability set of a Petri net. The new proof relies on an original invariant, is considerably shorter than that of [17], and allows to identify the property that the acceleration should meet to ensure the completeness of the algorithm.

As future work, we would like to extend MP Algorithm to more general classes of well-structured transition systems. To this end, we plan to rely on the representation of downward-closed sets using finite unions of ideals, as introduced in [6,7]. This setting has recently been used to develop an *ideal Karp and Miller* algorithm in [1], which should be a good basis for extending MP Algorithm to well-structured transition systems.

References

1. Blondin, M., Finkel, A., Goubault-Larrecq, J.: Forward analysis for wsts, part III: Karp-Miller trees. In: 37th IARCS Annual Conference on Foundations of Software Technology and Theoretical Computer Science, FSTTCS 2017, vol. 93 of LIPIcs, pp. 16:1–16:15. Leibniz-Zentrum fuer Informatik (2017)
2. Blondin, M., Finkel, A., Haase, C., Haddad, S.: Approaching the coverability problem continuously. In: Chechik, M., Raskin, J.-F. (eds.) TACAS 2016. LNCS, vol. 9636, pp. 480–496. Springer, Heidelberg (2016). https://doi.org/10.1007/978-3-662-49674-9_28
3. Cardoza, E., Lipton, R.J., Meyer, A.R.: Exponential space complete problems for petri nets and commutative semigroups: preliminary report. In: Proceedings of the 8th Annual ACM Symposium on Theory of Computing, Hershey, Pennsylvania, USA, 3–5 May 1976, pp. 50–54. ACM (1976)
4. Finkel, A.: A generalization of the procedure of Karp and Miller to well structured transition systems. In: Ottmann, T. (ed.) ICALP 1987. LNCS, vol. 267, pp. 499–508. Springer, Heidelberg (1987). https://doi.org/10.1007/3-540-18088-5_43
5. Finkel, A.: The minimal coverability graph for Petri nets. In: Rozenberg, G. (ed.) ICATPN 1991. LNCS, vol. 674, pp. 210–243. Springer, Heidelberg (1993). https://doi.org/10.1007/3-540-56689-9_45
6. Finkel, A., Goubault-Larrecq, J.: Forward analysis for WSTS, part I: completions. In: Proceedings of STACS 2009, vol. 3 of LIPIcs, pp. 433–444. Leibniz-Zentrum für Informatik (2009)
7. Finkel, A., Goubault-Larrecq, J.: Forward analysis for WSTS, part II: complete WSTS. Log. Methods Comput. Sci. **8**(3) (2012)
8. Geeraerts, G., Raskin, J., Begin, L.V.: Expand, enlarge and check: new algorithms for the coverability problem of WSTS. J. Comput. Syst. Sci. **72**(1), 180–203 (2006)

9. Geeraerts, G., Raskin, J.-F., Van Begin, L.: On the efficient computation of the coverability set for petri nets. Int. J. Found. Comput. Sci. **21**(2), 135–165 (2010)
10. Karp, R.M., Miller, R.E.: Parallel program schemata. J. Comput. Syst. Sci. **3**(2), 147–195 (1969)
11. Kloos, J., Majumdar, R., Niksic, F., Piskac, R.: Incremental, inductive coverability. In: Sharygina, N., Veith, H. (eds.) CAV 2013. LNCS, vol. 8044, pp. 158–173. Springer, Heidelberg (2013). https://doi.org/10.1007/978-3-642-39799-8_10
12. Li, Y., Deutsch, A., Vianu, V.: VERIFAS: a practical verifier for artifact systems. Proc. VLDB Endow. **11**(3), 283–296 (2017)
13. Lüttge, K.: Zustandsgraphen von Petri-Netzen. Master's thesis, Humboldt-Universität (1995)
14. Petri, C.A.: Kommunikation mit Automaten. Ph.D. thesis, Institut für Instrumentelle Mathematik, Bonn, Germany (1962)
15. Piipponen, A., Valmari, A.: Constructing minimal coverability sets. Fundam. Inf. **143**(3–4), 393–414 (2016)
16. Rackoff, C.: The covering and boundedness problems for vector addition systems. Theor. Comput. Sci. **6**, 223–231 (1978)
17. Reynier, P.-A., Servais, F.: Minimal coverability set for Petri nets: Karp and Miller algorithm with pruning. Fundam. Inf. **122**(1–2), 1–30 (2013)
18. Schmidt, K.: Model-checking with coverability graphs. Form. Methods Syst. Des. **15**(3), 239–254 (1999)

Deciding Reachability for Piecewise Constant Derivative Systems on Orientable Manifolds

Andrei Sandler and Olga Tveretina[(✉)]

Department of Computer Science, University of Hertfordshire, Hatfield, UK
{a.sandler,o.tveretina}@herts.ac.uk

Abstract. A hybrid automaton is a finite state machine combined with some k real-valued continuous variables, where k determines the number of the automaton dimensions. This formalism is widely used for modelling safety-critical systems, and verification tasks for such systems can often be expressed as the reachability problem for hybrid automata.

Asarin, Mysore, Pnueli and Schneider defined classes of hybrid automata lying on the boundary between decidability and undecidability in their seminal paper 'Low dimensional hybrid systems - decidable, undecidable, don't know' [9]. They proved that certain decidable classes become undecidable when given a little additional computational power, and showed that the reachability question remains unsolved for some 2-dimensional systems.

Piecewise Constant Derivative Systems on 2-dimensional manifolds (or PCD_{2m}) constitute a class of hybrid automata for which decidability of the reachability problem is unknown. In this paper we show that the reachability problem becomes decidable for PCD_{2m} if we slightly limit their dynamics, and thus we partially answer the open question of Asarin, Mysore, Pnueli and Schneider posed in [9].

Keywords: Hybrid systems · Reachability · Decidability

1 Introduction

A hybrid automaton is a formalism used to model dynamic systems that comprise both digital and analog components. Formally, it is a finite state machine combined with some k real-valued continuous variables, where k determines the number of the automaton dimensions. Examples of such systems can be found among others in avionics, robotics and bioinformatics, and most of them are safety-critical.

Verifying safety properties typically consists of construction of a set of reachable states and checking whether this set intersects with a set of unsafe states. Therefore, one of the most fundamental problems in the analysis of hybrid automata is the reachability problem. Formally, it is stated as follows: *for a*

© Springer Nature Switzerland AG 2019
E. Filiot et al. (Eds.): RP 2019, LNCS 11674, pp. 178–192, 2019.
https://doi.org/10.1007/978-3-030-30806-3_14

given automaton determine if there is a trajectory from some initial state to a target state.

Undecidability of reachability is usually proved by the simulation of a Turing machine or any other Turing-complete abstraction on the given hybrid automaton (see [3,9] for examples). This way, the existence of an algorithm deciding the reachability problem would solve the halting problem, which is a contradiction. On the other hand, decidability of reachability is typically shown by providing an algorithm which solves it, or by showing that the system admits finite bisimulation [13,14].

The reachability problem is undecidable even for simple classes of hybrid automata such as linear hybrid automata [1]. Nevertheless, there are classes of hybrid systems for which it is decidable. Examples of decidable systems include o-minimal systems [14] and initialized rectangular automata [11].

Despite the increasing interest in discovering new decidability results for hybrid automata, there is still no clear boundary between what is decidable and what is not for such systems [9].

Asarin et al. and Henzinger et al. presented hybrid automata that span the boundary between decidability and undecidability for the reachability problem in [9] and [11] respectively. Asarin et al. observed that certain decidable classes become undecidable, when given a little additional computational power. Thus, decidable 2-dimensional Piecewise Constant Derivative Systems (PCDs) become undecidable for three dimensions or higher [3,15].

Asarin and Schneider considered 2-dimensional Hierarchical Piecewise Constant Derivative Systems (HPCDs), an intermediate class lying between decidable 2-dimensional and undecidable 3-dimensional PCDs [4]. They proved that 2-dimensional HPCDs are equivalent to 1-dimensional Piecewise Affine Maps (PAMs), a class of dynamical systems for which reachability is a well-known open problem [8]. A 1-dimensional PAM is a piecewise function which is applied to the 1-dimensional real line, and the function within each interval of the real line is affine. It has been proven that a 1-dimensional PAM is equivalent to a 2-dimensional system called a planar pseudo-billiard system, also known as a 'strange billiards' model in bifurcation and chaos theory [12].

Variants of HPCDs called Restricted HPCDs (RHPCDs), have been considered in [7]. This class of systems has similarities with many well-known models such as rectangular automata and stopwatch automata. The authors show that 3-dimensional RHPCDs are undecidable by encoding a Minsky machine.

Mysore and Pnueli raised the following question [17]: Is there any class, simpler than 2-dimensional HPCDs, which is equivalent to 1-dimensional PAMs? Asarin et al. came up with further classes of hybrid automata, including PCDs on 2-dimensional manifolds (PCD$_{2m}$), which are equivalent to 1-dimensional PAMs [9].

We consider PCD$_{2m}$ with slightly limited dynamics by forbidding colliding and branching trajectories. We call such systems Regular PCD$_{2m}$ (PCD$_{r2m}$) and show that the reachability problem is decidable in this case.

As 'reference systems' we use dynamical systems on the closed orientable surfaces and rely on the topological properties of their trajectories [16]. Furthermore, we study the properties of the language generated by the trajectories of a PCD$_{r2m}$ by associating all generated words with a sequence of graphs, called Rauzy graphs or factor graphs [18].

The remainder of the paper is organised as follows. In Sect. 2 we recall the notion of 2-dimensional PCDs, and in Sect. 3 we extend it to orientable manifolds. Section 4 provides properties of dynamical systems on the closed orientable surfaces. In Sect. 5 we introduce the language of PCD$_{r2m}$, and in Sect. 6 we show that the reachability problem for PCD$_{r2m}$ is decidable. Section 7 contains concluding remarks.

2   Preliminaries

A Piecewise Constant Derivative System (PCD) can be viewed as a finite set of regions, where each region is associated with a vector field which determines the rate of change of the continuous variables. In this section we formally define them including the reachability problem for such systems.

2.1   Piecewise Constant Derivative Systems on a Plane

In this paper we deal with a 2-dimensional Euclidean space $X = \mathbb{R}^2$. An open half-space in X is the set of all points $x \in X$ satisfying $a \cdot x + b < 0$ for some rational a and b. A convex open polygonal set q is an intersection of a finite number of half-spaces. Let also $\mathsf{cl}(S)$ denote the topological closure of a set $S \subseteq X$.

A finite polygonal partition of X is a set $\mathsf{Q} = \{q_1, \ldots, q_k\}$ of polygonal sets, called regions, such that: (1) $q_i \neq \varnothing$ for all $1 \leq i \leq k$; (2) $q_i \cap q_j = \varnothing$ for all $1 \leq i, j \leq k$ such that $i \neq j$; (3) $\bigcup_{i=1}^{k} \mathsf{cl}(q_i) = X$

The boundary of each region $q \in \mathsf{Q}$ is $\mathsf{bd}(q) = \mathsf{cl}(q) \backslash q$. The interior $\mathsf{int}(X')$ of $X' \subseteq X$ is the set of points $x \in X'$ such that for some $\varepsilon > 0$ there exists an ε-neighbourhood $N_\varepsilon(x) \subseteq X'$ of x. If X' is 1-dimensional then ε-neighbourhood is assumed 1-dimensional too.

We use $\mathsf{E}(\mathsf{Q})$ to denote the set of edges of Q of the form $e = \mathsf{int}(\mathsf{cl}(q_i) \cap \mathsf{cl}(q_j))$, where $q_i, q_j \in \mathsf{Q}, i \neq j$, and $\mathsf{int}(\mathsf{cl}(q_i) \cap \mathsf{cl}(q_j)) \neq \varnothing$. Similarly, $\mathsf{V}(\mathsf{Q})$ denotes the set of vertices of Q of the form $v = \mathsf{cl}(e_i) \cap \mathsf{cl}(e_j)$, where $e_i, e_j \in \mathsf{E}(\mathsf{Q}), i \neq j$, and $\mathsf{cl}(e_i) \cap \mathsf{cl}(e_j) \neq \varnothing$.

We say that $\mathsf{Bd}(\mathsf{Q}) = \mathsf{E}(\mathsf{Q}) \cup \mathsf{V}(\mathsf{Q})$ is a set of border elements. Now the set $\mathsf{Q} \cup \mathsf{Bd}(\mathsf{Q})$ forms a partition of X. We define the border elements of a region $q \in \mathsf{Q}$ as $\mathsf{Bd}(q) = \{b \mid b \subseteq \mathsf{cl}(q)\} \cap \mathsf{Bd}(\mathsf{Q})$.

Definition 1 (2-PCD). *A 2-dimensional Piecewise Constant Derivative System (or 2-PCD) is a pair $\mathsf{H} = (\mathsf{Q}, \mathsf{F})$ with $\mathsf{Q} = \{q_i\}_{i \in I}$ a finite polygonal partition of $\mathbb{R}^2$ and $\mathsf{F} = \{v_i\}_{i \in I}$ a set of vectors from $\mathbb{R}^2$. The dynamics is determined by the equation $\dot{x} = v_i$ for $x \in q_i$.*

We can also define a 2-PCD on a convex subset $S \subset X$ and assume that the rest of $\mathbb{R}^2$ is split onto infinite convex parts with dynamics defined as a constant flow going in or out from S, depending on the flow on the boundary of S.

The set $\mathsf{Bd}(q)$ consists of all boundary elements of a region $q \in Q$ – edges and vertices. Now we define the *input* and *output* boundary elements of q.

Definition 2 (Input and output edges). *Assume $q \in Q$ with dynamics v, and an edge $e \in \mathsf{Bd}(q)$. We say that e in an input edge for q if for any $x \in e$ there is $t > 0$ such that $x + vt \in q$; and e in an output edge for q if for any $x \in e$ there is $t < 0$ such that $x + vt \in q$.*

By Definition 1, for every vertex $x \in \mathsf{bd}(q)$ with $q \in Q$ there are exactly two edges $e, e' \subseteq \mathsf{bd}(q)$ such that $x \in \mathsf{cl}(e) \cap \mathsf{cl}(e')$.

Definition 3 (Input and output vertices). *We say that x is an input vertex for q if both e and e' are input edges; x is an output vertex for q if both e and e' are output edges; and x is neutral with respect to q if e is an input edge and e' is an output edge.*

We denote by $\mathsf{In}(q) \subseteq \mathsf{Bd}(q)$ and $\mathsf{Out}(q) \subseteq \mathsf{Bd}(q)$ the sets of input and output border elements (edges and vertices) of some region q respectively. In the rest of the paper and similar to [5] we assume that $\mathsf{In}(q) \cap \mathsf{Out}(q) = \varnothing$.

2.2 Trajectories

In this section we define the notions of a trajectory, its discrete abstraction called an edge signature, and successor functions similar to [5,6][1] (Fig. 1).

Definition 4 (Trajectory). *A **trajectory segment** of $H = (Q, F)$ with the starting point x_0 is a continuous and almost-everywhere (except on finitely many points) derivable function $\tau : [0, T] \to \mathbb{R}^2$ such that $\tau(0) = x_0$ and for any $t \in [0, T]$, if $\tau(t) \in q_i$ then $\dot{\tau}(t) = v_i$. If $T = \infty$ then τ is called a **trajectory**.*

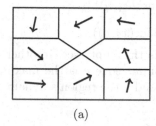 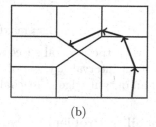

(a)   (b)

Fig. 1. (a) An example of a 2-PCD; (b) an example of a trajectory segment

In the following we consider the discrete abstraction of a trajectory called an *edge signature*.

---

[1] A PCD can be seen as a special case of Polygonal Differential Inclusion Systems (SPDIs).

Definition 5 (Edge signature). *The **edge signature** of a trajectory τ is the sequence $\sigma(\tau) = e_0 e_1 e_2 \dots$ of edges traversed by τ.*

The edge signature of any trajectory segment τ can be represented in the following form:
$$Sig(\tau) = r_1 s_1^{k_1} r_2 s_2^{k_2} \dots r_n s_n^{k_n} r_{n+1},$$
where $s_i^{k_i}$ denotes the cycles s_i of edges repeated k_i times, and r_i denotes the paths (sequence of edges) between cycles (see Theorem 4.1 in [6]). Cycles s_i are simple, that is, an edge can not appear twice in the cycle.

Definition 6 (Signature type). *The **signature type** of an edge signature $Sig(\tau) = r_1 s_1^{k_1} r_2 \dots r_n s_n^{k_n} r_{n+1}$ is the sequence $\mathsf{type}(\tau) = r_1 s_1 r_2 \dots r_n s_n r_{n+1}$.*

The following theorem defines the set of signature types which has to be examined to compute reachable states for a given 2-PCD.

Theorem 1 (Asarin, Schneider, Yovine, [6]). *Only those signature types having disjoint paths r_i and unique (as sets of edges) cycles s_i, could correspond to a trajectory starting in initial set S and ending in final set F. There are only finite number of such signature types on any given 2-PCD.*

For computing the successive interval images, it is convenient to introduce a one-dimensional coordinate system on each edge e, with zero (0) denoting one chosen vertex v_0 of e and one (1) denoting the other vertex v_1. Now each point has the coordinate $v_\lambda = \lambda v_0 + (1 - \lambda)v_1$ with $0 < \lambda < 1$. Then, a series of successor functions on edges of the 2-PCD can be defined.

- Let $x \in e$ for some $e \in \mathsf{In}(q)$. The successor $\mathsf{Succ}(x, q)$ of x is a point $x' \in e'$ for some $e' \in \mathsf{Out}(q)$ such that there is a trajectory segment that starts in x, ends in x' and goes only through q.
- Let $(x_1, x_2) \subseteq e$ for some $e \in \mathsf{In}(q)$. The successor $\mathsf{SuccInt}(x_1, x_2, e', q)$ of (x_1, x_2) on the interval $e' \in \mathsf{Out}(q)$ is $(x_1', x_2') \subseteq e'$ defined as follows:

$$x_1' = \min(1, \mathsf{Succ}(x_1, q))$$

$$x_2' = \max(0, \mathsf{Succ}(x_2, q))$$

If $x_1' > x_2'$, then the interval successor is the empty set. In other words, the successor of an interval (x_1, x_2) of an input edge e is the maximal interval (x_1', x_2') of an output edge e' reachable under the region's dynamics.

2.3   Reachability Problem

Reachability for PCD-like systems can typically be formulated as either point-to-point or edge-to-edge reachability. In this paper we are interested in edge-to-edge reachability.

Definition 7 (Point-to-point reachability). *Let $H = (Q, F)$ be a PCD. Then a point $b \in \mathsf{Bd}(Q)$ is reachable from a point $a \in \mathsf{Bd}(Q)$ if there is a trajectory segment that starts at a and ends at b.*

Definition 8 (Edge-to-edge reachability). *Let* $H = (Q, F)$ *be a PCD. Then an edge* $e_f \in E(Q)$ *is reachable from an edge* $e_s \in E(Q)$ *if there are points* $a_s \in e_s$ *and* $b_f \in e_f$ *such that there is a trajectory segment that starts at* a_s *and ends at* b_f.

3   Piecewise Constant Derivative Systems on Manifolds

All the definitions in this section are similar to the respective definitions in [9] and follow the combinatorial approach in [10].

Definition 9 (Triangulable space). *A topological space is triangulable if it is obtained from a set of triangles by the identification of edges and vertices, where any two triangles are identified either along a single edge or at a vertex, or are completely disjoint. The identification is done via an affine bijection.*

By a closed surface we mean a compact surface without boundary, and it is formally defined below.

Definition 10 (Closed surface). *A closed surface (or a 2-dimensional manifold) S is a compact triangulable space for which in addition the following holds:*

(1) Each edge is identified with exactly one other edge;
(2) The triangles identified at each vertex can always be arranged in a cycle $T_1, T_2, \ldots, T_k, T_1$ *so that adjacent triangles are identified along an edge.*

Examples of closed surfaces include a sphere, a torus (see Fig. 2) and projective planes. In the rest of the paper we only deal with orientable surfaces (a sphere and a connected sum of tori) even though we do not always state it explicitly.

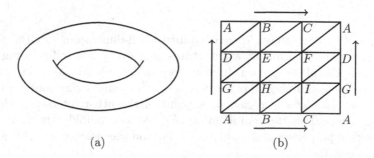

(a)   (b)

Fig. 2. Different representations of a torus: (a) A surface in $\mathbb{R}^3$; (b) A triangulated surface with identified edges

Definition 11 (PCD$_{2m}$). *We define a PCD on a 2-dimensional manifold (or PCD$_{2m}$)* $H = (Q, F)$ *as a 2-PCD on a closed orientable surface S.*

Below we introduce the subclass of PCD_{2m} which we will prove to be decidable. We define a Regular PCD_{2m} by imposing additional restrictions on the dynamics (the flow vectors do not collide or diverge on edges and vertices) to guarantee that any point of the trajectory has exactly one predecessor and one successor.

Definition 12 (PCD_{r2m}). *We say that a PCD_{2m} $H = (Q, F)$ is regular (or PCD_{r2m}) if the following holds:*

1. *For any $q, q' \in Q$, if $b \in Bd(q) \cap Bd(q')$ then either $b \in In(q)$ and $b \in Out(q')$, or $b \in Out(q)$ and $b \in In(q')$.*
2. *The vector field in any $q \in Q$ is not parallel to any $e \in cl(q) \cap E(Q)$.*

Examples of the dynamics forbidden by Definition 12 are given in Fig. 3. Now a manifold is a finite set of triangles with identified edges, and the constant vector field in each triangle defines a successor relation between edges of the triangle such that each triangle has either two input edges and one output edge or one input and two output edges.

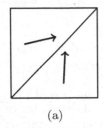 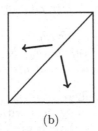 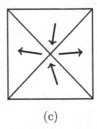 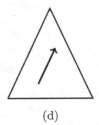

(a)   (b)   (c)   (d)

Fig. 3. Forbidden dynamics in PCD_{r2m}: (a) collision at an edge; (b) branching at an edge; (c) collision and branching on a vertex; (d) flow vector parallel to an edge

The proof of decidability of reachability for 2-dimensional PCDs is based on the Jordan theorem for $\mathbb{R}^2$ which states that every non-self-crossing closed curve divides the plane into an 'interior' region bounded by the curve and an 'exterior' region containing all other points of $\mathbb{R}^2$, so that every continuous path connecting a point of one region to a point of the other intersects with that curve somewhere. This theorem is not applicable to manifolds, therefore we use the topological properties of dynamical systems on the closed orientable surfaces defined in Sect. 4.

4   Dynamical Systems on the Closed Orientable Surfaces

Dynamical systems on the closed orientable surfaces and topological properties of their trajectories considered in [16] provide a formalism promising to serve as a 'reference system' for showing decidability of reachability problem on PCD_{r2m}.

Let S_g be a closed orientable 2-dimensional manifold (surface) with the genus $g \geqslant 0$, and R be a covering of this manifold by a finite number of regions r_i, $1 \leqslant i \leqslant n$, homeomorphic to a Euclidean disc such that every region r_i has its own coordinate system (φ_i, ψ_i). Let the dynamics in each r_i be defined by a system of differential equations:

$$\varphi_i' = \Phi(\varphi_i, \psi_i), \quad \psi_i' = \Psi(\varphi_i, \psi_i) \tag{1}$$

Furthermore, we assume the following:

1. Transformation of one coordinate system to another at the points which belong to two regions or more is done by continuous functions with continuous derivatives and nonzero Jacobian;
2. The right sides of the Eq. 1 are continuous functions and become zero only at a finite number of points;
3. Dynamics change between regions is done by the functions transforming one coordinate system to another.

For convenience we refer such systems as Regionwise Dynamical Systems on 2-dimensional manifolds (RDS_{2m}) in the rest of the paper.

In this study we are concerned with three different types of trajectories. Two of them, dense and orbital stable trajectories are defined below. The third type of interest are trajectories on a subset of $\mathbb{R}^2$ and their properties are described in Sect. 2.

Definition 13 (Dense trajectory). *A trajectory τ is dense on a set of intervals $e_1, \ldots, e_k$ if for any $x \in \tau$ and any interval $e_i' \subseteq e_i$, $1 \leqslant i \leqslant k$, there is $y \in e_i'$ such that y is reachable from x.*

Definition 14 (Orbital stable trajectory). *A trajectory τ is called orbital stable if for any $\varepsilon > 0$ there exists $\delta > 0$ such that if a trajectory τ' starts in the δ-neighbourhood of τ then it is also contained in the ε-neighbourhood of τ.*

Along with the covering of the given manifold by the regions as described above, we also consider another covering. Theorem 2 below shows that any RDS_{2m} can be decomposed into components consisting of trajectories which are equivalent topologically.

Theorem 2 (Mayer, [16]). *A RDS_{2m} S_g is a disjoint union of a finite number of areas $M_1, \ldots, M_k$ (referred later as **dynamical components**) of the following types:*

1. ***Type A**: Any trajectory inside the area is orbital stable and non-closed. Furthermore, all the trajectories have the same set of limit points; the area is flat and at most 2-connected;*
2. ***Type B**: Any trajectory inside the area is closed; the area is either flat and at most 2-connected or equals to the whole manifold in case of $g = 1$ (only for a torus);*

3. **Type C:** *Any trajectory inside the area is everywhere dense; the area is not flat and the number of areas of this type does not exceed g.*

All other trajectories, called **separatrices**, form boundaries between the areas of the above types.

Proposition 1. *Any PCD$_{r2m}$ is a RDS$_{2m}$, and therefore it is a disjoint union of a finite number of areas of Types A, B and C as defined in Theorem 2 and separatrices.*

Proof. By definition of PCD$_{r2m}$ the dynamics change between regions is non-degenerate and local coordinates change with nonzero Jacobian. The flow in each region is constant, hence there is a local coordinate system where both φ'_i and ψ'_i are nonzero.   □

Example 1. Let us consider the PCD$_{r2m}$ represented in Fig. 4. The dynamical component highlighted with grey colour is of Type A and its borders (separatrices) are depicted with dotted lines. The regions outside the grey area can form dynamical components of Types A, B and C depending on their dynamics.

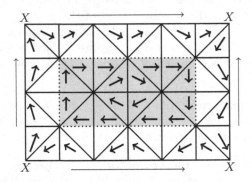

Fig. 4. The dynamical component of Type A is highlighted with grey colour

Proposition 2. *Any separatrix of a PCD$_{r2m}$ H $=$ (Q, F) starts and ends at a vertex.*

Proof. Let us assume that there is an infinite (or half-infinite) separatrix τ_s. Then there is an ε-tube around τ_s (see Fig. 6) such that the trajectories from different sides of τ_s belong to different dynamic components but never split, which contradicts to the definition of a separatrix (a trajectory separating areas of different types).   □

Note that some trajectories which start and end at vertices are not separatrices in the above sense. Such trajectories can divide, for example, a dynamical component of Type B into several smaller components, each of which is of Type B (see Example 2). For convenience and simplicity we still refer such trajectories as separatrices.

Example 2. The PCD_{r2m} depicted in Fig. 5 consists of one dynamical component of Type B, which is divided into two areas by the trajectories starting and ending at the vertices X and Y.

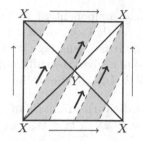

Fig. 5. A PCD_{r2m} consisting of one dynamical component of Type B

5   Language of PCD_{r2m}

In the following we assume a finite non-empty set A called an alphabet, and the elements of A are called letters. By A^* we denote the set of all finite sequences of A called words. A language L over A is a subset of A^*.

Definition 15 (Factorial language). *A language L over an alphabet A is* **factorial** *if $u_0 u_1 \ldots u_n \in L$ implies $u_1 u_2 \ldots u_n \in L$ and $u_0 u_1 \ldots u_{n-1} \in L$ for arbitrary $u_0, \ldots, u_n \in A$.*

Definition 16 (Prolongable language). *A language L over an alphabet A is* **prolongable** *if for any $u \in L$ there exist $a, b \in A$ such that $au \in L$ and $ub \in L$.*

In the following we consider the words induced by the trajectories of a PCD_{r2m} in the following sense: $e_1 \ldots e_k$ is a word if it is a signature of some trajectory segment.

Proposition 3. *Let $H = (Q, F)$ be a PCD_{r2m}, and $L(H)$ be the set of all finite words generated by the trajectories of H. Then $L(H)$ is a factorial and prolongable language over the alphabet $Bd(Q)$.*

Proof. By definition, $L(H)$ is a language over $Bd(Q)$. It is obviously factorial and prolongable because for every boundary point x there are boundary points x_{succ} and x_{pred} such that x_{succ} is reachable from x and x is reachable from x_{pred}. □

A recurrent word is an infinite word over A in which every finite subword occurs infinitely often.

Definition 17 (Uniformly recurrent language). *A language L is called* ***uniformly recurrent*** *if for any $n \in \mathbb{N}$ there exists $\eta_n \in \mathbb{N}$ such that every word from L of length η_n contains all of the words from L of length n as subwords.*

The language of a dynamic component of PCD_{r2m} can be naturally seen as a constraint of a language of PCD_{r2m} on a set of words produced by the trajectories of this component. It is a language with the same properties as in Proposition 3. The following lemma gives the sufficient condition for the language of a dynamic component of a PCD_{r2m} to be uniformly recurrent.

Lemma 1 (Uniformity condition). *Density of the trajectories of a dynamical component of Type C of a PCD_{r2m} H = (Q, F) implies the uniformal recurrence of its language L_C.*

Proof. Let $w = e_1 e_2 \dots e_n$ be a word in L_C, and $s_w \subseteq e_1$ be a maximal subinterval such that w is generated by a trajectory starting at some $x \in s_w$.

The density of trajectories in C implies that the first-return map $f : s_w \to s_w$ is defined for any $x \in s_w$.

To prove the uniformal recurrence of L_C it is sufficient to show that there exists a constant $C(s_w)$ such that $r(x) < C(s_w)$ for any $x \in s_w$, where $r(x)$ is the length of the word corresponding to the trajectory segment connecting x and $f(x)$. Then by choosing $\eta_n = \max_{w:|w|=n} C(s_w)$ we guarantee that any word of length η_n contains any word of length n as a subword.

Fig. 6. An 'ε-tube' around the trajectory starting at x

We observe that for any inner point $x \in s_w$ there is a *returning interval* $i_x \subseteq s_w$ around x such that $r(y) = r(x)$ for any $y \in i_x$. This is because a trajectory of x does not meet any vertex and there is always an "ε-tube" around the subsequent images of x (see Fig. 6).

We also observe that any two adjacent returning intervals $i(x)$ and $i(y)$ are divided by a trajectory of some vertex (see Fig. 7). As C contains a finite number of vertices, we conclude that there is $k \in \mathbb{N}$ such that s_w is a disjoint union of k returning intervals corresponding to some $x_1, \dots, x_k \in s_w$.

We define $C(s_w) = \max_{i \in \{1,\dots,k\}} r(i_i)$ and this concludes the proof. □

Corollary 1. *The language of any dynamical component of Type B of any PCD_{r2m} is uniformly recurrent.*

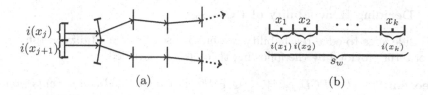

Fig. 7. An example illustrating the proof of Lemma 1

6   Decidability of Reachability for PCD_{r2m}

In this section we show that the reachability problem is decidable for PCD_{r2m}. Moreover, we provide an algorithm which decides in a finite number of steps if a target edge is reachable from an initial edge.

6.1   Rauzy Graphs

A Rauzy graph of power k for a factorial language L is a directed graph formally defined below.

Definition 18 (Rauzy graph). *Rauzy graph of power $k \geqslant 1$ for a language L is a directed graph $R^k(L) = (V^k, E^k)$ defined as follows:*

- $V^k = \{w \in L \mid |w| = k\}$;
- *For any two vertices $u = u_1 u_2 \ldots u_k \in V^k$ and $v = v_1 v_2 \ldots v_k \in V^k$ there is an edge $(u, v) \in E^k$ if $u_2 = v_1, u_3 = v_2, \ldots, u_k = v_{k-1}$ and $u_1 u_2 \ldots u_k v_k \in L$.*

In other words, any two words of length k are connected in the Rauzy graph of power k if they are a prefix and a suffix of some word of length $k + 1$.

Example 3. Let $L = \{a, b, c, ab, bc, ba, ca, abc, aba, bab, bca, cab\}$ be a language over the alphabet $A = \{a, b, c\}$. The Rauzy graphs of power one and two for L, $R^1(L)$ and $R^2(L)$ respectively, are depicted in Fig. 8.

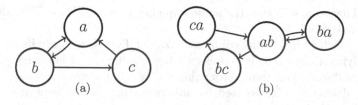

Fig. 8. (a) Rauzy graph of power one for the language in Example 3; (b) Rauzy graph of power two for the same language

6.2   Deciding Reachability of PCD_{r2m}

To decide edge-to-edge reachability, we build a sequence of Rauzy graphs until the criteria provided by the following theorem are satisfied.

Theorem 3. *Let a PCD_{r2m} $H = (Q, F)$ have n dynamical components, and L be the language of H. Then there exists $t_{stop} \in \mathbb{N}$ such that any $R^i(L)$, $i \geqslant t_{stop}$, consists of $k \geqslant n$ disconnected components such that at least one of the following conditions holds for each component $K_j^i = (V_j^i, E_j^i)$:*

(1) All vertices in V_j^i contain the same set of letters;
(2) There is $t' < t_{stop}$ such that the set of signature types of V_j^i equals to the set of signature types of some component in the graph $R^{t'}(L)$.

Proof. By Theorem 2, H is divided into a finite number of regions with different dynamics (regions of Types A, B and C, and the separatrices forming the boundary of the regions). By Proposition 2 each separatrix starts and ends at a vertex. Let m be the maximal length of a separatrix in H. Then any $R^i(L)$, $i > m$, contains at least n components as there could be no words of length greater than m (see Fig. 9).

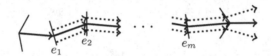

Fig. 9. An example illustrating the proof of Theorem 3

Each dynamical component D_A of Type A is flat and therefore can be seen as a 2-PCD. Then by Theorem 1, D_A has a finite number of signature types. All the signature types of D_A will be constructed at some step in a sequence of Rauzy graphs and thus no new signature type can be later discovered. Hence, Condition (2) will eventually hold.

Each trajectory inside any component D_B of Type B is closed and therefore periodic. That is it will visit the same sequence of edges. Hence, Condition (1) will eventually hold.

Each trajectory inside any component D_C of Type C is dense in D_C. Hence, the underlying component is connected. By Lemma 1, the language generated by D_C is uniformly recurrent. From this follows that there exists η_C such that any word of length η_C generated by any trajectory of D_C contains all words of length 1 (singular letters) as subwords. Hence, Condition (1) will eventually hold.

By Proposition 2, each separatrix starts and ends at a vertex and thus there is a finite number of them. Let us assume that some separatrix ends at a vertex v. Then, another separatrix starts at v, hence, all separatrices form a finite number of cyclic trajectories on H. Depending on which vertex belongs to which region of H, a cyclic trajectory s of any set of separatrices can generate either the same words as the dynamical components or a finite set of words corresponding to a special component K_s in $R^i(L)$. Because each K_s is generated by a cyclic trajectory, Condition (1) will eventually hold for each K_s. □

Theorem 4. *Edge-to-edge reachability for a PCD_{r2m} is decidable.*

Proof. First, let us prove that any $R^k(L)$ can be constructed in finite time. For any PCD_{r2m} H = (Q, F), the set V^k of vertices of $R^k(L)$ consists of all the words of length k over the finite alphabet on the edges of H. This set can be constructed by applying the successor function $k-1$ times to each edge. The edges of $R^k(L)$ can be computed in a finite time by checking if any two vertices have common suffix and prefix of length $k-1$.

To check reachability on H, it is sufficient to build a sequence $R^k(L)$ until a finite t_{stop}, defined by Theorem 3. An edge $e_f \in$ E(Q) is reachable from an edge $e_s \in$ E(Q) if and only if $R^{t_{stop}}(L)$ contains a component with a vertex labelled by a word $(\ldots e_s \ldots e_f \ldots)$. □

7 Conclusions

In this paper we introduced a class of hybrid systems which we called Regular PCD_{2m} (or PCD_{r2m}). It is a subclass of introduced earlier so-called PCDs on manifolds (or PCD_{2m}) [9]. While the reachability problem for the whole class PCD_{2m} is still an open question, we proved that under certain limitations on the systems dynamics it becomes decidable.

As future work we consider to extend the current results to non-orentable manifolds using properties of trajectories on non-orientable manifolds presented in [2].

Acknowledgements. The authors thank Vincent Delecroix, Alexey Kanel-Belov, Alexey Klimenko, Alexandra Skripchenko and Eugene Asarin for their kind help and consultations.

References

1. Alur, R., et al.: The algorithmic analysis of hybrid systems. Theor. Comput. Sci. **138**(1), 3–34 (1995)
2. Aranson, S.H.: Trajectories on nonorientable two-dimensional manifolds. Math. USSR-Sbornik **9**(3), 297–313 (1969)
3. Asarin, E., Maler, O., Pnueli, A.: Reachability analysis of dynamical systems having piecewise-constant derivatives. Theor. Comput. Sci. **138**(1), 35–65 (1995)
4. Asarin, E., Schneider, G.: Widening the boundary between decidable and undecidable hybrid systems*. In: Brim, L., Křetínský, M., Kučera, A., Jančar, P. (eds.) CONCUR 2002. LNCS, vol. 2421, pp. 193–208. Springer, Heidelberg (2002). https://doi.org/10.1007/3-540-45694-5_14

5. Asarin, E., Schneider, G., Yovine, S.: On the decidability of the reachability problem for planar differential inclusions. In: Di Benedetto, M.D., Sangiovanni-Vincentelli, A. (eds.) HSCC 2001. LNCS, vol. 2034, pp. 89–104. Springer, Heidelberg (2001). https://doi.org/10.1007/3-540-45351-2_11
6. Asarin, E., Schneider, G., Yovine, S.: Algorithmic analysis of polygonal hybrid systems, part I: reachability. Theor. Comput. Sci. **379**(1–2), 231–265 (2007)
7. Bell, P.C., Chen, S., Jackson, L.: On the decidability and complexity of problems for restricted hierarchical hybrid systems. Theor. Comput. Sci. **652**, 47–63 (2016)
8. Bournez, O., Kurganskyy, O., Potapov, I.: Reachability problems for one-dimensional piecewise affine maps. Int. J. Found. Comput. Sci. **29**(4), 529–549 (2018)
9. Asarin, E., Mysore, V., Pnueli, A., Schneider, G.: Low dimensional hybrid systems - decidable, undecidable, don't know. Inf. Comput. **211**, 138–159 (2012)
10. Henle, M.: A Combinatorial Introduction to Topology. Dover Publications Inc., New York City (1979)
11. Henzinger, T.A., Kopke, P.W., Puri, A., Varaiya, P.: What's decidable about hybrid automata? J. Comput. Syst. Sci. **57**(1), 94–124 (1998)
12. Kurganskyy, O., Potapov, I., Caparrini, F.S.: Computation in one-dimensional piecewise maps. In: Bemporad, A., Bicchi, A., Buttazzo, G. (eds.) HSCC 2007. LNCS, vol. 4416, pp. 706–709. Springer, Heidelberg (2007). https://doi.org/10.1007/978-3-540-71493-4_66
13. Lafferriere, G., Pappas, G.J., Yovine, S.: A new class of decidable hybrid systems. In: Vaandrager, F.W., van Schuppen, J.H. (eds.) HSCC 1999. LNCS, vol. 1569, pp. 137–151. Springer, Heidelberg (1999). https://doi.org/10.1007/3-540-48983-5_15
14. Lafferriere, G., Pappas, G.J., Sastry, S.: O-minimal hybrid systems. Math. Control Sig. Syst. **13**(1), 1–21 (2000)
15. Maler, O., Pnueli, A.: Reachability analysis of planar multi-linear systems. In: Courcoubetis, C. (ed.) CAV 1993. LNCS, vol. 697, pp. 194–209. Springer, Heidelberg (1993). https://doi.org/10.1007/3-540-56922-7_17
16. Mayer, A.: Trajectories on the closed orientable surfaces. Rec. Math. [Mat. Sbornik] N.S. **12**(54), 1, 71–84 (1943)
17. Mysore, V., Pnueli, A.: Refining the undecidability frontier of hybrid automata. In: Sarukkai, S., Sen, S. (eds.) FSTTCS 2005. LNCS, vol. 3821, pp. 261–272. Springer, Heidelberg (2005). https://doi.org/10.1007/11590156_21
18. Rauzy, G.: Suites à termes dans un alphabet fini. In: Seminar on Number Theory (1982–1983). University of Bordeaux I, Talence, vol. 25, pp. 1–16 (1983)

Coverability Is Undecidable in One-Dimensional Pushdown Vector Addition Systems with Resets

Sylvain Schmitz[1,2]([✉]) [iD] and Georg Zetzsche[3]([✉]) [iD]

[1] LSV, ENS Paris-Saclay & CNRS, Université Paris-Saclay, Cachan, France
sylvain.schmitz@lsv.fr
[2] IUF, Paris, France
[3] MPI-SWS, Kaiserslautern and Saarbrücken, Germany
georg@mpi-sws.org

Abstract. We consider the model of pushdown vector addition systems with resets. These consist of vector addition systems that have access to a pushdown stack and have instructions to reset counters. For this model, we study the coverability problem. In the absence of resets, this problem is known to be decidable for one-dimensional pushdown vector addition systems, but decidability is open for general pushdown vector addition systems. Moreover, coverability is known to be decidable for reset vector addition systems without a pushdown stack. We show in this note that the problem is undecidable for one-dimensional pushdown vector addition systems with resets.

Keywords: Pushdown vector addition systems · Decidability

1 Introduction

Vector addition systems with states (VASS) play a central role for modelling systems that manipulate discrete resources, and as such provide an algorithmic toolbox applicable in many different fields. Adding a pushdown stack to vector addition systems yields so-called *pushdown VASS* (PVASS), which are even more versatile: one can model for instance recursive programs with integer variables [2] or distributed systems with a recursive server and multiple finite-state clients, and PVASS can be related to decidability issues in logics on data trees [9]. However, this greater expressivity comes with a price: the *coverability problem* for PVASS is only known to be decidable in dimension one [12]. This problem captures most of the decision problems of interest and in particular safety properties, and is the stumbling block in a classification for a large family of models combining pushdown stacks and counters [16].

Another viewpoint on one-dimensional PVASS [12] is to see those systems as extensions of two-dimensional VASS, where one of the two counters is replaced

Work partially funded by ANR-17-CE40-0028 BraVAS.

E. Filiot et al. (Eds.): RP 2019, LNCS 11674, pp. 193–201, 2019.
https://doi.org/10.1007/978-3-030-30806-3_15

by a pushdown stack. In this context, a complete classification with respect to decidability of coverability, and of the more difficult *reachability problem*, was provided by Finkel and Sutre [6], whether one uses plain counters ($\mathbb{N}$), counters with resets ($\mathbb{N}_r$), counters whose contents can be transferred to the other counter ($\mathbb{N}_t$), or counters with zero tests ($\mathbb{N}_z$); see Table 1. In particular, two-dimensional VASS with one counter extended to allow resets and one extended to allow zero tests have a decidable reachability problem [6]: put differently, the coverability problem for one-dimensional PVASS *with resets* (1-PRVASS) is decidable if the stack alphabet is of the form $\{a, \bot\}$ where $\bot$ is a distinguished bottom-of-stack symbol.

Table 1. Decidability status of the coverability and reachability problems in extensions of two-dimensional VASS; our contribution is indicated in bold.

(a) Coverability problem.

$\mathbb{N}$	$\mathbb{N}_r$	$\mathbb{N}_t$	$\mathbb{N}_z$	PD	
D [8]	D [1]	D [4]	D [15]	D [12]	$\mathbb{N}$
	D [1]	D [4]	D [6]	**U**	$\mathbb{N}_r$
		D [4]	U [6]	U [6]	$\mathbb{N}_t$
			U [14]	U [14]	$\mathbb{N}_z$
				U	PD

(b) Reachability problem.

$\mathbb{N}$	$\mathbb{N}_r$	$\mathbb{N}_t$	$\mathbb{N}_z$	PD	
D [7]	D [15]	D [15]	D [15]	??	$\mathbb{N}$
	D [6]	D [6]	D [6]	**U**	$\mathbb{N}_r$
		D [6]	U [6]	U [6]	$\mathbb{N}_t$
			U [14]	U [14]	$\mathbb{N}_z$
				U	PD

Contributions. In this note, we show that Finkel and Sutre's decidability result does not generalise to one-dimensional pushdown VASS with resets over an arbitrary finite stack alphabet.

Theorem 1. *The coverability problem for* 1-*PRVASS is undecidable.*

As far as the coverability problem is concerned, this fully determines the decidability status in extensions of two-dimensional VASS where one may also replace counters by pushdown stacks (PD); see Table 1a.

Technically, the proof of Theorem 1 presented in Sect. 3 reduces from the reachability problem in two-counter Minsky machines. The reduction relies on the ability to *weakly implement* [13] basic operations—like multiplication by a constant—and their inverses—like division by a constant. This in itself would not bring much; for instance, plain two-dimensional VASS can already weakly implement multiplication and division by constants. The crucial point here is that, in a 1-PRVASS, we can also weakly implement the inverse of a *sequence* of basic operations performed by the system, by using the pushdown stack to record a sequence of basic operations and later replaying it in reverse, and relying on resets to "clean-up" between consecutive operations. Note that without resets, while PVASS are known to be able to weakly implement Ackermannian functions already in dimension one [11], they cannot weakly compute sublinear functions [10]—like iterated division by two, i.e., logarithms.

2   Pushdown Vector Addition Systems with Resets

For an alphabet Σ, we use Σ^* to denote the set of words over Σ and ε for the empty word. A *(1-dimensional) pushdown vector addition system with resets (1-PRVASS)* is a tuple $\mathcal{V} = (Q, \Gamma, A)$, where Q is a finite set of *states*, Γ is a finite set of *stack symbols*, and $A \subseteq Q \times I^* \times Q$ is a finite set of *actions*. Here, transitions are labelled by finite sequences of *instructions* from $I \stackrel{\text{def}}{=} \Gamma \cup \bar{\Gamma} \cup \{\boxplus, \boxminus, r\}$ where $\bar{\Gamma} \stackrel{\text{def}}{=} \{\bar{z} \mid z \in \Gamma\}$ is a disjoint copy of Γ.

A 1-PRVASS defines a (generally infinite) transition system acting over *configurations* $(q, w, n) \in Q \times \Gamma^* \times \mathbb{N}$. For an instruction $x \in I$, $w, w' \in \Gamma^*$, and $n, n' \in \mathbb{N}$, we write $(w, n) \xrightarrow{x} (w', n')$ in the following cases:

push if $x = z$ for $z \in \Gamma$, then $w' = wz$ and $n' = n$,
pop if $x = \bar{z}$ for $z \in \Gamma$, then $w = w'z$ and $n' = n$,
increment if $x = \boxplus$, then $w' = w$ and $n' = n + 1$.
decrement if $x = \boxminus$, then $w' = w$ and $n' = n - 1$, and
reset if $x = r$, then $w' = w$ and $n' = 0$.

Moreover, for a sequence of instructions $u = x_1 \cdots x_k$ with $x_1, \dots, x_k \in I$, we have $(w_0, n_0) \xrightarrow{u} (w_k, n_k)$ if for some $(w_1, n_1), \dots, (w_{k-1}, n_{k-1}) \in \Gamma^* \times \mathbb{N}$, we have $(w_i, n_i) \xrightarrow{x_i} (w_{i+1}, n_{i+1})$ for all $0 \leq i < k$. For two configurations $(q, w, n), (q', w', n') \in Q \times \Gamma^* \times \mathbb{N}$, we write $(q, w, n) \rightarrow_{\mathcal{V}} (q', w', n')$ if there is an action $(q, u, q') \in A$ such that $(w, n) \xrightarrow{u} (w', n')$. Finally, $\rightarrow_{\mathcal{V}}^*$ denotes the reflexive transitive closure of $\rightarrow_{\mathcal{V}}$.

The *coverability problem* for 1-PRVASS is the following decision problem.

given a 1-PRVASS $\mathcal{V} = (Q, \Gamma, A)$, states $s, t \in Q$.
question are there $w \in \Gamma^*$ and $n \in \mathbb{N}$ with $(s, \varepsilon, 0) \rightarrow_{\mathcal{V}}^* (t, w, n)$?

3   Reduction from Minsky Machines

We present in this section a reduction from reachability in two-counter Minsky machines to coverability in 1-PRVASS.

3.1   Preliminaries

Recall that a *two-counter (Minsky) machine* is a tuple $\mathcal{M} = (Q, A)$, where Q is a finite set of states and $A \subseteq Q \times \{0, 1\} \times \{\boxplus, \boxminus, z\} \times Q$ a set of actions. A *configuration* is now a triple (q, n_0, n_1) with $q \in Q$ and $n_0, n_1 \in \mathbb{N}$. We write $(q, n_0, n_1) \rightarrow_{\mathcal{M}} (q', n_0', n_1')$ if there is an action $(q, c, x, q') \in A$ such that $n_{1-c}' = n_{1-c}$ and

increment if $x = \boxplus$, then $n_c' = n_c + 1$,
decrement if $x = \boxminus$, then $n_c' = n_c - 1$, and
zero test if $x = z$, then $n_c' = n_c = 0$.

Finally, $\to_{\mathcal{M}}^*$ denotes the reflexive transitive closure of $\to_{\mathcal{M}}$.

The *reachability problem* for two-counter machines is the following undecidable decision problem [14].

given a two-counter machine $\mathcal{M} = (Q, A)$, and states $s, t \in Q$.
question does $(s, 0, 0) \to_{\mathcal{M}}^* (t, 0, 0)$ hold?

Gödel Encoding. The first ingredient of the reduction is to use the well-known encoding of counter values $(n_0, n_1) \in \mathbb{N} \times \mathbb{N}$ as a single number $2^{n_0} 3^{n_1}$; for instance, the pair $(0, 0) \in \mathbb{N} \times \mathbb{N}$ is encoded by $2^0 3^0 = 1$. In this encoding, incrementing the first counter means multiplying by 2, decrementing the second counter means dividing by 3, and testing the second counter for zero means verifying that the encoding is not divisible by 3, etc. Note that, in each case, we encode the instruction as a partial function $g \colon \mathbb{N} \nrightarrow \mathbb{N}$; let us define its *graph* as the binary relation $\{(m, n) \in \mathbb{N} \times \mathbb{N} \mid g \text{ is defined on } m \text{ and } g(m) = n\}$. Thus the encoded instructions are the partial functions with the following graphs:

$$R_{m_f} \stackrel{\text{def}}{=} \{(n, f \cdot n) \mid n \in \mathbb{N}\} \qquad \text{for multiplication,}$$
$$R_{d_f} \stackrel{\text{def}}{=} \{(f \cdot n, n) \mid n \in \mathbb{N}\} \qquad \text{for division, and}$$
$$R_{t_f} \stackrel{\text{def}}{=} \{(n, n) \mid n \not\equiv 0 \bmod f\} \qquad \text{for the non-divisibility test,}$$

for a factor $f \in \{2, 3\}$. This means that we can equivalently see

- a two-counter machine with distinguished source and target states s and t as a regular language $M \subseteq \Delta^*$ over the alphabet $\Delta \stackrel{\text{def}}{=} \{m_f, d_f, t_f \mid f \in \{2, 3\}\}$, and
- reachability as the existence of a word $u = x_1 \cdots x_\ell$ in the language M, with $x_1, \ldots, x_\ell \in \Delta$, such that the pair $(1, 1)$ belongs to the composition $R_{x_1} R_{x_2} \cdots R_{x_\ell}$.

Weak Relations. Here, the problem is that it does not seem possible to implement these operations (multiplication, division, divisibility test) directly in a 1-PRVASS. Therefore, a key idea of our reduction is to perform the instructions of u *weakly*—meaning that the resulting value may be smaller than the correct result—but *twice*: once forward and once backward. More precisely, for any relation $R \subseteq \mathbb{N} \times \mathbb{N}$, we define the *weak forward* and *backward* relations $\overrightarrow{R}$ and $\overleftarrow{R}$ by

$$\overrightarrow{R} \stackrel{\text{def}}{=} \{(m, n) \in \mathbb{N} \times \mathbb{N} \mid \exists \tilde{n} \geq n \colon (m, \tilde{n}) \in R\}$$
$$\overleftarrow{R} \stackrel{\text{def}}{=} \{(m, n) \in \mathbb{N} \times \mathbb{N} \mid \exists \tilde{m} \geq m \colon (\tilde{m}, n) \in R\}.$$

Let us call a relation $R \subseteq \mathbb{N} \times \mathbb{N}$ *strictly monotone* if for $(m, n) \in R$ and $(m', n') \in R$, we have $m < m'$ if and only if $n < n'$. We shall rely on the following proposition, which is proven in Appendix A.

Proposition 2. *If $R_1, \ldots, R_\ell \subseteq \mathbb{N} \times \mathbb{N}$ are strictly monotone relations, then* $R_1 R_2 \cdots R_\ell = \overrightarrow{R_1} \overrightarrow{R_2} \cdots \overrightarrow{R_\ell} \cap \overleftarrow{R_1} \overleftarrow{R_2} \cdots \overleftarrow{R_\ell}$.

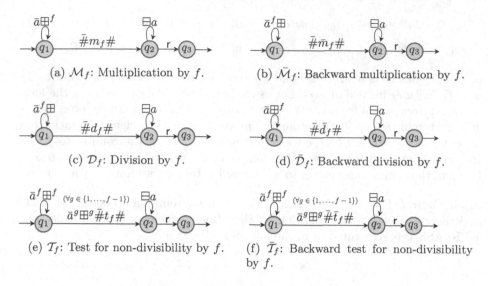

(a) $\mathcal{M}_f$: Multiplication by f.

(b) $\bar{\mathcal{M}}_f$: Backward multiplication by f.

(c) $\mathcal{D}_f$: Division by f.

(d) $\bar{\mathcal{D}}_f$: Backward division by f.

(e) $\mathcal{T}_f$: Test for non-divisibility by f.

(f) $\bar{\mathcal{T}}_f$: Backward test for non-divisibility by f.

Fig. 1. Gadgets used in the reduction.

We shall thus construct in Sect. 3.2 a 1-PRVASS $\mathcal{V}$ in which a particular state is reachable if and only if there exists a word $u \in M$ with $u = x_1 \cdots x_\ell$ and $x_1, \ldots, x_\ell \in \Delta$, such that $(1, 1) \in \overrightarrow{R_{x_1}} \cdots \overrightarrow{R_{x_\ell}}$ and $(1, 1) \in \overleftarrow{R_{x_1}} \cdots \overleftarrow{R_{x_\ell}}$. Since the relations R_{m_f}, R_{d_f}, and R_{t_f} for $f \in \{2, 3\}$ are strictly monotone, Proposition 2 guarantees that this is equivalent to $(1, 1) \in R_{x_1} \cdots R_{x_\ell}$. Intuitively, if we make a mistake in the forward phase $\overrightarrow{R_{x_1}} \cdots \overrightarrow{R_{x_\ell}}$, then at some point, we produce a number n that is smaller than the correct result $\tilde{n} > n$. Then, the backward phase cannot compensate for that, because it can only make the results even smaller, and cannot reproduce the initial value.

3.2 Construction

We now describe the construction of our 1-PRVASS $\mathcal{V}$. Its stack alphabet is $\Gamma \stackrel{\text{def}}{=} \Delta \cup \{\bot, \#, a\}$, where $\Delta = \{m_f, d_f, t_f \mid f \in \{2, 3\}\}$ as before. In $\mathcal{V}$, each configuration will be of the form $(q, \bot w \# a^n, k)$, where $w \in \Delta^*$, and $n, k \in \mathbb{N}$. In the forward phase, we simulate the run of the two-counter machine so that n is the Gödel encoding of the two counters. In order to perform the backward phase, the word w records the instruction sequence of the forward phase. The resettable counter is used as an auxiliary counter in each weak computation step.

Gadgets. For each weak computation step, we use one of the gadgets from Fig. 1. Note that there, for instance, "$\boxplus^f$" denotes the sequence of instructions $\boxplus \cdots \boxplus$ of length f. Moreover, "$\boxminus a$" decrements the counter and pushes a on the stack. Observe that we have:

$$(q_1, \bot u\#a^m, 0) \to^*_{\mathcal{M}_f} (q_3, \bot v\#a^n, 0) \quad \text{iff} \quad v = um_f \text{ and } (m, n) \in \overrightarrow{R_{m_f}} \quad (1)$$

$$(q_1, \bot u\#a^m, 0) \to^*_{\mathcal{M}_f} (q_3, \bot v\#a^n, 0) \quad \text{iff} \quad u = vm_f \text{ and } (n, m) \in \overleftarrow{R_{m_f}} \quad (2)$$

and analogous facts hold for $\mathcal{D}_f$ and $\bar{\mathcal{D}}_f$ (with d_f instead of m_f) and also for $\mathcal{T}_f$ and $\bar{\mathcal{T}}_f$ (with t_f instead of m_f). Let us explain this in the case $\mathcal{M}_f$. In the loop at q_1, $\mathcal{M}_f$ removes a from the stack and adds f to the auxiliary counter. When $\#$ is on top of the stack the automaton moves to q_2 and changes the stack from $\bot u\#$ to $\bot um_f\#$. Therefore, once $\mathcal{M}_f$ is in q_2, it has set the counter to $f \cdot m$. In the loop at q_2, it decrements the counter and pushes a onto the stack before it resets the counter and moves to q_3. Thus, in state q_3, we have $0 \le n \le f \cdot m$.

Main Control. Let $\mathcal{A} = (\Delta, Q, A, s, t)$ be a finite automaton accepting $M \subseteq \Delta^*$ and thereby encoding the behaviour of the Minsky machine. Schematically, our 1-PRVASS $\mathcal{V}$ is structured as in the following diagram:

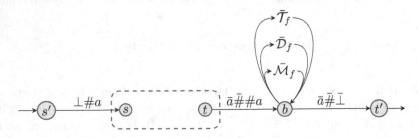

The part in the dashed rectangle is obtained from $\mathcal{A}$ as follows. Whenever there is an action (q, m_f, q') in $\mathcal{A}$, we glue in a fresh copy of $\mathcal{M}_f$ between q and q', including ε-actions from q to q_1 and from q_3 to q'. The original action (q, m_f, q') is removed. We proceed analogously for actions (q, d_f, q') and (q, t_f, q'), where we glue in fresh copies of $\mathcal{D}_f$ and $\mathcal{T}_f$, respectively. Clearly, the part in the dashed rectangle realizes the forward phase as described above.

Once $\mathcal{V}$ reaches t, it can check whether the current number stored on the stack equals 1 and if so, move to state b. In b, the backward phase is implemented. The 1-PRVASS $\mathcal{V}$ contains a copy of $\bar{\mathcal{M}}_f$, $\bar{\mathcal{D}}_f$, and $\bar{\mathcal{T}}_f$ for each $f \in \{2, 3\}$. Each of these copies can be entered from b and goes back to b when exited.

Finally, the stack is emptied by an action from b to t', which can be taken if and only if the stack content is $\bot\#a$. We can check that from $(s', \varepsilon, 0)$, one can reach a configuration (t', w, m) with $w \in \Gamma^*$ and $m \in \mathbb{N}$, if and only if there exists $u \in M$ with $u = x_1 \cdots x_\ell$ and $x_1, \ldots, x_\ell \in \Delta$ such that $(1, 1) \in \overrightarrow{R_{x_1}} \cdots \overrightarrow{R_{x_\ell}} \cap \overleftarrow{R_{x_1}} \cdots \overleftarrow{R_{x_\ell}}$. According to Proposition 2, the latter is equivalent to $(1, 1) \in R_{x_1} \cdots R_{x_\ell}$.

4   Concluding Remarks

In this note, we have proven the undecidability of coverability in one-dimensional pushdown VASS with resets (c.f. Theorem 1). The only remaining open question in Table 1 regarding extensions of two-dimensional VASS is a long-standing one, namely the reachability problem for one-dimensional PVASS. Another fruitful research avenue is to pinpoint the exact complexity in the decidable cases of Table 1. Here, not much is known except regarding coverability and reachability in two-dimensional VASS: these problems are PSPACE-complete if counter updates are encoded in binary [3] and NL-complete if updates are encoded in unary [5].

A   Proof of Proposition 2

It remains to prove Proposition 2. We will use the following lemma.

Lemma 1. *Let* $R_1, \ldots, R_\ell \subseteq \mathbb{N} \times \mathbb{N}$ *be strictly monotone relations and* $(m, n) \in \overrightarrow{R_1} \cdots \overrightarrow{R_\ell}$ *and* $(m', n') \in \overleftarrow{R_1} \cdots \overleftarrow{R_\ell}$. *If* $n' \leq n$, *then* $m' \leq m$. *Moreover, if* $n' < n$, *then* $m' < m$.

Proof. It suffices to prove the lemma in the case $\ell = 1$: then, the general version follows by induction. Let $(m, n) \in \overrightarrow{R_1}$ and $(m', n') \in \overleftarrow{R_1}$. Then there are $\tilde{n} \geq n$ with $(m, \tilde{n}) \in R_1$ and $\tilde{m} \geq m'$ with $(\tilde{m}, n') \in R_1$. If $n' < n$, then we have the following relationships:

$$
\begin{array}{c}
m \ R_1 \ \tilde{n} \\
\text{IV} \\
n \\
\text{V} \\
\tilde{m} \ R_1 \ n' \\
\text{IV} \\
m'
\end{array}
$$

Since R_1 is strictly monotone, this implies $\tilde{m} < m$ and thus $m' < m$. The case $n' \leq n$ follows by the same argument.   □

We are now ready to prove Proposition 2.

Proposition 2. *If* $R_1, \ldots, R_\ell \subseteq \mathbb{N} \times \mathbb{N}$ *are strictly monotone relations, then* $R_1 R_2 \cdots R_\ell = \overrightarrow{R_1} \overrightarrow{R_2} \cdots \overrightarrow{R_\ell} \cap \overleftarrow{R_1} \overleftarrow{R_2} \cdots \overleftarrow{R_\ell}$.

Proof. Of course, for any relation $R \subseteq \mathbb{N} \times \mathbb{N}$, one has $R \subseteq \overrightarrow{R}$ and $R \subseteq \overleftarrow{R}$. In particular, $R_1 R_2 \cdots R_\ell$ is included in both $\overrightarrow{R_1} \overrightarrow{R_2} \cdots \overrightarrow{R_\ell}$ and $\overleftarrow{R_1} \overleftarrow{R_2} \cdots \overleftarrow{R_\ell}$.

For the converse inclusion, suppose $(m, n) \in \overrightarrow{R_1} \overrightarrow{R_2} \cdots \overrightarrow{R_\ell} \cap \overleftarrow{R_1} \overleftarrow{R_2} \cdots \overleftarrow{R_\ell}$. Then there are $p_0, \ldots, p_\ell \in \mathbb{N}$ with $p_0 = m$, $p_\ell = n$, and $(p_{i-1}, p_i) \in \overrightarrow{R_i}$ for $0 < i \leq \ell$. There are also $q_0, \ldots, q_\ell \in \mathbb{N}$ with $q_0 = m$, $q_\ell = n$, and $(q_{i-1}, q_i) \in \overleftarrow{R_i}$ for $0 < i \leq \ell$.

Towards a contradiction, suppose that $(p_{i-1}, p_i) \notin R_i$ for some $0 < i \leq \ell$. Then there is a $\tilde{p}_i > p_i$ with $(p_{i-1}, \tilde{p}_i) \in R_i$. With this, we have

$$
\begin{array}{c}
m = p_0 \ \overrightarrow{R_1} \cdots \overrightarrow{R_i} \ \tilde{p}_i \\
\text{V} \\
p_i \ \overrightarrow{R_{i+1}} \cdots \overrightarrow{R_\ell} \ p_\ell \\
\| \\
m = q_0 \ \overleftarrow{R_1} \cdots \overleftarrow{R_i} \ q_i \ \overleftarrow{R_{i+1}} \cdots \overleftarrow{R_\ell} \ q_\ell
\end{array}
$$

Since $p_\ell = q_\ell$, Lemma 1 applied to $R_{i+1}, \ldots, R_\ell$ implies $q_i \leq p_i$ and thus $q_i < \tilde{p}_i$. Applying Lemma 1 to $R_1, \ldots, R_i$ then yields $q_0 < p_0$, a contradiction. Therefore, we have $(p_{i-1}, p_i) \in R_i$ for every $0 < i \leq \ell$ and thus $(m, n) \in R_1 \cdots R_\ell$. $\qquad \square$

References

1. Arnold, A., Latteux, M.: Récursivité et cônes rationnels fermés par intersection. CALCOLO **15**(4), 381–394 (1978). https://doi.org/10.1007/BF02576519
2. Atig, M.F., Ganty, P.: Approximating Petri net reachability along context-free traces. In: FSTTCS 2011. Leibniz International Proceedings in Informatics, vol. 13, pp. 152–163 (2011). https://doi.org/10.4230/LIPIcs.FSTTCS.2011.152
3. Blondin, M., Finkel, A., Göller, S., Haase, C., McKenzie, P.: Reachability in two-dimensional vector addition systems with states is PSPACE-complete. In: LICS 2015, pp. 32–43. IEEE (2015). https://doi.org/10.1109/LICS.2015.14
4. Dufourd, C., Finkel, A., Schnoebelen, P.: Reset nets between decidability and undecidability. In: Larsen, K.G., Skyum, S., Winskel, G. (eds.) ICALP 1998. LNCS, vol. 1443, pp. 103–115. Springer, Heidelberg (1998). https://doi.org/10.1007/BFb0055044
5. Englert, M., Lazić, R., Totzke, P.: Reachability in two-dimensional unary vector addition systems with states is NL-complete. In: LICS 2016, pp. 477–484. ACM (2016). https://doi.org/10.1145/2933575.2933577
6. Finkel, A., Sutre, G.: Decidability of reachability problems for classes of two counters automata. In: Reichel, H., Tison, S. (eds.) STACS 2000. LNCS, vol. 1770, pp. 346–357. Springer, Heidelberg (2000). https://doi.org/10.1007/3-540-46541-3_29
7. Hopcroft, J.E., Pansiot, J.J.: On the reachability problem for 5-dimensional vector addition systems. Theor. Comput. Sci. **8**, 135–159 (1979). https://doi.org/10.1016/0304-3975(79)90041-0
8. Karp, R.M., Miller, R.E.: Parallel program schemata. J. Comput. Syst. Sci. **3**(2), 147–195 (1969). https://doi.org/10.1016/S0022-0000(69)80011-5
9. Lazić, R.: The reachability problem for vector addition systems with a stack is not elementary. arXiv:1310.1767 [cs.LO] (2013). Presented orally at RP 2012
10. Leroux, J., Praveen, M., Schnoebelen, Ph., Sutre, G.: On functions weakly computable by pushdown Petri nets and related systems (2019). arXiv:1904.04090 [cs.FL]. extended abstract published in: RP 2014. Lect. Notes in Comput. Sci., vol. 8762, pp. 190–202. Springer (2014)
11. Leroux, J., Praveen, M., Sutre, G.: Hyper-Ackermannian bounds for pushdown vector addition systems. In: CSL-LICS 2014. IEEE (2014). https://doi.org/10.1145/2603088.2603146
12. Leroux, J., Sutre, G., Totzke, P.: On the coverability problem for pushdown vector addition systems in one dimension. In: Halldórsson, M.M., Iwama, K., Kobayashi, N., Speckmann, B. (eds.) ICALP 2015. LNCS, vol. 9135, pp. 324–336. Springer, Heidelberg (2015). https://doi.org/10.1007/978-3-662-47666-6_26
13. Mayr, E.W., Meyer, A.R.: The complexity of the finite containment problem for Petri nets. J. ACM **28**(3), 561–576 (1981). https://doi.org/10.1145/322261.322271
14. Minsky, M.L.: Recursive unsolvability of Post's problem of "tag" and other topics in theory of Turing machines. Ann. Math. **74**(3), 437–455 (1961). https://doi.org/10.2307/1970290

15. Reinhardt, K.: Reachability in Petri nets with inhibitor arcs. In: RP 2008. Electronic Notes in Theoretical Computer Science, vol. 223, pp. 239–264 (2008). https://doi.org/10.1016/j.entcs.2008.12.042. Based on a manuscript already available from the author's webpage in 1995
16. Zetzsche, G.: The emptiness problem for valence automata over graph monoids. Inform. Comput. (2018) arXiv:1710.07528 [cs.FL]. To appear; extended abstract published in: RP 2015. Lect. Notes in Comput. Sci., vol. 9328, pp. 166–178. Springer (2015)

Synthesis of Structurally Restricted b-bounded Petri Nets: Complexity Results

Ronny Tredup[✉]

Institut für Informatik, Theoretische Informatik, Universität Rostock,
Albert-Einstein-Straße 22, 18059 Rostock, Germany
ronny.tredup@uni-rostock.de

Abstract. Let $b \in \mathbb{N}^+$. A b-bounded Petri net (b-net) *solves* a transition system (TS) if its reachability graph and the TS are isomorphic. Synthesis (of b-nets) is the problem of finding for a TS A a b-net N that solves it. This paper investigates the computational complexity of synthesis, where the searched net is structurally restricted in advance. The restrictions relate to the cardinality of the preset and the postset of N's transitions and places. For example, N is *choice-free* (CF) if the postset-cardinality of its places do not exceed one. If additionally the preset-cardinality of N's transitions is at most one then it is fork-attribution. This paper shows that deciding if A is solvable by a pure or test-free b-net N which is *choice-free*, *fork-attribution*, *free-choice*, *extended free-choice* or *asymmetric-choice*, respectively, is NP-complete. Moreover, we show that deciding if A is solvable by a b-bounded weighted (m,n)-*T-systems*, $m, n \in \mathbb{N}$, is NP-complete if m, n belong to the input. On the contrary, synthesis for this class becomes tractable if $m, n \in \mathbb{N}$ are chosen *a priori*. We contrast this result with the fact that synthesis for weighted (m,n)-*S-systems*, being the T-systems's dual class, is NP-complete for any fixed $m, n \geq 2$.

1 Introduction

Examining the behaviour of a system and deducing its behavioral properties is the task of system *analyses*. Its counterpart, *synthesis*, is the task to find for a given behavioral specification an implementing system. A valid synthesis procedure computes systems which are correct by design. However, the chances for obtaining an (efficient) algorithm for both analyses and synthesis, depend drastically on the given specification and the searched system: In [8] it has been shown that deciding liveness (the behavioral property) is EXPSPACE-hard for bounded Petri nets (the system), while it is NP-complete for free-choice Petri nets and polynomial for 1-safe free-choice nets. Similarly, the reachability problem is EXPSPACE-hard for bounded Petri nets, PSPACE-complete for free-choice 1-safe nets, NP-complete for acyclic 1-safe and conflict-free nets and polynomial for 1-safe conflict-free nets [8,10].

© Springer Nature Switzerland AG 2019
E. Filiot et al. (Eds.): RP 2019, LNCS 11674, pp. 202–217, 2019.
https://doi.org/10.1007/978-3-030-30806-3_16

In [12] it has been shown that it is impossible to decide if a *modal* transition system (the specification) can be implemented by a bounded Petri net, while synthesis of bounded Petri nets can be done in polynomial time if the specification is a transition system (TS, for short) [1]. An even better procedure for synthesis from TS is possible if the searched bounded Petri net is to be choice-free or a marked graph [4,7]. Moreover, restricting the searched system to b-bounded Petri nets makes synthesis from modal TSs decidable for every fixed integer b [13].

In this paper, we investigate the following instance of synthesis: The specification is a TS A and the searched system is a b-bounded Petri net N (b-net, for short). We demand that N implements A up to isomorphism, that is, N's reachability graph and A are isomorphic. Recently, in [15] we have shown that deciding the existence of N is NP-complete for every fixed $b \geq 1$. However, the former examples provide several results where restricting the system makes the corresponding analyses and synthesis problems easier. Encouraged by these results, we continue our work of [15] in this paper and address whether structurally restricting a searched b-net N influences positively the computational complexity of synthesis. The restrictions relate to the preset- and postset-cardinality of N's transitions and places and correspond to well-known subclasses of Petri nets [3,6,9,14]. Surprisingly, it turns out that almost all applied net restrictions do not bring the synthesis down to polynomial time. More exactly, we show that synthesis remains intractable if N is pure or test-free and satisfies one of the following properties: *choice-free* [6,14], *fork-attribution* [14], *free-choice*, *extended free-choice* or *asymmetric-choice* [3]. Moreover, we adapt the classes of *(weighted) T-systems* and *(weighted) marked graphs* [9] for b-nets and introduce for $m, n \in \mathbb{N}$ their extension of *weighted (m, n)-T-systems* restricting the cardinality of the preset and the postset of N's places by m and n, respectively. We show that synthesis of weighted (m, n)-T-systems is hard if m, n are part of the input and becomes tractable for every fixed m, n. In particular, synthesis of b-bounded weighted T-systems is polynomial which answers partly a question from [5, p.144]. Furthermore, we introduce their dual class of *weighted (m, n)-S-systems* which restricts the cardinality of the preset and postset of N's transitions by m and n, respectively. In contrast to the result of its dual class, deciding if A is implementable by a pure or test-free b-net, being a weighted (m, n)-S-system, is NP-complete for every fixed $m, n \geq 2$. We get all intractability results by a reduction of the cubic monotone one-in-three-3-sat-problem and partly apply our methods from [15]. However, the reductions here are extremely specialized and tailored to synthesis of restricted nets.

The next Sect. 2 introduces all necessary preliminary notions, Sect. 3 presents our main result and Sect. 4 closes the paper.

2   Preliminaries

This section introduces all necessary preliminary notions and Fig. 1 gives corresponding examples. In the remainder of this paper, if not stated explicitly otherwise then $b \in \mathbb{N}^+$ is assumed to be arbitrary but fixed.

Transition Systems. An *initialized transition system* (TS, for short) $A = (S, E, \delta, s_0)$ consists of a finite disjoint set S of states, E of events, a partial *transition function* $\delta : S \times E \to S$ and an *initial state* $s_0 \in S$. A can be interpreted as edge-labeled directed graph where every triple $\delta(s, e) = s'$ is an e-labeled edge $s \xrightarrow{e} s'$, called *transition*. An event e *occurs* at state s, denoted by $s \xrightarrow{e}$, if $\delta(s, e) = s'$ for some state s'. This notation is extended to words $w' = wa$, $w \in E^*, a \in E$, by inductively defining $s \xrightarrow{\varepsilon} s$ for all $s \in S$ and $s \xrightarrow{w'} s''$ if and only if there is a state $s' \in S$ satisfying $s \xrightarrow{w} s'$ and $s' \xrightarrow{a} s''$. If $w \in E^*$ then $s \xrightarrow{w}$ denotes that there is a state $s' \in S$ such that $s \xrightarrow{w} s'$. We assume all TSs to be *reachable*: $\forall s \in S, \exists w \in E^* : s_0 \xrightarrow{w} s$.

b-bounded Petri Nets. A *b-bounded Petri net* (b-net, for short) $N = (P, T, f, M_0)$ consists of finite and disjoint sets of *places* P and *transitions* T, a (total) *flow function* $f : P \times T \to \{0, \ldots, b\}^2$ and an *initial marking* $M_0 : P \to \{0, \ldots, b\}$. If $f(p, t) = (m, n)$ then $f^-(p, t) = m$ defines the *consuming effect* of t on p. Similarly, $f^+(p, t) = n$ defines t's *producing effect* on p. The *preset* of a place p is defined by $^\bullet p = \{t \in T \mid f^+(p, t) > 0\}$, the set of transitions that produce on p. Accordingly, p's *postset* is defined by $p^\bullet = \{t \in T \mid f^-(p, t) > 0\}$ and contains the transitions that consume from p. Similarly, the *preset* $^\bullet t = \{p \in P \mid f^-(p, t) > 0\}$ of a transition t is defined by the places from which t consumes and its *postset* $t^\bullet = \{p \in P \mid f^+(p, t) > 0\}$ by the places on which t produces. Notice that neither $^\bullet p \cap p^\bullet$ nor $^\bullet t \cap t^\bullet$ is necessarily empty. A transition $t \in T$ can *fire* or *occur* in a marking $M : P \to \{0, \ldots, b\}$, denoted by $M \xrightarrow{t}$, if $M(p) \geq f^-(p, t)$ and $M(p) - f^-(p, t) + f^+(p, t) \leq b$ for all places $p \in P$. The firing of t in marking M leads to the marking $M'(p) = M(p) - f^-(p, t) + f^+(p, t)$ for $p \in P$, denoted by $M \xrightarrow{t} M'$. Again, this notation extends to sequences $\sigma \in T^*$ and the *reachability set* $RS(N) = \{M \mid \exists \sigma \in T^* : M_0 \xrightarrow{\sigma} M\}$ contains all of N's reachable markings. The firing rule preserves the *b-boundedness* of N by definition: $M(p) \leq b$ for all places p and all $M \in RS(N)$. The *reachability graph* of N is the TS $A_N = (RS(N), T, \delta, M_0)$, where for every reachable marking M of N and transition $t \in T$ with $M \xrightarrow{t} M'$ the transition function δ of A_N is defined by $\delta(M, t) = M'$.

Structurally Restricted Subclasses of b-nets. A b-net N is *pure* if $\forall (p, t) \in P \times T : f^-(p, t) = 0$ or $f^+(p, t) = 0$, that is, $\forall p \in P : {}^\bullet p \cap p^\bullet = \emptyset$; *test-free* if $\forall (p, t) \in P \times T : f(p, t) \neq (0, 0) \Rightarrow f^-(p, t) \neq f^+(p, t)$; *choice-free* (CF) or *place-output-nonbranching* if $\forall p \in P : |p^\bullet| \leq 1$; *fork-attribution* (FA) if it is CF and, additionally, $\forall t \in T : |{}^\bullet t| \leq 1$; *free-choice* (FC) if $\forall p, \tilde{p} \in P : p^\bullet \cap \tilde{p}^\bullet \neq \emptyset \Rightarrow |p^\bullet| = |\tilde{p}^\bullet| = 1$; *extended-free-choice* (EFC) if $\forall p, \tilde{p} \in P : p^\bullet \cap \tilde{p}^\bullet \neq \emptyset \Rightarrow p^\bullet = \tilde{p}^\bullet$; *asymmetric-choice* (AC) if $\forall p, \tilde{p} \in P : p^\bullet \cap \tilde{p}^\bullet \neq \emptyset \Rightarrow (p^\bullet \subseteq \tilde{p}^\bullet$ or $\tilde{p}^\bullet \subseteq p^\bullet)$; for $m, n \in \mathbb{N}$ a *weighted (m, n)-T-system* if $\forall p \in P : |{}^\bullet p| \leq m, |p^\bullet| \leq n$; for $m, n \in \mathbb{N}$ a *weighted (m, n)-S-system* if $\forall t \in T : |{}^\bullet t| \leq m, |t^\bullet| \leq n$.

b-bounded Regions. For the purpose of finding a b-net N implementing a TS A, we want to synthesize N's components purely from the input A. Demanding A

$$s_0 \xrightarrow{k} s_1 \xrightarrow{k} s_2 \xrightarrow{z} s_3 \xrightarrow{o} s_4 \xrightarrow{k} s_5 \xrightarrow{k} s_6$$

sup	s_0	s_1	s_2	s_3	s_4	s_5	s_6	sig	k	z	o
sup_1	0	1	2	2	0	1	2	sig_1	$(0,1)$	$(0,0)$	$(2,0)$
sup_2	2	1	0	2	2	1	0	sig_2	$(1,0)$	$(0,2)$	$(0,0)$
sup_3	2	2	2	0	2	2	2	sig_3	$(0,0)$	$(2,0)$	$(0,2)$
sup_4	0	0	0	2	2	2	2	sig_4	$(0,0)$	$(0,2)$	$(0,0)$
sup_5	0	1	2	1	0	1	2	sig_5	$(0,1)$	$(1,0)$	$(1,0)$

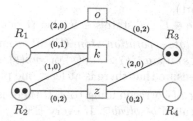

Fig. 1. Top: Input TS A. Middle: For $i \in \{1,2,3,4,5\}$ pure 2-regions $R_i = (sup_i, sig_i)$ of A, where $R_1, \ldots, R_4$ already solve all of A's (E)SSP atoms. For example, the region R_1 solves (k, s_i), $\forall i \in \{2,3,6\}$ and (o, s_i), $\forall i \in \{0,1,4,5\}$. Bottom: Pure 2-net $N_A^{\mathcal{R}}$, built by $\mathcal{R} = \{R_1, R_2, R_3, R_4\}$, where, for example, ${}^\bullet R_3 = \{o\}$ and $R_3{}^\bullet = \{z\}$ and ${}^\bullet o = \{R_1\}$ and $o^\bullet = \{R_3\}$. Moreover, $N_A^{\mathcal{R}}$ is FA because of $|R^\bullet| \leq 1$ and $|{}^\bullet e_{\mathcal{R}}| \leq 1$ for all $R \in \mathcal{R}$ and $e \in E(A)$. The net $N_A^{\mathcal{R}}$ origins from $N_A^{\mathcal{R}'}$, where $\mathcal{R}' = \mathcal{R} \cup \{ R_5 \}$, by removing R_5. Both $\mathcal{R}$ and $\mathcal{R}'$ are b-admissible sets. Thus, the reachability graphs of their synthesized nets are both isomorphic to A. However, because $z \in R_3{}^\bullet \cap R_5{}^\bullet$ and $R_5{}^\bullet = \{z, o\}$, the net $N_A^{\mathcal{R}'}$ is not even free-choice.

and A_N to be isomorphic suggests that A's events correspond to N's transitions. However, the notion of a *place* is not known for TSs. A *b-bounded region R* (*b-region*, for short) of a TS $A = (S, E, \delta, s_0)$ is a pair $R = (sup, sig)$ of *support* $sup : S \to \{0, \ldots, b\}$ and *signature* $sig : E \to \{0, \ldots, b\}^2$, where $sig^-(e) = m$ and $sig^+(e) = n$ for $sig(e) = (m, n)$, such that for every edge $s \xrightarrow{e} s'$ of A holds $sup(s) \geq sig^-(e)$ and $sup(s') = sup(s) - sig^-(e) + sig^+(e)$. A region (sup, sig) models a place p and the corresponding part of the flow function f: $sig^+(e)$ models $f^+(e)$, $sig^-(e)$ models $f^-(e)$ and $sup(s)$ models $M(p)$ in the marking $M \in RS(N)$ corresponding to $s \in S(A)$. Accordingly, a region R is *test-free* if $sig(e) \neq (0, 0)$ implies $sig^-(e) \neq sig^+(e)$. The *preset* of R is defined by ${}^\bullet R = \{e \in E \mid sig^+(e) > 0\}$ and its *postset* by $R^\bullet = \{e \in E \mid sig^-(e) > 0\}$. The Region R is *pure* if ${}^\bullet R \cap R^\bullet = \emptyset$. For a set $\mathcal{R}$ of b-regions and $e \in E$ we define by ${}^\bullet e_{\mathcal{R}} = \{(sup, sig) \in \mathcal{R} \mid sig^-(e) > 0\}$ the *preset* and by $e_{\mathcal{R}}^\bullet = \{(sup, sig) \in \mathcal{R} \mid sig^+(e) > 0\}$ the *postset* of e (in accordance to $\mathcal{R}$). Every set $\mathcal{R}$ of b-regions of A defines the *synthesized b-net* $N_A^{\mathcal{R}} = (\mathcal{R}, E, f, M_0)$ with flow function $f((sup, sig), e) = sig(e)$ and initial marking $M_0((sup, sig)) = sup(s_0)$ for all $(sup, sig) \in \mathcal{R}, e \in E$. We emphasize once again that a *region R of $\mathcal{R}$ becomes a place of $N_A^{\mathcal{R}}$ with the preset ${}^\bullet R$ and the postset $R^\bullet$. Moreover, every event $e \in E$ becomes a *transition* of $N_A^{\mathcal{R}}$ with preset ${}^\bullet e = {}^\bullet e_{\mathcal{R}}$ and postset $e^\bullet = e_{\mathcal{R}}^\bullet$. It is well known that $A_{N_A^{\mathcal{R}}}$ and A are isomorphic if and only if $\mathcal{R}$'s regions solve certain separation atoms [2], to be introduced next.

A pair (s, s') of distinct states of A define a *state separation atom* (SSP atom, for short). A b-region $R = (sup, sig)$ *solves* (s, s') if $sup(s) \neq sup(s')$. The meaning of R is to ensure that $N_A^{\mathcal{R}}$ contains at least one place R such that $M(R) \neq M'(R)$ for the markings M and M' corresponding to s and s', respectively. If there is a b-region that solves (s, s') then s and s' are called b-*solvable*. If every SSP atom of A is b-solvable then A has the b-*state separation property* (b-SSP, for short).

A pair (e, s) of event $e \in E$ and state $s \in S$ where e does not occur at s, that is $\neg s \xrightarrow{e}$, define an *event state separation atom* (ESSP atom, for short). A b-region $R = (sup, sig)$ *solves* (e, s) if $sig^-(e) > sup(s)$ or $sup(s) - sig^-(e) + sig^+(e) > b$. The meaning of R is to ensure that there is at least one place R in $N_A^{\mathcal{R}}$ such that $\neg M \xrightarrow{e}$ for the marking M corresponding to s. If there is a b-region that solves (e, s) then e and s are called b-*solvable*. If every ESSP atom of A is b-solvable then A has the b-*event state separation property* (b-ESSP, for short).

A set $\mathcal{R}$ of b-regions of A is called b-*admissible* if for every of A's (E)SSP atoms there is a b-region R in $\mathcal{R}$ that solves it. The following lemma, borrowed from [2, p.163], summarizes the already implied connection between the existence of b-admissible sets of A and (the solvability of) synthesis:

Lemma 1. ([2]). *A b-net N has a reachability graph isomorphic to a given TS A if and only if there is a b-admissible set $\mathcal{R}$ of A such that $N = N_A^{\mathcal{R}}$.*

We say a b-net N *solves* A if A_N and A are isomorphic. By Lemma 1, searching for a restricted b-net reduces to finding a b-admissible set of accordingly restricted regions. The following two examples illustrate this fact.

Example 1. If $\mathcal{R}$ is a b-admissible set of pure regions of A satisfying $\forall R \in \mathcal{R} : |R^\bullet| \leq 1$ and $\forall e \in E(A) : |{}^\bullet e_{\mathcal{R}}| \leq 1$ then $N_A^{\mathcal{R}}$ is a pure FA b-net solving A.

Example 2. If $\mathcal{R}$ is a b-admissible set of pure regions of A and $\forall e \in E(A) : |{}^\bullet e_{\mathcal{R}}| \leq 2, |e_{\mathcal{R}}^\bullet| \leq 2$ then $N_A^{\mathcal{R}}$ is a pure solving b-net, being a weighted $(2, 2)$-S-system.

3 Our Contribution

Theorem 1. *For a given TS A the following conditions are true:*

1. *If $P \in \{CF, FA, FC, EFC, AC\}$ then to decide if A is solvable by a pure or a test-free b-net which is P is NP-complete.*
2. *Given integers $\ell, \ell' \in \mathbb{N}$, deciding if A is solvable by a pure or a test-free b-net, being a weighted (ℓ, ℓ')-T-System, is NP-complete.*
3. *For any fixed $\ell, \ell' \geq 2$, deciding if A is solvable by a pure or a test-free b-net, being a weighted (ℓ, ℓ')-S-system, is NP-complete.*

4. *For any fixed $\ell, \ell' \in \mathbb{N}$, one can decide in polynomial time if A is solvable by a b-net, being a weighted (ℓ, ℓ')-T-System.*

To prove Theorem 1.1–Theorem 1.3 we show that the corresponding decision problems are in NP and NP-hard. Membership in NP can be seen as follows: By Lemma 1, if N is a b-net that solves A then there is a b-admissible set $\mathcal{R}'$ of A such that $N_A^{\mathcal{R}'} = N$. By definition, A has at most $|S|^2$ SSP atoms and at most $|E| \cdot |S|$ ESSP atoms. Thus, there is a b-admissible subset $\mathcal{R} \subseteq \mathcal{R}'$ with $|\mathcal{R}| \leq |S|^2 + |E| \cdot |S|$. In particular, $N_A^{\mathcal{R}}$ originates from $N_A^{\mathcal{R}'} = N$ by (possibly) removing places, which can not increase any preset- or postset cardinality. Consequently, removing places preserves property $P \in \{CA, FA, FC, EFC, AC\}$, the weighted (m, n)-T-system property and the weighted (m, n)-S-system property. This makes $N_A^{\mathcal{R}}$ a searched net. A non-deterministic Turing machine can guess in polynomial time a corresponding set $\mathcal{R}$, check its b-admissibility, build $N_A^{\mathcal{R}}$ and check its structural properties in accordance to the regarded decision problem.

To show hardness we use the NP-complete problem CUBIC MONOTONE ONE-IN-THREE-3-SAT (CM 1-IN-3 3SAT) from [11] which is defined as follows: The input for CM 1-IN-3 3SAT is a negation-free boolean expression $\varphi = \{\zeta_0, \dots, \zeta_{m-1}\}$ of three-clauses $\zeta_0, \dots, \zeta_{m-1}$ with set of variables $V(\varphi)$ where every variable occurs in exactly three clauses. Notice that this implies $|V(\varphi)| = m$. The question is whether there is a subset $M \subseteq V(\varphi)$ satisfying $|M \cap \zeta_i| = 1$, $\forall i \in \{0, \dots, m-1\}$.

For Theorem 1.(1–2) we reduce an input instance φ with m clauses (in polynomial time) to a TS A_φ^b satisfying the following condition:

Condition 1. *1. If a test-free b-net solves A_φ^b then φ is one-in-three satisfiable.*
2. If φ is one-in-three satisfiable then there is a b-admissible set $\mathcal{R}$ of pure regions of A_φ^b satisfying $\forall R \in \mathcal{R} : |R^\bullet| \leq 1 \wedge |{}^\bullet R| \leq 7m+4$ and $\forall c \in E(A) : |{}^\bullet e_\mathcal{R}| \leq 1$.

A reduction that satisfies Condition 1 proves Theorem 1.(1–2) as follows: By definition of test-freeness, every b-net of Theorem 1.(1–2) is *at least* test-free, although possibly further restricted. Hence, Condition 1.1 ensures that if A_φ^b is solvable by such a net then φ has a one-in-three model. Moreover, a b-admissible set $\mathcal{R}$ that satisfies Condition 1.2 implies that $N_{A_\varphi^b}^{\mathcal{R}}$ is a pure b-net that is FA and solves A, cf. Example 1. Every pure FA b-net is test-free (by $f^+(p, t) = 0$ or $f^-(p, t) = 0$) and CF (by definition). By $N_{A_\varphi^b}^{\mathcal{R}}$ being CF, all of its places p satisfy $|p^\bullet| \leq 1$. Thus, the net is also FC, EFC and AC. Finally, by $\ell = 7m + 4$ and $\ell' = 1$, the net $N_{A_\varphi^b}^{\mathcal{R}}$ is a weighted (ℓ, ℓ')-T-system. Altogether, Condition 1 ensures that A_φ^b is solvable by a b-net of Theorem 1.(1–2) if and only if φ is one-in-three satisfiable.

For Theorem 1.3 we reduce φ to a TS B_φ^b that satisfies the following condition:

Condition 2. *1. If a test-free b-net solves B_φ^b then φ is one-in-three satisfiable.*
2. If φ is one-in-three satisfiable then there is a b-admissible set $\mathcal{R}$ of pure regions such that $|{}^\bullet e_\mathcal{R}| \leq 2$ and $|e_\mathcal{R}^\bullet| \leq 2$ for all $e \in E(A)$.

A reduction satisfying Condition 2 proves Theorem 1.3 as follows: By the definition of test-freeness and weighted (m, n)-S-systems, a pure weighted $(2, 2)$-S-system is a test-free weighted (m, n)-S-System for all $m, n \geq 2$. Moreover, a b-admissible set that satisfies Condition 2.2 implies that $N^{\mathcal{R}}_{B^b_\varphi}$ is a pure weighted $(2, 2)$-S-system solving B^b_φ, cf. Example 2. Thus, Condition 2 ensures that B^b_φ is solvable by a b-net of Theorem 1.3 if and only if φ is one-in-three satisfiable.

3.1 The Reduction and the Proof of Condition 1.1 and Condition 2.2

In accordance to Condition 1.1 and Condition 2.1, our goal is to combine the existence of a b-admissible set $\mathcal{R}$, the b-solvability of A^b_φ and B^b_φ, with the one-in-three satisfiability of φ. For this purpose, both TSs (among others) apply gadgets that represent φ's clauses and use their variables as events. Moreover, both A^b_φ and B^b_φ have a certain separation atom and the signature of a solving region (sup, sig) defines a one-in-three model of φ via the variable events. So far, this approach is like that of [15]. However, the main difference and the biggest challenge is to consider the restrictions of Condition 1.1 and Condition 2.2. To master this challenge, we apply refined, specialized and different gadgets. Particularly noteworthy in this context is the representation of φ's clauses by $\{0, \ldots, b\}^3$-grids instead of simple sequences, as it has been done in [15].

We proceed by introducing the gadgets of A^b_φ and B^b_φ that represent φ's clauses. In particular, the clause-gadgets' functionality will serve as motivation for the remaining parts of A^b_φ and B^b_φ, which are presented afterwards.

Let $i \in \{0, \ldots, m - 1\}$. The TSs A^b_φ and B^b_φ have for the clause $C_i = \{X_{i,0}, X_{i,1}, X_{i,2}\}$ the $\{0, \ldots, b\}^3$-grid C^b_i with transitions that use the variables of C_i as events. More exactly, the $\{0, \ldots, b\}^3$-grid C^b_i is built by the following sequences $P^{i,0}_{\alpha,\beta}, P^{i,1}_{\alpha,\beta}, P^{i,2}_{\alpha,\beta}$, where $\alpha, \beta \in \{0, \ldots, b\}$. Figure 2 shows C^2_i.

$$P^{i,0}_{\alpha,\beta} = \quad t^i_{0,\alpha,\beta} \xrightarrow{X_{i,0}} t^i_{1,\alpha,\beta} \xrightarrow{X_{i,0}} \cdots \xrightarrow{X_{i,0}} t^i_{b-1,\alpha,\beta} \xrightarrow{X_{i,0}} t^i_{b,\alpha,\beta}$$

$$P^{i,1}_{\alpha,\beta} = \quad t^i_{\alpha,\beta,0} \xrightarrow{X_{i,1}} t^i_{\alpha,\beta,1} \xrightarrow{X_{i,1}} \cdots \xrightarrow{X_{i,1}} t^i_{\alpha,\beta,b-1} \xrightarrow{X_{i,1}} t^i_{\alpha,\beta,b}$$

$$P^{i,2}_{\alpha,\beta} = \quad t^i_{\alpha,0,\beta} \xrightarrow{X_{i,2}} t^i_{\alpha,1,\beta} \xrightarrow{X_{i,2}} \cdots \xrightarrow{X_{i,2}} t^i_{\alpha,b-1,\beta} \xrightarrow{X_{i,2}} t^i_{\alpha,b,\beta}$$

Among others, C^b_i provides the following sequence P_i where each of $X_{i,0}, X_{i,1}$ and $X_{i,2}$ occur b times in a row:

$$P_i = t^i_{0,0,0} \xrightarrow{X_{i,0}} \cdots \xrightarrow{X_{i,0}} t^i_{b,0,0} \xrightarrow{X_{i,1}} \cdots \xrightarrow{X_{i,1}} t^i_{b,0,b} \xrightarrow{X_{i,2}} \cdots \xrightarrow{X_{i,2}} t^i_{b,b,b}$$

Notice that, except for $t^i_{b,b,b}$, every variable of C_i occur at every state of C^b_i. This has the advantage that we never have to solve an ESSP atom (X, s) such that $X \in \{X_{i,0}, X_{i,1}, X_{i,2}\}$ and s occur in the same grid and s is a source of another variable event $Y \in \{X_{i,0}, X_{i,1}, X_{i,2}\} \setminus \{X\}$. This property is crucial to ensure Condition 1.2 and Condition 2.2. In particular, it prevents atoms like $(X_{i,1}, t^i_{b-1,0,0})$ which would be unsolvable for $b \geq 2$.

The TSs A^b_φ and B^b_φ use the grid C^b_i as follows: Both TSs have at least one separation atom such that a corresponding b-solving region (sup, sig) satisfies either $sup(t^i_{0,0,0}) = 0$ and $sup(t^i_{b,b,b}) = b$ or $sup(t^i_{0,0,0}) = b$ and $sup(t^i_{b,b,b}) = 0$. In the following, we assume $sup(t^i_{0,0,0}) = 0$ and $sup(t^i_{b,b,b}) = b$ and argue that this implies that there is exactly one $X \in \{X_{i,0}, X_{i,1}, X_{i,2}\}$ with $sig(X) \neq (0,0)$. If $X \in \{X_{i,0}, X_{i,1}, X_{i,2}\}$ then, by $sup(t^i_{0,0,0}) = 0$ and $t^i_{0,0,0} \xrightarrow{X}$, we have immediately $sig^-(X) = 0$ (no consuming is possible). Moreover, by the definition of regions, we have $sup(s') = sup(s) - sig^-(e) + sig^+(e)$ for every $s \xrightarrow{e} s' \in P_i$. We use all this together and obtain inductively that $b = sup(t^i_{b,b,b}) = b \cdot (sig^+(X_{i,0}) + sig^+(X_{i,1}) + sig^+(X_{i,2})) > 0 = sup(t^i_{0,0,0})$. It is easy to see that this expression is satisfied if and only if there is exactly one variable event with a positive value sig^+ (and this value equals 1). Thus, there is exactly one event $X \in \{X_{i,0}, X_{i,1}, X_{i,2}\}$ with $sig(X) \neq (0,0)$. By the arbitrariness of i this is simultaneously true for all grids $C^b_0, \ldots, C^b_{m-1}$. Consequently, the set $M = \{X \in V(\varphi) \mid sig(X) \neq (0,0)\}$ selects exactly one element of every clause C_i which makes it a one-in-three model of φ. Similarly, if $sup(t^i_{0,0,0}) = b$ and $sup(t^i_{b,b,b}) = 0$ then M yields also a one-in-three model of φ.

With the just presented functionality of C^b_i in mind, in what follows, we introduce A^b_φ's and B^b_φ's remaining parts. In particular, we explain how they collaborate to ensure the existence of a region satisfying $sup(t^i_{0,0,0}) = 0$ and $sup(t^i_{b,b,b}) = b$ or $sup(t^i_{0,0,0}) = b$ and $sup(t^i_{b,b,b}) = 0$. Before we start, the following lemma provides a basic result, to be used in the sequel, and shows how to connect the signature of some events with the solvability of an ESSP atom.

Lemma 2. Let $q_0 \xrightarrow{e_1} \ldots \xrightarrow{e_1} q_b \xrightarrow{e_2} q_{b+1} \xrightarrow{e_3} q_{b+2} \xrightarrow{e_1} \ldots \xrightarrow{e_1} q_{2b+2}$ be a sequence of a TS $A = (S, E, \delta, s_0)$, where e_1, e_2, e_3 are pairwise distinct events, which starts and ends with e_1 b-times in a row. A test-free b-region solves the ESSP atom (e_1, q_{b+1}) if and only if $sig(e_1) = (0,1)$, $sig^-(e_2) = sig^+(e_2)$ and $sig(e_3) = (b,0)$ or $sig(e_1) = (1,0)$, $sig^-(e_2) = sig^+(e_2)$ and $sig(e_3) = (0,b)$.

We start by introducing the parts of A^b_φ. Figure 2 sketches a snippet of A^2_φ. The initial state of A^b_φ is s. Firstly, the TS A^b_φ has the sequence Q^b:

$$Q^b = s \xrightarrow{a} q_0 \xrightarrow{k} \cdots \xrightarrow{k} q_b \xrightarrow{z} q_{b+1} \xrightarrow{o} q_{b+2} \xrightarrow{k} \cdots \xrightarrow{k} q_{2b+2}$$

The sequence Q^b provides the ESSP-atom (k, q_{b+1}). If A^b_φ is b-solvable then, by Lemma 1, there is a b-admissible set $\mathcal{R}$ of (test-free) regions such that $N = N^{\mathcal{R}}_{A^b_\varphi}$. As $\mathcal{R}$ is b-admissible, there is a test-free b-region $(sup, sig) \in \mathcal{R}$ that solves (k, q_{b+1}). By Lemma 2, we have either $sig^-(z) = sig^+(z)$ and $sig(o) = (b,0)$ or $sig^-(z) = sig^+(z)$ and $sig(o) = (0,b)$. Let's discuss the former case. The region R implies for transitions $s \xrightarrow{o} s'$ and $s'' \xrightarrow{z} s'''$ (of A^b_φ) that $sup(s) = b$, $sup(s') = 0$ and $sup(s'') = sup(s''')$. The TS A^b_φ uses this to ensure a particular

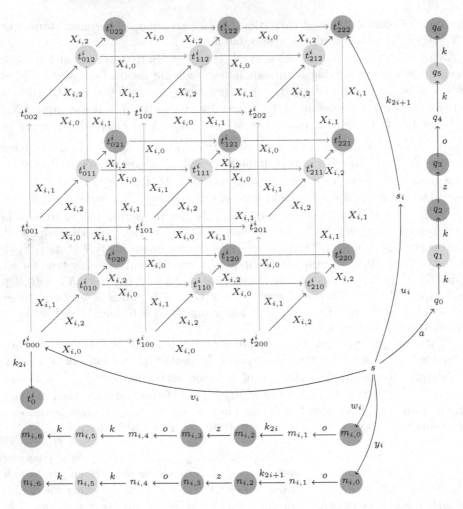

Fig. 2. A snippet of A_φ^2 showing the sequences Q^2, M_i^2, N_i^2, the $\{0,1,2\}^3$-grid C_i^2 for the clause $C_i = \{X_{i,0}, X_{i,1}, X_{i,2}\}$ and the paths $L_{i,0}$ and $L_{i,1}$. For clarity, edges labeled by the same variable event have the same color. The coloring of the states corresponds to the 2-region R_1 which is defined in Table 1 and where $X_{i,0} \in M$: Light (dark) red colored states are mapped to 1 (2) and the others are mapped to 0. (Color figure online)

signature of the events k_{2i}, k_{2i+1} that are provided by the following sequences N_i^b and M_i^b, for all $i \in \{0, \ldots, m-1\}$:

$$M_i^b = s \xrightarrow{w_i} m_{i,0} \xrightarrow{o} m_{i,1} \xrightarrow{k_{2i}} m_{i,2} \xrightarrow{z} m_{i,3} \xrightarrow{o} m_{i,4} \xrightarrow{k} \cdots \xrightarrow{k} m_{i,b+4}$$

$$N_i^b = s \xrightarrow{y_i} n_{i,0} \xrightarrow{o} n_{i,1} \xrightarrow{k_{2i+1}} n_{i,2} \xrightarrow{z} n_{i,3} \xrightarrow{o} n_{i,4} \xrightarrow{k} \cdots \xrightarrow{k} n_{i,b+4}$$

The TS A_φ^b uses M_i^b, N_i^b, R and the occurrences of z and o for the announced goal as follows: By $sig(o) = (b, 0)$, we have $sup(m_{i,1}) = sup(n_{i,1}) = 0$ and $sup(m_{i,3}) = sup(n_{i,3}) = b$ which, by $sig^-(z) = sig^+(z)$, implies $sup(m_{i,2}) = sup(n_{i,2}) = b$. By $m_{i,1} \xrightarrow{k_{2i}} m_{i,2}$, $n_{i,1} \xrightarrow{k_{2i+1}} n_{i,2}$ this leads to $sig(k_{2i}) = sig(k_{2i+1}) = (0, b)$. In particular, for all edges $s \xrightarrow{k_{2i}} s'$ and $s'' \xrightarrow{k_{2i+1}} s'''$ of A_φ^b holds $sup(s) = sup(s'') = 0$ and $sup(s') = sup(s''') = b$. Finally, A_φ^b uses other occurrences of k_{2i} and k_{2i+1} to ensure $sup(t_{0,0,0}^i) = 0$ and $sup(t_{b,b,b}^i) = b$. More exactly, A_φ^b installs the paths $L_{i,0} = s \xrightarrow{v_i} t_{0,0,0}^i \xrightarrow{k_{2i}} t_0^i$ and $L_{i,1} = s \xrightarrow{u_i} s_i \xrightarrow{k_{2i+1}} t_{b,b,b}^i$. On the one hand, $L_{i,0}$ ensures reachability of A_φ^b. On the other hand, by $t_{0,0,0}^i \xrightarrow{k_{2i}} t_0^i$, $s_i \xrightarrow{k_{2i+1}} t_{b,b,b}^i$ and the discussion above, $L_{i,0}, L_{i,1}$ ensure that $sup_0(t_{0,0,0}^i) = 0$ and $sup_0(t_{b,b,b}^i) = b$.

Similarly, one argues that $sig(o) = (0, b)$ and $sig^-(z) = sig^+(z)$ yields $sig(k_{2i}) = sig(k_{2i+1}) = (b, 0)$, implying $sup_1(t_{0,0,0}^i) = b$ and $sup_1(t_{b,b,b}^i) = 0$. By the discussed functionality of the grids, this proves that A_φ^b satisfies Condition 1.1.

We proceed by presenting the remaining gadgets of B_φ^b. The TS B_φ^b has the initial state s and it has for every $i \in \{0, \ldots, m-1\}$ the following six sequences:

$$F_i^b = s \xrightarrow{b_{2m+5}^i} a_{2m+5}^i \cdots a_1^i \xrightarrow{b_0^i} f_0^i \xrightarrow{k} \cdots \xrightarrow{k} f_b^i \xrightarrow{z_{2i}} f_{b+1}^i \xrightarrow{o} f_{b+2}^i \xrightarrow{k} \cdots \xrightarrow{k} f_{2b+2}^i$$

$$G_i^b = s \xrightarrow{d_{2m+5}^i} c_{2m+5}^i \cdots c_1^i \xrightarrow{d_0^i} g_0^i \xrightarrow{k} \cdots \xrightarrow{k} g_b^i \xrightarrow{z_{2i+1}} g_{b+1}^i \xrightarrow{o} g_{b+2}^i \xrightarrow{k} \cdots \xrightarrow{k} g_{2b+2}^i$$

$$M_i^b = s \xrightarrow{w_{2m+5}^i} r_{2m+5}^i \cdots r_1^i \xrightarrow{w_0^i} m_0^i \xrightarrow{o} m_1^i \xrightarrow{k_{2i}} m_2^i \xrightarrow{z_{2i}} m_3^i \xrightarrow{o} m_4^i \xrightarrow{k} \cdots \xrightarrow{k} m_{2b+2}^i$$

$$N_i^b = s \xrightarrow{y_{2m+5}^i} s_{2m+5}^i \cdots s_1^i \xrightarrow{y_0^i} n_0^i \xrightarrow{o} n_1^i \xrightarrow{k_{2i+1}} n_2^i \xrightarrow{z_{2i+1}} n_3^i \xrightarrow{o} n_4^i \xrightarrow{k} \cdots \xrightarrow{k} n_{2b+2}^i$$

$$L_{i,0} = s \xrightarrow{v_{2m+5}^i} q_{2m+5}^i \cdots q_1^i \xrightarrow{v_0^i} t_0^i \xrightarrow{k_{2i}} t_0^i \qquad L_{i,1} = s \xrightarrow{u_{2m+5}^i} p_{2m+5}^i \cdots p_2^i \xrightarrow{u_1^i} p_1^i \xrightarrow{k_{2i+1}} t_{b,b,b}^i$$

In terms of Condition 2.2, the gadgets $M_i^b, N_i^b, L_{i,0}$ and $L_{i,1}$ work similar to the corresponding ones of A_φ^b. However, Condition 2.2 requires to distribute the task of one event to multiple events. For example, the events $z_0, \ldots, z_{2m-1}$ of B_φ^b play the same role as z of A_φ^b. This is achieved by F_i^b and G_i^b. More exactly, if B_φ^b is b-solvable then, by Lemma 1, every atom (k, f_{b+1}^i) is too. By Lemma 2, if (sup, sig) is a solving test-free b-region then $sig(k) = (0, 1)$ and $sig(o) = (b, 0)$ or $sig(k) = (1, 0)$ and $sig(o) = (0, b)$. If $sig(k) = (0, 1)$ then, by $sup(f_b^i) = sup(g_b^i) = b \cdot sig^+(k) = b$ and $sup(f_{b+1}^i) = sup(f_{b+1}^i) = b$, we get $sig^+(z_i) = sig^-(z_i)$ and, thus, $sig(k_i) = (0, b)$, $\forall i \in \{0, \ldots, 2m-1\}$. Similarly, if $sig(k) = (1, 0)$ then $sig(k_i) = (b, 0)$, $\forall i \in \{0, \ldots, 2m-1\}$. Thus, by the grids' functionality, the set $M = \{X \in V(\varphi) \mid sig(X) \neq (0, 0)\}$ is a sought model.

3.2   The Proof of Condition 1.2 and Condition 2.2

In this section, we provide b-admissible sets of A_φ^b and B_φ^b in accordance to Condition 1.2 and Condition 2.2, respectively. For the sake of simplicity, we present for every region (sup, sig) only its signature sig and the value $sup(s)$ of the initial state s. Because A_φ^b and B_φ^b are reachable and $sup(s'') = sup(s') - sig^-(e) + sig^+(e)$ for every transition $s' \xrightarrow{e} s''$, this completely defines the region. In the remainder of this section, unless stated explicitly otherwise, let $i \in \{0, \ldots, m-1\}$ and M be a one-in-three model of φ. Moreover, for $\alpha \in \{0, 1, 2\}$ let $\beta_\alpha, \gamma_\alpha \in \{0, \ldots, m-1\} \setminus \{i\}$ be the distinct indices such that $X_{i,\alpha} \in C_i \cap C_{\beta_\alpha} \cap C_{\gamma_\alpha}$, that is, $\beta_\alpha, \gamma_\alpha$ choose the other two clauses of φ containing $X_{i,\alpha}$.

We start with Condition 1.2 and provide a b-admissible set $\mathcal{R}$ of pure regions of A_φ^b such that $|R^\bullet| \leq 1$ and $|{}^\bullet e_\mathcal{R}| \leq 1$ for all $R \in \mathcal{R}$ and $e \in E(A_\varphi^b)$. Moreover, because A_φ^b has exactly $7m+4$ events, every region R of A_φ^b satisfies $|{}^\bullet R| \leq 7m + 4$. For abbreviation, we define $U = \{u_0, \ldots, u_{m-1}\}, V = \{v_0, \ldots, v_{m-1}\}, W = \{w_0, \ldots, w_{m-1}\}, Y = \{y_0, \ldots, y_{m-1}\}$ and $K = \{k_0, \ldots, k_{2m-1}\}$. We solve all atoms concerning the events of $\{a\} \cup U \cup V \cup W \cup Y$ with the region $R = (sup, sig)$, defined by $sup(s) = 0$ and $sig(e) = (0, b)$ if $e \in \{a\} \cup U \cup V \cup W \cup Y$ and, otherwise, $sig(e) = (0, 0)$. This region satisfies $|R^\bullet| = 0$ (no event consumes). Moreover, none of the subsequently presented regions of A_φ^b is in the preset of any of $\{a\} \cup U \cup V \cup W \cup Y$, thus, $|{}^\bullet e_\mathcal{R}| \leq 1$ for $e \in \{a\} \cup U \cup V \cup W \cup Y$. We proceed with presenting for every event $k, z, o, v, k_{2i}, k_{2i+1}$ and $X_{i,0}, X_{i,1}, X_{i,2}$ corresponding regions that solves it. Every row of Table 1 (below) defines a region $R = (sup_R, sig_R)$ with $sup_R(s) = 0$ as follows: For every $e \in E(A_\varphi^b)$ we have either $sig_R(e) = (0, 0)$ or $sig_R(e) \in \{(1, 0), (0, 1), (b, 0), (0, b)\}$. In the latter case, e occurs according to its signature in the corresponding column either as a single event or as member of the event set shown. For example, for R_1 we have $sig_{R_1}(k) = (0, 1)$ and $sig_{R_1}(e) = (0, 1)$ for $e \in M$.

Table 1. Pure regions of A_φ^b that solve $k, z, o, k_{2i}, k_{2i+1}$ and $X_{i,0}, X_{i,1}, X_{i,2}$.

R	$(1,0)$	$(0,1)$	$(b,0)$	$(0,b)$
R_1		k, M	o	W, Y, K
R_2	k			z, a
R_3			z	a, o, U, V
R_4				z, U, V
$R_{k_{2i}}^z$			k_{2i}	z, u_i, v_i, w_i
$R_{k_{2i+1}}^z$			k_{2i+1}	z, u_i, y_i
$R_{k_{2i}}^\alpha$ for $X_{i,\alpha} \notin M$		$X_{i,\alpha}$		$a, Y, \ell \in \{i, \beta_\alpha, \gamma_\alpha\} : u_\ell, k_{2\ell},$ $W \setminus \{w_\ell \mid \ell \in \{i, \beta_\alpha, \gamma_\alpha\}$
$R_{k_{2i+1}}$				$k_{2i+1}, a, W, V, U \setminus \{u_{2i+1}\}, Y \setminus \{y_{2i+1}\}$
$R_{X_{i,\alpha}}$	$X_{i,\alpha}$			$v_i, v_{\beta_\alpha}, v_{\gamma_\alpha}$

The regions of Table 1 solve the events $k, z, o, k_{2i}, k_{2i+1}$ and $X_{i,0}, X_{i,1}, X_{i,2}$ as follows. (k): R_1 solves k at the sinks of z and R_2 solves k at the remaining states. (z): R_2 solves z at the sources of k and R_3 solves z at o's sources and at s. $R^z_{k_{2i}}$ and $R^z_{k_{2i+1}}$, where $i \in \{0, \ldots, m-1\}$, solve z at the sources of $k_0, \ldots, k_{2m-1}$. Finally, R_4 solves z at the remaining states. (o): R_1 solves o at the sources of k, $k_0, \ldots, k_{2m-1}$ and at s and R_3 solves o at the remaining states. (k_{2i}): R_1 solves k_{2i} at all sources of o and all sources of $X_{i,\alpha}$ in C^b_i, where $X_{i,\alpha} \in M$. $R^z_{k_{2i}}$ solves k_{2i} at all sources of k_j, where $2i \neq j \in \{0, \ldots, 2m-1\}$ and at s. The remaining atoms are solved by (the two regions defined by) $R^\alpha_{k_{2i}}$, where $\alpha \in \{0, 1, 2\}$ such that $X_{i,\alpha} \notin M$. (k_{2i+1}): R_1 solve k_{2i+1} at $n_{i,0}$ and $R^z_{k_{2i+1}}$ at s and $R_{k_{2i+1}}$ at all remaining states. $(X_{i,\alpha})$: If $X_{i,\alpha} \in M$ then the region R_1 solves it at t^i_0, otherwise, $X_{i,\alpha}$ is solved at t^i_0 by $R^\alpha_{k_{2i}}$. The remaining atoms are solved by $R_{X_{i,\alpha}}$.

In the following we argue that A^b_φ has the SSP, too: To separate $S(Q_b)$ from $S(A^b_\varphi) \setminus S(Q_b)$ we use the region $R_Q = (sup_Q, sig_Q)$ where $sup_Q(s) = 0$, $sig_Q(a) = (0, b)$ and $sig_Q(e) = (0, 0)$ for the other events. Moreover, the states of Q_b are pairwise separated by R_1, R_2 and R_4. To separate the states $S(M^b_i)$ from $S(A^b_\varphi) \setminus S(M^b_i)$ we define the region $R_{M_i} = (sup_{M_i}, sig_{M_i})$ where $sup_{M_i}(s) = 0$, $sig_{M_i}(w_i) = (0, b)$ and $sig_{M_i}(e) = (0, 0)$ for the other events. The states of M^b_i are pairwise separated by R_1, R_2, R_3 and R_4. Similarly, the states $S(N^b_i)$ are separated by R_1, R_2, R_3, R_4 and $R_{N_i} = (sup_{N_i}, sig_{N_i})$ where $sup_{N_i}(s) = 0$, $sig_{N_i}(y_i) = (0, b)$ and $sig_{N_i}(e) = (0, 0)$ for the other events. To separate the states of $S(C^b_i) \cup \{t^i_0, s_i\}$ from all the other states we use the region $R_{C_i} = (sup_{C_i}, sig_{C_i})$ where $sup(C_i)(s) = 0$, $sig_{C_i}(u_i) = sig_{C_i}(v_i) = (0, b)$ and $sig_{C_i}(e) = (0, 0)$ for the other events. Moreover, the states of $S(C^b_i) \cup \{t^i_0, s_i\}$ are pairwise separated by $R_1, R_{k_{2i+1}}$ and $R^\alpha_{X_{i,\alpha}}$, where $X_{i,\alpha} \notin M$.

Altogether, the set $\mathcal{R} = \mathcal{R}_1 \cup \mathcal{R}_2 \cup \mathcal{R}_3 \cup \mathcal{R}_4$ where $\mathcal{R}_1 = \{R_1, R_2, R_3, R_4\}$, $\mathcal{R}_2 = \{R^z_{k_{2i}}, R^z_{k_{2i+1}}, R^\alpha_{k_{2i}}, R_{k_{2i+1}} \mid i \in \{0, \ldots, m-1\}, \alpha \in \{0, 1, 2\}, X_{i,\alpha} \notin M\}$, $\mathcal{R}_3 = \{R_{X_{i,\alpha}} \mid i \in \{0, \ldots, m-1\}, \alpha \in \{0, 1, 2\}\}$ and $\mathcal{R}_3 = \{R_Q, R_{M_i}, R_{N_i}, R_{C_i} \mid i \in \{0, \ldots, m-1\}\}$, is an admissible set of A^b_φ. We briefly argue that it is FA: It is easy to see that every presented region $R \in \mathcal{R}$ satisfy $|R^\bullet| \leq 1$. Moreover, $|^\bullet e_{\mathcal{R}}| \leq 1$ is also true for $e \in E(A^b_\varphi)$: The regions $R_1 \in {}^\bullet o$, $R_2 \in {}^\bullet k$, $R_3 \in {}^\bullet z$ and $R^z_{k_{2i}} \in {}^\bullet k_{2i}$ and $R^z_{k_{2i+1}} \in {}^\bullet k_{2i+1}$ are unique. Furthermore, if $X_{i,\alpha} = X_{j,\beta} = X_{\ell,\gamma}$ then $R_{X_{i,\alpha}} = R_{X_{j,\beta}} = R_{X_{\ell,\gamma}}$ where $i, j, \ell \in \{0, \ldots, m-1\}, \alpha, \beta, \gamma \in \{0, 1, 2\}$. As $\mathcal{R}$ is a set, this region is the only element in ${}^\bullet X_{i,\alpha}$. No other region $(sup, sig) \in \mathcal{R}$ satisfies $sig^-(e) > 0$ for any $e \in E(A^b_\varphi)$. Thus, A^b_φ satisfies Condition 1.2.

To prove Condition 2.2 we provide a b-admissible set $\mathcal{R}$ of pure regions of B^b_φ such that $|e_{\mathcal{R}}^\bullet| \leq 2$ and $|{}^\bullet e_{\mathcal{R}}| \leq 2$ for all $e \in E(B^b_\varphi)$. For brevity, we define for $j \in \{0, \ldots, m-1\}$ the following sets: $B_j = \{b^i_j \mid i \in \{0, \ldots, m-1\}\}$, $D_j = \{d^i_j \mid i \in \{0, \ldots, m-1\}\}$, $U_j = \{u^i_j \mid i \in \{0, \ldots, m-1\}\}$, $V_j = \{v^i_j \mid i \in \{0, \ldots, m-1\}\}$, $W_j = \{w^i_j \mid i \in \{0, \ldots, m-1\}\}$, $Y_j = \{y^i_j \mid i \in \{0, \ldots, m-1\}\}$, $K = \{k_i \mid i \in \{0, \ldots, 2m-1\}\}$ and $Z = \{z_i \mid i \in \{0, \ldots, 2m-1\}\}$. By a little abuse of notation, we let $\mathcal{C}_i = F^b_i \cup G^b_i \cup M^b_i \cup N^b_i \cup F^b_i \cup C^b_i \cup L_{i,0} \cup L_{i,1}$ and $\delta_i = 2m + 5 - i$. Table 2 (below) defines a regions R of B^b_φ with $sup_R(s) = 0$.

Table 2. Pure b-regions of B_φ^b that solve several separation atoms.

R	$(1,0)$	$(0,1)$	$(b,0)$	$(0,b)$
R_1		k, M	o	W_0, Y_0, K
R_2	k			Z, B_0, D_0
R_3			o	Z, W_3, Y_3
$R_{z_{2i}}$			z_{2i}	b_1^i, w_1^i
$R_{z_{2i+1}}^0$			z_{2i+1}	v_5^i, k_{2i+1}, d_1^i
$R_{z_{2i+1}}^1$			z_{2i+1}	d_0^i, y_1^i
$R_{k_{2i+1}}$			k_{2i+1}	b_1^i, w_1^i
R_{2i}^2			z_{2i}	$k_{2i}, b_0^i, (V_{\delta_i} \cup U_{\delta_i} \cup B_{\delta_i} \cup D_{\delta_i} \cup W_{\delta_i} \cup Y_{\delta_i}) \setminus E(\mathcal{C}_i)$

$(k), (o)$: The regions R_1 and R_2 solve k and the regions R_1 and R_3 solve o. $(z_{2i}), (z_{2i+1})$: The region R_2 solves z_{2i}, z_{2i+1} at k's sources and R_3 solves them at o's sources, at $s_{i,1}, s_{i,2}, s_{i,3}$ and at $r_{i,1}, r_{i,2}, r_{i,3}$. R_{2i}^2 solves z_{2i} at the remaining states of $\mathcal{C}_i \setminus \{t_0^i\}$ and $R_{z_{2i}}$ solves z_{2i} at the remaining states of B_φ^b. $R_{z_{2i+1}}^0$ solves z_{2i+1} at n_0^i, n_1^i and $s_{i,1}$ and $R_{z_{2i+1}}^1$ solves it at the remaining states.

(k_{2i}): For a correct referencing, we need the following definitions: If $j \in \{0, \ldots, m-1\}$ then let $\alpha_j \in \{0, 1, 2\}$ be the index such that $X_{j,\alpha_j} \in M$ and let by $\beta_j < \gamma_j \in \{0, 1, 2\} \setminus \{\alpha_j\}$ the other variable events of C_j^b be chosen. Moreover, let $\ell \neq j \in \{0, \ldots, m-1\}$ such that $X_{i,\beta_i} \in C_i \cap C_\ell, \cap C_j$ and let $\ell' \neq j' \in \{0, \ldots, m-1\}$ such that $X_{i,\gamma_i} \in C_i \cap C_{\ell'}, \cap C_{j'}$. That is, ℓ, j and ℓ', j' choose the other two clauses where $X_{i,\beta_i}, X_{i,\gamma_i}$ occur. We use this to define the region $R_{2i}^0 = (sup_{2i}^0, sig_{2i}^0)$ where $sup_{2i}^0(s) = 0$, $sig(X_{i,\beta_i}) = (1,0)$ and for $\delta \in \{i, \ell, j\}$ it is $sig_{2i}^0(k_{2\delta}) = (b,0)$ and $sig_{2i}^0(w_0^\delta) = (0,b)$ if $X_{i,\beta_i} = X_{\delta,\beta_\delta}$ and $sig_{2i}^0(w_2^\delta) = (0,b)$ if $X_{i,\beta_i} = X_{\delta,\gamma_\delta}$. Similarly, we define the region $R_{2i}^1 = (sup_{2i}^1, sig_{2i}^1)$ by $sup_{2i}^1(s) = 0$, $sig(X_{i,\gamma_i}) = (1,0)$ and for $\delta \in \{i, \ell', j'\}$ it is $sig_{2i}^1(k_{2\delta}) = (b,0)$ and $sig_{2i}^1(w_{\delta,2}) = (0,b)$ if $X_{i,\gamma_i} = X_{\delta,\gamma_\delta}$ and $sig_{2i}^1(w_{\delta,0}) = (0,b)$ if if $X_{i,\gamma_i} = X_{\delta,\beta_\delta}$. Notice that if $X_{i,\beta_i} = X_{\delta,\gamma_\delta}$ then $R_{2i}^0 = R_{2\delta}^1$ and if $X_{i,\gamma_i} = X_{\delta,\beta_\delta}$ then $R_{2i}^1 = R_{2\delta}^0$. This is our way to correctly, restrict the postset of the events $w_0^{\cdots}$ and $w_2^{\cdots}$. The region R_1 solves k_{2i} at m_0^i and the sinks of X_{i,α_i}. R_{2i}^0 and R_{2i}^1 solve k_{2i} at all states of $C_i^b \cup \{s\}$ and $\bigcup_{j=1}^{2m+5} \{q_j^\ell, p_j^\ell, a_j^\ell, c_j^\ell, r_j^\ell, s_j^\ell \mid \ell \in \{0, \ldots, m-1\} \setminus \{i\}\}$. Finally, to solve k_{2i} at the remaining states we use the region R_{2i}^2 defined as follows: If $\alpha = 2m + 5 - i$ then $R_{2i}^2 = (sup_{2i}^2, sig_{2i}^2)$ is defined by $sup_{2i}^2(s) = 0$, $sig_{2i}^2(k_{2i}) = sig_{2i}^2(b_{i,0}) = sig_{2i}^2(e)$, where $e \in \{v_{j,\alpha}, u_{j,\alpha}, b_{j,\alpha}, d_{j,\alpha}, w_{j,\alpha}, y_{j,\alpha} \mid j \in \{0, \ldots, m-1\} \setminus \{i\}\}$ and $sig_{2i}^2(z_{2i}) = (b,0)$.

(k_{2i+1}): R_1 and $R_{k_{2i+1}}$ solve k_{2i+1} at all states of B_φ^b.

$(X_{i,0}, X_{i,1}, X_{i,2})$: Let $\alpha_i, \beta_i, \gamma_i$ be defined as above. To separate $X_{i,\alpha_i} = X_{\ell,\alpha_\ell} = X_{j,\alpha_j}$, i, j, ℓ pairwise distinct, from $q_1^i, q_2^i, q_1^\ell, q_2^\ell, q_1^j, q_2^j$, respectively, we use the region $R_q^i = R_q^\ell = R_q^j$ that maps s to 0, X_{i,α_1} to $(0,b)$, v_0^i, v_0^ℓ, v_0^j to $(b,0)$, v_2^i, v_2^ℓ, v_2^j to $(0,b)$ and the other events to $(0,0)$. This region is necessary as the presets ${}^\bullet v_0^i, {}^\bullet v_0^\ell, {}^\bullet v_0^j$ have already two elements. To separate X_{i,α_i} from the remaining states, we use $R^i_{\alpha_i} = (sup^i_{\alpha_i}, sig^i_{\alpha_i})$, where $sup^i_{\alpha_i}(s) = 0$, $sig^i_{\alpha_i}(X_{i\alpha_i}) = (1,0)$ $sig^i_{\alpha_i}(v^i_1) = sig^i_{\alpha_i}(v^\ell_1) = sig^i_{\alpha_i}(v^j_1) = (0,b)$ and $X_{i,\alpha_i} \in C_i \cap C_\ell \cap C_j$.

The regions $R^i_{\beta_i}$ for X_{i,β_i} and $R^i_{\gamma_i}$ for X_{i,γ_i} are defined accordingly, where we use $v_{\ddot{3}}$ and $v_{\ddot{4}}$ (without repetition or confusion) as preset events, respectively. Notice that, so far, $X_{i,\beta_i}, X_{i,\gamma_i}$ are already separated from $q_1, \ldots, q_{2m+5}$ by R^0_{2i} and R^1_{2i}, respectively.

$(u^i_j, v^i_j, b^i_j, d^i_j, w^i_j, y^i_j, j \in \{1, \ldots, 2m - 5\})$: So far, for all of these events e holds $|{}^{\bullet}e_{\mathcal{R}}| = 0$ and, even more, if $j \neq 1$ then $|e_{\mathcal{R}}{}^{\bullet}| \leq 1$. Hence, for $e, e' \in \{u^i_j, v^i_j, b^i_j, d^i_j, w^i_j, y^i_j, j \in \{1, \ldots, 2m - 4\}\}$ with $\xrightarrow{e'} x \xrightarrow{e} \in B^b_\varphi$ we use the region (sup_e, sig_e) where $sup_e(s) = 0$, $sig_e(e') = (0,b)$ and $sig_e(e) = (b,0)$ and $sig_e(e'') = (0,0)$ for $E(B^b_\varphi) \setminus \{e, e'\}$. Notice that e, e' are unique and that this region also separates x. For the $2m + 5$-indexed events we use the region where all these (and only these) events are mapped to $(b, 0)$ and s is mapped to b.

So far, the presented regions justify B^b_φ's b-ESSP. It remains to justify its b-SSP: One verifies that all distinct states $s, s' \in C_i$ are separated by the already presented regions. If $e \in \{u^i_j, v^i_j, b^i_j, d^i_j, w^i_j, y^i_j \mid i \in \{0, \ldots, m - 1\}, j \in \{1, \ldots, 2m - 5\}\}$ and $s \xrightarrow{e}$ then s is separated by the region defined for the separation of e. Moreover, so far, if $e \in \{u^i_j, v^i_j, b^i_j, d^i_j, w^i_j, y^i_j \mid i \in \{0, \ldots, m - 1\}, j \in \{m, \ldots, 2m + 6\}\}$ then $|e_{\mathcal{R}}{}^{\bullet}| = 1$. Hence, we choose for every $i \in \{0, \ldots, m - 1\}$ the region $R_{C_i} = (sup_{C_i}, sig_{C_i})$ where $sup_{C_i}(s) = 0$, $sig_{C_i}(e) = (0,b)$ if $e \in \{u^i_j, v^i_j, b^i_j, d^i_j, w^i_j, y^i_j \mid j = 2m + 5 - i\}$ and, otherwise, $sig_{C_i}(e) = (0,0)$. Clearly, R_{C_i} separates the remaining states in question from $S(B^b_\varphi) \setminus C_i$. Moreover, the regions $R_{C_0}, \ldots, R_{C_{m-1}}$ preserve the $(2,2)$-S-system property.

Altogether, the union of all introduced regions yields a b-admissible set $\mathcal{R}$ of pure regions that has the $(2, 2)$-S-system property.

3.3   The Proof of Theorem 1.4

By Lemma 1, a b-net N, being a weighted (m, n)-T-system, solves A if and only if there is a b-admissible set $\mathcal{R}$ with $N = N^{\mathcal{R}}_A$. By definition, every $R = (sup, sig) \in \mathcal{R}$ satisfies $|{}^{\bullet}R| = |\{e \in E(A) \mid sig^+(e) > 0\}| \leq m$ and $|R^{\bullet}| = |\{e \in E(A) \mid sig^-(e) > 0\}| \leq n$. The maximum set $\mathcal{R}$ of A's b-regions that satisfy the (m, n)-condition is computable in polynomial time: To define $R = (sup, sig) \in \mathcal{R}$ we have for $\ell \in \{1, \ldots, m\}$ and $\ell' \in \{1, \ldots, n\}$ at most $\binom{|E|}{\ell}$ and $\binom{|E|}{\ell'}$ events for ${}^{\bullet}R$ and $R^{\bullet}$, respectively. This makes at most $\binom{|E|}{\ell} \cdot \binom{|E|}{\ell'} \cdot (b+1)^{\ell+\ell'}$ possibilities for sig, each of it is to combine with the at most $b+1$ values for $sup(s_0)$. As b, m and n are not part of the input, altogether, there are at most $\mathcal{O}(|E|^{m+n})$ b-regions. Moreover, one can decide in polynomial time if $sup(s_0)$ and sig define actually a fitting b-region as follows: Firstly, compute a spanning tree A' of A, having at most $|S(A)|$ paths, in time $\mathcal{O}(|E(A)| \cdot |S(A)|^3)$ [16]. Secondly, use $sup(s_0)$ and sig to determine $sup(s_j)$ for all $s_j \in S(A)$ by the unique path $s_0 \xrightarrow{e_1} \ldots \xrightarrow{e_j} s_j \in A'$. Thirdly, check for the at most $|S|^2 \cdot |E|$ edges $s \xrightarrow{e} s' \in A$ if both $sup(s) \geq sig^-(e)$ and $sup(s') = sup(s) + sig^-(e) + sig^+(e) \leq b$ are satisfied.

Having computed the (maximum) set $\mathcal{R}$, it remains to check (in polynomial time) whether the at most $|S|^2 + |S| \cdot |E|$ separation atoms of A are solved by $\mathcal{R}$.

4   Conclusion

This paper shows that deciding if a TS is solvable by a b-net which is CF, FA, FC, EFC or AC remains NP-complete. Moreover, our proof imply that synthesis is also hard if the searched net is to be *behaviorally free choice*, *behaviorally asymmetric choice* or *reducedly asymmetric choice* [3]. Furthermore, we show that synthesis of (m, n)-S-systems is NP-complete for every fixed $m, n \geq 2$. While synthesis of weighted (m, n)-T-systems, being dual to the S-systems, is also hard if m, n are part of the input, it becomes tractable for any fixed m, n. In particular, fixing m, n puts the problem into the complexity class XP. Consequently, for future work, it remains to be investigated whether the synthesis of weighted (m, n)-T-systems parameterized by $m + n$ is fixed parameter tractable.

Acknowledgements. I would like to thank the reviewers for their helpful comments.

References

1. Badouel, E., Bernardinello, L., Darondeau, P.: Polynomial algorithms for the synthesis of bounded nets. In: Mosses, P.D., Nielsen, M., Schwartzbach, M.I. (eds.) CAAP 1995. LNCS, vol. 915, pp. 364–378. Springer, Heidelberg (1995). https://doi.org/10.1007/3-540-59293-8_207
2. Badouel, E., Bernardinello, L., Darondeau, P.: Petri Net Synthesis. An EATCS Series. Springer, Heidelberg (2015). https://doi.org/10.1007/978-3-662-47967-4
3. Best, E.: Structure theory of petri nets: the free choice hiatus. In: Brauer, W., Reisig, W., Rozenberg, G. (eds.) ACPN 1986. LNCS, vol. 254, pp. 168–205. Springer, Heidelberg (1987). https://doi.org/10.1007/978-3-540-47919-2_8
4. Best, E., Devillers, R.: Characterisation of the state spaces of live and bounded marked graph petri nets. In: Dediu, A.-H., Martín-Vide, C., Sierra-Rodríguez, J.-L., Truthe, B. (eds.) LATA 2014. LNCS, vol. 8370, pp. 161–172. Springer, Cham (2014). https://doi.org/10.1007/978-3-319-04921-2_13
5. Best, E., Devillers, R.R.: State space axioms for t-systems. Acta Inf. **52**(2–3), 133–152 (2014). https://doi.org/10.1007/s00236-015-0219-0
6. Best, E., Devillers, R.R.: Synthesis and reengineering of persistent systems. Acta Inf. **52**(1), 35–60 (2015). https://doi.org/10.1007/s00236-014-0209-7
7. Best, E., Devillers, R.R.: Synthesis of bounded choice-free petri nets. In: CONCUR. LIPIcs, vol. 42, pp. 128–141. Schloss Dagstuhl - Leibniz-Zentrum fuer Informatik (2015). https://doi.org/10.4230/LIPIcs.CONCUR.2015.128
8. Cheng, A., Esparza, J., Palsberg, J.: Complexity results for 1-safe nets. Theor. Comput. Sci. **147**(1&2), 117–136 (1995). https://doi.org/10.1016/0304-3975(94)00231-7
9. Devillers, R., Hujsa, T.: Analysis and synthesis of weighted marked graph petri nets. In: Khomenko, V., Roux, O.H. (eds.) PETRI NETS 2018. LNCS, vol. 10877, pp. 19–39. Springer, Cham (2018). https://doi.org/10.1007/978-3-319-91268-4_2

10. Howell, R.R., Rosier, L.E.: Completeness results for conflict-free vector replacement systems. J. Comput. Syst. Sci. **37**(3), 349–366 (1988). https://doi.org/10.1016/0022-0000(88)90013-X
11. Moore, C., Robson, J.M.: Hard tiling problems with simple tiles. Discret. Comput. Geom. **26**(4), 573–590 (2001). https://doi.org/10.1007/s00454-001-0047-6
12. Schlachter, U.: Bounded petri net synthesis from modal transition systems is undecidable. In: CONCUR. LIPIcs, vol. 59, pp. 15:1–15:14. Schloss Dagstuhl - Leibniz-Zentrum fuer Informatik (2016). https://doi.org/10.4230/LIPIcs.CONCUR.2016.15
13. Schlachter, U., Wimmel, H.: k-bounded petri net synthesis from modal transition systems. In: CONCUR. LIPIcs, vol. 85, pp. 6:1–6:15. Schloss Dagstuhl - Leibniz-Zentrum fuer Informatik (2017). https://doi.org/10.4230/LIPIcs.CONCUR.2017.6
14. Teruel, E., Colom, J.M., Suárez, M.S.: Choice-free petri nets: a model for deterministic concurrent systems with bulk services and arrivals. IEEE Trans. Syst. Man Cybern. Part A **27**(1), 73–83 (1997). https://doi.org/10.1109/3468.553226
15. Tredup, R.: Hardness results for the synthesis of b-bounded petri nets. In: Donatelli, S., Haar, S. (eds.) PETRI NETS 2019. LNCS, vol. 11522, pp. 127–147. Springer, Cham (2019). https://doi.org/10.1007/978-3-030-21571-2_9
16. Turau, V.: Algorithmische Graphentheorie, (2. Aufl). Oldenbourg (2004)

Reachability of Five Gossip Protocols

Hans van Ditmarsch[1] , Malvin Gattinger[2] , Ioannis Kokkinis[3(✉)] ,
and Louwe B. Kuijer[4]

[1] CNRS, LORIA, University of Lorraine, Nancy, France
`hans.van-ditmarsch@loria.fr`
[2] University of Groningen, Groningen, The Netherlands
`malvin@w4eg.eu`
[3] TU Dortmund, Dortmund, Germany
`ioannis.kokkinis@tu-dortmund.de`
[4] University of Liverpool, Liverpool, UK
`l.b.kuijer@gmail.com`

Abstract. Gossip protocols use point-to-point communication to spread
information within a network until every agent knows everything. Each
agent starts with her own piece of information ('secret') and in each call
two agents will exchange all secrets they currently know. Depending on
the protocol, this leads to different distributions of secrets among the
agents during its execution. We investigate which distributions of secrets
are *reachable* when using several distributed epistemic gossip protocols
from the literature. Surprisingly, a protocol may reach the distribution
where all agents know all secrets, but not all other distributions. The five
protocols we consider are called ANY, LNS, CO, TOK, and SPI. We find
that TOK and ANY reach the same distributions but all other protocols
reach different sets of distributions, with some inclusions. Additionally,
we show that all distributions are *subreachable* with all five protocols:
any distribution can be reached, if there are enough additional agents.

Keywords: Gossip · Networks · Reachability

1 Introduction

Let each of a set of agents $\{a, b, c, \ldots\}$ know a single secret $\{A, B, C, \ldots\}$, respectively. The agents can communicate via telephone calls. When they call, they share all the secrets they know at the moment the call takes place. An agent who knows all secrets is an *expert*. The goal is to turn all agents into experts. A protocol to achieve this state of knowledge is called a *gossip protocol* [10,11].

Here we consider five gossip protocols of a distributed nature [1–4]:

ANY *Any call is allowed, i.e., for every two agents a and b, a is allowed to call b.*
CO *Short for "call once". An agent a may call b iff they have not spoken before, i.e., if a has not called b before and b has not called a before.*

© Springer Nature Switzerland AG 2019
E. Filiot et al. (Eds.): RP 2019, LNCS 11674, pp. 218–231, 2019.
https://doi.org/10.1007/978-3-030-30806-3_17

LNS *Short for "learn new secrets". Agent a may call b iff a doesn't yet know B.*
During the execution of this protocol after every call at least one new secret
is learned, hence the protocol name.

TOK *Short for "token". Agent a may call b iff a has a token. The caller passes
her token to the callee.*
Every agent starts with a token. After a call all tokens held by an agent merge
to one. In this protocol an agent who lost her token can get it back when she
receives a new call.
Equivalently, we can say that a may call b iff a has either not been involved
in any calls, or a was the callee in the last call a was involved in.

SPI *Short for "spider". Agent a may call b iff a has a token. The callee passes
her token to the caller.*
Every agent starts with a token. After a call all tokens held by an agent
merge to one. In this protocol an agent who has been called loses her token
permanently and can never initiate a call again. This protocol tends to lead
to a small number of agents making many calls. When drawn as a graph, this
looks like a spider web with the agent making the calls at the centre, hence
the name "spider".
Equivalently, we can say that a may call b iff a has never received any calls.

All protocols run in a sequential manner as follows: starting from the situation
where each agent only knows her own secret, each moment in time a single call
satisfying the protocol condition is selected and executed. The selection of calls
continues until all agents are experts. Here we investigate which distributions
of secrets may be reached during the protocol execution (under *any sequence* of
calls), hence we do not have to fix a specific algorithm for call selection.

Knowing which distribution can be reached by which protocol can help the
agents (or an external observer) understand which protocol is being used during
the exchange of information. Moreover, reachability can be of importance for
security or privacy reasons.

Let us illustrate the topic of reachability by an example with three agents.
We represent a distribution of secrets by listing the secrets known by each agent.
Given initial distribution (A, B, C), the call ab (the call from a to b) results in
(AB, AB, C). (Strictly, we go from $(\{A\}, \{B\}, \{C\})$ to $(\{A, B\}, \{A, B\}, \{C\})$.)
This is therefore ANY-reachable. After the call sequence $ab; bc; ac$, which is per-
mitted in ANY, LNS, and CO, all three agents are experts. But already for three
agents there is a difference between the five protocols. The sequence $ab; bc; ca$ is
CO-permitted but not LNS-permitted: as c already knows A, the call ca is not
allowed in LNS. The sequence $ab; bc; ab$ is not CO-permitted (repeating ab is not
allowed); and clearly if a call is not CO-permitted it is also not LNS-permitted.
Call sequence $ab; ba; ab$ is TOK-permitted but not SPI-permitted, whereas call
sequence $ab; ab$ is SPI-permitted but not TOK-permitted.

We assume that communication between all agents is possible, i.e., a complete
network topology. Unreachability results in this setting are very strong: if one
of the five protocols cannot reach a distribution s assuming a complete network
topology, then this protocol also cannot reach s assuming any other topology. It

is not difficult to see that already for two agents, unreachable distributions can occur: the distribution (AB, A) cannot be reached by any of the five protocols.

We also study a less restrictive notion called *subreachability*. Given agents a, b, c, if b calls c, and then c calls a, the resulting distribution is (ABC, BC, ABC). The restriction of that distribution to the agents a and b only is (AB, B). We say that distribution (AB, B), although not reachable, is subreachable. Knowing the knowledge situations that can be subreached by a protocol is particularly interesting when the number of agents is not common knowledge among the agents, or when the agents have limited reasoning power and cannot reason like "there are two agents beside me and a call has taken place, so these agents now know each other's secrets". In such a situation the agents should not only consider the reachable but also the subreachable distributions possible.

We further investigate reachability under *unordered distributions* ("given n agents and n subsets of the set of all secrets, is there a bijection between these sets of agents and subsets?"). Unordered distributions should be taken into consideration by an observer who is uncertain about which agent holds which sets of secrets in a distribution.

Our Contributions. For up to three agents all five protocols can reach the same distributions. Thus, with at most three agents present, an observer cannot tell which protocol is currently used, by simply observing the distributions of secrets. But with four or more agents there is a difference in the reachability strength. In Fig. 1 we give a complete overview of each protocol's reachability strength. Figure 1 (together with the relevant theorems) can serve as guide for an (internal or external observer) that wants to know which protocol the agents are using for information exchange. For example, if the observer finds out that the distribution of secrets $(ABCD, ABCD, ABC, ABD)$ has appeared, then she can be certain that agents are not using the CO-protocol since this distribution is not CO-reachable (see Theorem 3).

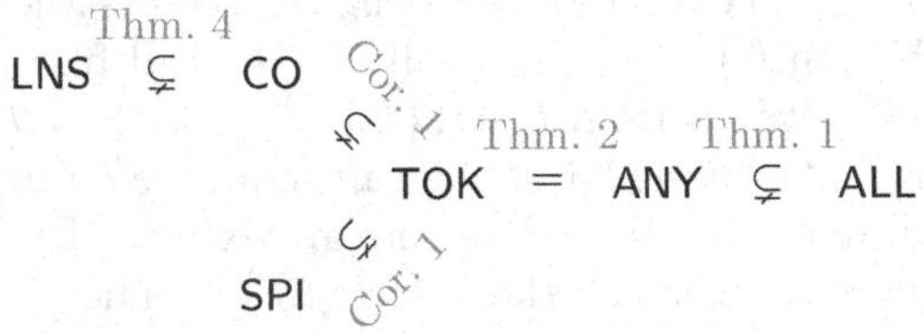

$$\text{LNS} \overset{\text{Thm. 4}}{\subsetneq} \text{CO}$$
$$\text{TOK} \overset{\text{Thm. 2}}{=} \text{ANY} \overset{\text{Thm. 1}}{\subsetneq} \text{ALL}$$
$$\text{SPI}$$

Fig. 1. Overview of results, in which a protocol's name stands for the set of distributions reachable by it and ALL stands for the set of all distributions. Besides transitivity no other inclusions hold, i.e., SPI and CO properly intersect (see Theorem 3 and Corollary 1), and so do SPI and LNS (see Theorems 4 and 5).

In Theorem 6 we show that all distributions are subreachable by all five protocols. As we mentioned before this has the consequence that an observer cannot infer which of the five protocols is being used if, for example, she does not know how many agents there are, since every distribution of secrets can possibly occur among a subset of the agents using any of the five protocols. Finally in Theorem 7 we show that SPI, ANY and TOK reach the same set of unordered distributions. The consequence of the latter theorem is that the observer cannot distinguish between the protocols SPI, ANY and TOK if she is uncertain about which agent holds which set of secrets in a given distribution.

Related Work. The combinatorial properties of gossip protocols have been investigated several times in the literature. In [1] the focus is on distributed gossip, including information change in one direction only, and termination. The extension (permitted call sequences of the protocols) and the characterization of the classes of graphs where the (dynamic versions) of our protocols terminate were investigated in [5], where their main result is for LNS (in [9] the same question was answered for the (static) protocol CO). In [2,3] the focus is on the logical dynamics of call exchange. In [6,7] the gossip protocols were treated as random processes and it was shown that TOK and SPI have the same expectation. As simulations (some of which where theoretically corroborated) in [6,7] indicate, the expected duration of all protocols considered here is of the order $n \log n$, the 'usual' suspect in the gossiping community, but the constant factor may be different.

Organization of the Paper. In Sect. 2 we present all the definitions and relevant notions that are necessary for understanding our results. In Sect. 3 we present our main result which is the comparison of the reachability strength of the 5 protocols. In Sect. 4 we study the subreachability strength of our protocols and their reachability strength in unordered distributions. Finally in Sect. 5 we give directions for further work, mainly on parallel calls.

2   Terminology for Gossip Protocols and Reachability

In this section we give formal definitions for the notions of secret (sub)distributions and (sub)reachability. We always assume a complete network topology. A set of agents is represented by $\mathcal{A}$. We use the lower-case letters $a, b, c, d, \ldots$ for agents. At the start of any gossip protocol each agent has a unique secret. We denote the secrets by the corresponding upper-case letters $A, B, C, D, \ldots$ and there are no other secrets.

Definition 1 (Distribution of Secrets). *An n-distribution of secrets for a set of agents $\mathcal{A} = \{a_1, \ldots, a_n\}$ is an ordered n-tuple $(S_{a_1}, \ldots, S_{a_n})$ where each S_{a_i} is a subset of the set of all secrets $\{A_1, \ldots, A_n\}$. In the* initial *distribution every agent knows only her own secret, i.e. $S_{a_i} = \{A_i\}$ for all a_i. An agent a_i is an* expert *iff she knows all secrets, i.e. iff $S_{a_i} = \{A_1, \ldots, A_n\}$. In the* final *distribution every agent is an expert.*

In general, a distribution $(S_{a_1}, \ldots, S_{a_n})$ represents the situation in which each agent a_i knows exactly the secrets in S_{a_i}. We drop the references to $\mathcal{A}$, n

and secrets if this causes no confusion. We write (ABC, AB, ABC) instead of $(\{A, B, C\}, \{A, B\}, \{A, B, C\})$. We use the letters s, t (possibly primed or with subscripts) to represent a distribution. Finally, we observe that a distribution of secrets implicitly assumes an ordering on the agents.

Definition 2 (Call). *A* call *is an ordered pair* (a, b), *where* $a \neq b$ *for some agents* a, b. *We write* ab *instead of* (a, b). *A* call sequence *is a (possibly empty) finite or infinite sequence of calls. We write* $ab; cd; \ldots$ *for a call sequence. If* ab *occurs in a call sequence* σ, *we also write* $ab \in \sigma$, *slightly abusing language. By* (ab) *we mean the call* ab *or the call* ba. *Let* $s = (S_{a_1}, \ldots, S_{a_n})$ *be a distribution and consider* $(a_i a_j)$ *for some* $i < j$. *We apply any of the two calls* $a_i a_j$ *and* $a_j a_i$ *to* $(S_{a_1}, \ldots, S_{a_n})$ *as follows and obtain the new distribution*

$$s^{a_i a_j} := s^{a_j a_i} :=$$

$$(S_{a_1}, \ldots, S_{a_{i-1}}, S_{a_i} \cup S_{a_j}, S_{a_{i+1}}, \ldots, S_{a_{j-1}}, S_{a_i} \cup S_{a_j}, S_{a_{j+1}}, \ldots, S_{a_n}) .$$

We apply a finite call sequence σ *to a distribution* s *as follows:*

$$s^\sigma := \begin{cases} s, & \text{if } \sigma = \epsilon \\ (s^{ab})^\tau, & \text{if } \sigma = ab; \tau . \end{cases}$$

For example, we have $(A, B, C)^{ab;bc} = (AB, AB, C)^{bc} = (AB, ABC, ABC)$.

A call sequence σ is P-*permitted* if the restrictions of P allow every call in σ to be executed in the order given in σ. A P-permitted call sequence will also be called P-call sequence. A call sequence σ is called *successful* if the application of σ to an initial distribution leads to the final distribution where all agents know all secrets. If the applications of either σ or τ to the initial distribution lead to the same distribution we write $\sigma \approx \tau$.

Definition 3 (Reachability). *A distribution* s *is* P-reachable *if* s *can be obtained by applying a* P-*permitted call sequence on the initial distribution.*

The ANY-permitted calls are also called the *possible* calls and an ANY-reachable distribution is also called a *possible* distribution.

From a given n-distribution we can derive the set of possible calls that could have contributed to reaching that distribution, including an order on their execution. It is defined as follows.

Definition 4. *Let* s *be a distribution. The set of potential calls for* s *is* $PC(s) := \{ab \mid A \in S_b \text{ and } B \in S_a, \text{ for some agents } a, b\}$. *The order* $<_s$ *on* $PC(s)$ *is defined as follows. For any* $ab, cd \in PC(s)$:

$$\begin{aligned} ab < cd \ \ &\text{if } a = c \text{ and } D \notin S_b, \text{ or} \\ &b = c \text{ and } D \notin S_a, \text{ or} \\ &a = d \text{ and } C \notin S_b, \text{ or} \\ &b = d \text{ and } C \notin S_a. \end{aligned}$$

The pair $(PC(s), <_s)$ *is called the* set of potential call sequences *(for* s).

A call sequence σ consisting of calls from $PC(s)$ respects the order $<_s$ if, for every $ab <_s cd$, no occurrence of (ab) in σ is after any occurrence of (cd) in σ.

Let $(ab) <_s (cd)$ denote: $ab <_s cd$, $ba <_s cd$, $ab <_s dc$, and $ba <_s dc$. Now let σ and τ be call sequences. By $\sigma <_s \tau$ we mean that for every $xy \in \sigma$ and every $zw \in \tau$ if xy is related to zw then $xy <_s zw$; and that no pair of calls in σ are comparable and that the same holds for τ. We may additionally employ $(\sigma) <_s (\tau)$ meaning that for every $xy \in \sigma$ and every $zw \in \tau$, if xy and zw are comparable then $(xy) <_s (zw)$.

The proof of the next proposition is obvious.

Proposition 1. *Each distribution s uniquely determines a pair $(PC(s), <_s)$. Distribution s can only be obtained by a call sequence in which only calls in $PC(s)$ occur, and that respects the order $<_s$.*

We note (*i*) that a pair $(PC(s), <_s)$ does *not* uniquely determine a given distribution s, (*ii*) that calls may occur more than once (for example, in both directions, and as long as the order $<_s$ is respected), and (*iii*) that not all calls in $PC(s)$ need occur in a sequence reaching s. The proof of Theorems 5 demonstrates (*i*) and (*iii*). Concerning (*ii*), note that any call ab can be followed by (if the protocol so permits) the dual call ba as long as neither a nor b have been involved in other calls, without the second call ba affecting the distribution at that time.

Example 1. Consider the 4-distribution $s = (ABCD, ABCD, ABCD, ABCD)$. We have $PC(s) = \{ab, ac, ad, bc, bd, cd, ba, ca, da, cb, db, dc\}$ and $<_s = \emptyset$. Two different call sequences reaching s are $ab; cd; ac; bd$ and $ac; bd; ab; cd$. There are also call sequences that respect $<_s$ and do not reach s (e.g. $ab; ac; bd; cd$).

As a second example, consider the 3-distribution $t = (AB, ABC, BC)$. Then $PC(t) = \{ab, bc, ba, cb\}$ and $(ab) <_t (bc)$ and $(bc) <_t (ab)$. No call sequence respecting $<_t$ reaches t. Indeed, t is not ANY-reachable.

Finally we present the notion of *subreachability* that uses that of the *restriction* of a distribution.

Definition 5. *Suppose we have $\mathcal{A}' \subseteq \mathcal{A}$ with $m = |\mathcal{A}'|$ and $n = |\mathcal{A}|$. The $\mathcal{A}'$-restriction of an n-distribution $(S_{a_1}, \ldots, S_{a_n})$ for $\mathcal{A}$ is the m-distribution $(S_{a'_1}, \ldots, S_{a'_m})$ such that for all $a'_j \in \mathcal{A}'$, if $a_i = a'_j$ then $S_{a'_j} = \{A_k \in S_{a_i} \mid a_k \in \mathcal{A}'\}$.*

Definition 6 (Subreachability). *A distribution s for a set of agents $\mathcal{A}'$ is P-subreachable if there is a distribution t for an extended set of agents $\mathcal{A} \supseteq \mathcal{A}'$ such that t is P-reachable and s is the $\mathcal{A}'$-restriction of t.*

Note that P-reachable implies P-subreachable, namely when $\mathcal{A}' = \mathcal{A}$ above.

3   Reachability

In this section we provide an answer to the question: "are all P_1-reachable distributions also P_2-reachable?" for any P_1 and P_2 from the five protocols. It is interesting that although all five protocols can reach the final distribution on complete graphs [5], their reachability strength on intermediate distributions varies.

Theorem 1.

1. *There is a distribution that is not reachable by any of the five protocols.*
2. *Every* CO-, LNS-, SPI- *and* TOK-*distribution is* ANY-*reachable.*
3. *Every* LNS-*reachable distribution is* CO-*reachable.*

Proof. This follows from the protocol definitions and because (AB, A) is not reachable by any of the protocols.

Our next, rather unexpected, result is that, although TOK has a stricter calling condition than ANY, these two protocols reach the same set of distributions. Recall that TOK can be thought of as demanding that, in order to make a call, an agent has to possess a token. Every agent starts out with a token, and in a call ab the token of a is given to b. In the following lemma we use the fact that a call ab can be followed by a call ba in which the token is returned to a.

Lemma 1 (Token Density Lemma). *Let s be a* TOK-*reachable distribution and let a, b be two agents. Then s can be reached by a* TOK-*call sequence σ such that after the execution of σ at least one of a and b have a token.*

Proof. The Lemma follows easily from the following more general claim.

Claim. Let σ be any TOK sequence, let $k \in \mathbb{N}$, $I = \{1, \ldots, k\}$ and let $f, g : I \to \mathcal{A}$ be injections such that $f(I) \cap g(I) = \emptyset$ for some set of agents $\mathcal{A}$. Then there is a TOK sequence σ' such that (i) $\sigma \approx \sigma'$ and (ii) for every $1 \leq i \leq k$ at least one of $f(i)$ and $g(i)$ has a token after σ'.

Proof (of the Claim and the Lemma). By induction on the length of σ. If σ is of length 1 the claim is trivial. Assume then as induction hypothesis that the claim holds for all sequences shorter than σ. Now, let $\sigma = \tau; ab$. We distinguish whether the agents of the final call in σ are in the images of f and g.

- Suppose $a, b \notin f(I) \cup g(I)$. Then let $f', g' : I \cup \{k + 1\} \to \mathcal{A}$ be extensions of f and g with $f'(k + 1) = a, g'(k + 1) = b$. By the induction hypothesis, there is τ' such that $\tau \approx \tau'$ and for every $1 \leq i \leq k + 1$ either $f(i)$ or $g(i)$ has a token after τ'. Then $\tau'; (ab) \approx \sigma$ and for every $1 \leq i \leq k$, either $f(i)$ or $g(i)$ has a token after $\tau'; (ab)$.
- Suppose $a \in f(I) \cup g(I)$ and $b \notin f(I) \cup g(I)$. Without loss of generality, suppose that $f(1) = a$. Now, let f', g' be as f, g except $g'(1) = b$. By the induction hypothesis, there is a τ' such that $\tau \approx \tau'$ and either $f'(i)$ or $g'(i)$

ends up with a token. In particular, either a or b has a token after τ'. If a has the token, let $\sigma' = \tau'; ab; ba$, otherwise let $\sigma' = \tau'; ba$. In either case, (i) $\sigma' \approx \sigma$, (ii) for $i > 1$ either $f(i)$ or $g(i)$ has a token because they had it after τ' and (iii) a has a token so either $f(1)$ or $g(1)$ has a token.

- Suppose $a = f(i)$ and $b = g(i)$. By the induction hypothesis τ' exists, and $\sigma' = \tau'; (ab)$ suffices.
- Suppose $a = f(i)$ and $b \in f(I) \cup g(I) \setminus g(i)$. Without loss of generality, $b = f(j)$. Let f', g' be as f, g except $g'(i) = b$ and $f'(j) = g(i)$. Let τ' be such that $\tau \approx \tau'$ and for every l either $f'(l)$ or $g'(l)$ ends up with a token. Since $f'(i) = a$ and $g'(i) = b$, the sequence $\tau'; (ab)$ is TOK. Note furthermore that $f'(j) = g(i)$ and $g'(j) = g(j)$, so one of the pairs $(a, g(i))$ and $(f(j), b)$ has at least one token. By inverting the (ab) call if necessary, we can ensure that the other pair keeps the token of the (ab) call. As such, either $\tau'; (ab); (ba)$ or $\tau'; (ab)$ satisfies the conditions of the claim. $\square$

Theorem 2. *Every* ANY-*reachable distribution is* TOK-*reachable.*

Proof. We will show that for every ANY sequence σ there is a TOK sequence σ' such that $\sigma \approx \sigma'$. The proof proceeds by induction on the length of the call sequence σ and by repeatedly applying Lemma 1.

If σ is of length 1, then σ is a TOK sequence. Assume then as induction hypothesis that the theorem holds for all sequences shorter than σ, and let $\sigma = \tau; ab$. By the induction hypothesis, there is a TOK sequence τ' such that $\tau \approx \tau'$. Because τ' is a TOK sequence it follows from Lemma 1 that there is a TOK sequence τ'' such that (i) $\tau' \approx \tau''$ and (ii) either a or b has a token after τ''. It follows that $\sigma' = \tau''; (ab)$ is a TOK sequence, and $\sigma \approx \sigma'$. $\square$

We continue to compare the sets of distributions reachable by all other protocols. Theorems 3 and 4 are generalized versions of [6, Theorems 3 and 4].

Theorem 3. *There is a* SPI-*reachable distribution that is not* CO-*reachable.*

Proof. Consider the 4-distribution $t = (ABCD, ABCD, ABC, ABD)$. We show that in order to reach t one has to choose the same call twice.

- The initial 4-distribution is (A, B, C, D).
- Since c and d must not learn each others secret, the first call cannot be cd. Furthermore, if the first call is ac then, when d learns a's secret, she will also learn c's secret. With similar arguments we can show that the first call cannot be ad, bc or bd. Thus in order to reach t we have to select ab which leads to (AB, AB, C, D).
- Now, d has to learn A and B. So, without loss of generality the next call is ad which leads to (ABD, AB, C, ABD).
- Now, c has to learn A and B. The only way of achieving this is by selecting cb which leads to (ABD, ABC, ABC, ABD).
- Until now we have made the CO-permitted call sequence: $ab; ad; bc$. The only way of reaching t is by selecting call ab again, which is a violation of CO.

The call sequence that reached t is: $\sigma = ab; ad; cb; ab$. No agent who has been called initiates a call, hence σ is SPI-permitted.   □

Theorem 4. *There is a* CO- *and* SPI-*reachable distribution that is not* LNS-*reachable.*

Proof. Consider the 6-distribution:

$$t = (ABCDEF, ABC, ABCDE, ABCDEF, DEF, ABDEF) .$$

We will show that we can reach t without violating CO or SPI, but at the price of having to make a call between agents that already know each other's secrets.

- The initial 6-distribution is (A, B, C, D, E, F).
- Agent b has to learn A and C and nothing else and e has to learn d and f and nothing else. So, without loss of generality, the first four calls can be $ab; cb; ed; ef$, which are clearly both SPI-and CO-permitted and lead to

$$(AB, ABC, ABC, DE, DEF, DEF) .$$

- Now c has to learn everything but F. The only way of achieving this is by selecting the call cd. Similarly in order for f to learn everything but c we need to select call af. So, until now we have made the LNS- and SPI-permitted call sequence $ab; cb; ed; ef; cd; af$ which leads to

$$(ABDEF, ABC, ABCDE, ABCDE, DEF, ABDEF) .$$

- Only the CO- and SPI-permitted call ad will now lead to t. But ad is not LNS-permitted.   □

Theorem 5. *There is an* LNS-*reachable distribution that is not* SPI-*reachable.*

Proof. We will show that there is a 16-distribution reachable by LNS but not by SPI. Recall that (ab) represents a call between a and b, which can be instantiated as either ab or ba. Consider the following call sequence $\sigma := \sigma_1; \sigma_2; \sigma_3$, where

$$\sigma_1 = \quad (12); (34); (56); (78); (ab); (cd); (ef); (gh)$$
$$\sigma_2 = \quad (23); (45); (67); (81); (bc); (de); (fg); (ha)$$
$$\sigma_3 = \quad (1a); (4c); (7h); (6f).$$

This sequence has three phases σ_1, σ_2, σ_3, as shown on different lines. We can represent this sequence visually as in Fig. 2, where the solid lines are calls that happen in σ_1, dashed lines happen in σ_2, and dotted lines in σ_3.

We will show that σ is not SPI-permitted (nor any of its order variants). Suppose towards a contradiction that it is.

In the first stage, the callee member of each pair loses its token. In the second stage, every agent is involved in one more call. If agent 1 still has a token, then 2 does not. So 3 must have a token, otherwise (23) could not take place. But then 4 does not have a token, so 5 must have it, and so on. It follows that in

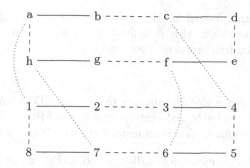

Fig. 2. The call sequence $\sigma = \sigma_1; \sigma_2; \sigma_3$.

both blocks, either all even agents have lost their token or all odd agents have lost their token (where a, c, e, g are "odd" and b, d, f, h are "even").

Now, consider the third stage. Here, calls (1a), (4c), (7h) and (6f) are supposed to happen. Note that these include every combination of even/odd from both groups: odd number and odd letter (1a), even number and odd letter (4c), odd number and even letter (7h), and even number and even letter (6f). So at least one of these calls is between two agents that do not have their token any more. It follows that the sequence σ is not SPI-permitted.

We still need to show that there is no sequence of other calls that is SPI-permitted and reaches the same distribution of secrets. However, this is fairly straightforward. The distribution s produced by the sequence σ is as follows (let the secret by agent named $i \in \mathbb{N}$ be also i):

$1: 1278ABGH$	$5: 3456$	$a: 1278ABGH$	$e: CDEF$
$2: 1234$	$6: 5678EFGH$	$b: ABCD$	$f: 5678EFGH$
$3: 1234$	$7: 5678ABGH$	$c: 3456ABCD$	$g: EFGH$
$4: 3456ABCD$	$8: 1278$	$d: CDEF$	$h: 5678ABGH$

Given this s, we now compute the set of *potential call sequences* $(PC(s), <_s)$. It is easy to show that the set $PC(s)$ consists of all calls in σ plus (7a) and (6h).

Our first observation is that since $A \notin S_6$ and $6 \notin S_a$ it holds that $(67) <_s$ (7a) and that (7a) $<_s$ (67). Thus (67) and (7a) cannot exist in the same call sequence leading to s. It is not difficult to see that (67) is necessary in order to produce s since give the order constraints there is no other way for 6 and 7 to exchange their secrets. Hence, (7a) cannot be used to a call sequence leading to s. In a similar fashion we obtain that $(ha) <_s$ (6h) and that (6h) $<_s$ (ha) and since (ha) is again necessary we conclude that (6h) cannot be used to a call sequence leading to s.

Additional we observe that $(\sigma_1) <_s (\sigma_2)$ and that $(\sigma_2) <_s (\sigma_3)$. Therefore, except for the order of calls within σ_1, σ_2, and σ_3, and the call directions, only σ leads to s. Finally, one can easily verify that σ is also LNS-permitted. □

Theorems 1, 3 and 5 lead to the following corollary. Together with some already discussed inclusions this completes the comparison of the reachability strength between the five protocols (see also Fig. 1).

Corollary 1.

1. *There is a* TOK-*reachable distribution that is not* CO-*reachable.*
2. *There is a* TOK-*reachable distribution that is not* SPI-*reachable.*
3. *There is a* CO-*reachable distribution that is not* SPI-*reachable.*

We presented several examples of distributions that are reachable by some of the protocols and unreachable by others. A natural question to ask is "are these distributions optimal counter-examples?", i.e., "did we use the smallest possible number of agents?". We implemented an algorithm that counts the reachable distributions for all five protocols modulo isomorphism (i.e., modulo renaming the agents) [13]. The results of this implementation can be found in Table 1.

Given the inclusions of Theorem 1, Table 1 tells us that all protocols reach the same set of distributions for up to 3 agents. This implies that the 4 distribution in Theorem 3 is optimal. We also see that LNS and CO reach the same set of distributions for up to 5 agents, which implies that the 6-distribution in Theorem 4 is optimal. We do not know whether the non-SPI reachable 16-distribution in the proof of Theorem 5 is optimal (due to a huge combinatorial explosion the implementation in [13] can only count distributions up to at most 7 agents).

Table 1. Number of non-isomorphic reachable distributions for up to 5 agents. For LNS and ANY these numbers are also in the On-Line Encyclopedia of Integer Sequences (OEIS) as https://oeis.org/A307085 and https://oeis.org/A318154, respectively.

n	LNS	CO	SPI	TOK = ANY
2	2	2	2	2
3	4	4	4	4
4	15	15	16	16
5	97	97	111	111

4 Subreachability, Unorderded Distributions

Subreachability in Ordered Distributions. While there are distributions that are not even ANY-reachable, all possible distributions are subreachable by any of the five protocols we consider. In [8] this was shown for a more general setting using incomplete network topologies that change dynamically when agents exchange 'phone numbers', but only for the protocol ANY.

Theorem 6. *All distributions are* ANY-, CO-, LNS-, SPI-, TOK-*subreachable.*

Proof. We adapt the proof of [8, Section 6.2]. Given a distribution s for agents $\mathcal{A}$, let the *number of secrets known by the agents in s* be defined as $\mathsf{sec}(s) = \Sigma_{a \in \mathcal{A}} |S_a|$, where S_a is the set of secrets known by a in s.

For any protocol P and for any distribution s, we prove by induction on $m = \mathsf{sec}(s)$ that s is P-subreachable. In the base case $m = 1$ the distribution must have shape (A) for a single agent a. This distribution is clearly (sub)reachable by all protocols and the empty call sequence.

Assuming that the result holds for m secrets we will show that it holds for $m + 1$ secrets. We need to distinguish two subcases: either there is an agent a who knows a single secret (i.e., an agent who has not made any call yet), or not.

In the first subcase, as $\Sigma_{b \in \mathcal{A} \setminus \{a\}} |S_b| = m$, by induction hypothesis there is a call sequence σ such that the $(\mathcal{A} \setminus \{a\})$-restriction of s is P-subreachable by σ from the initial distribution for the set of agents $\mathcal{A} \setminus \{a\}$. Clearly, s is then P-subreachable by the same call sequence σ from the initial distribution for the set of agents $\mathcal{A}$, as agent a has not been involved in any call. This holds for all five protocols ANY, CO, LNS, SPI, TOK.

In the second subcase, there must be an agent a who knows at least one other secret B than its own secret A. As $|S_a \setminus \{B\}| + \Sigma_{b \in \mathcal{A}, b \neq a} |S_b| = m$, by induction there is a call sequence σ such that s' is P-subreachable by σ, where s' is as s (and defined for the same set of agents) except that $S'_a = S_a \setminus \{B\}$.

First, assume that P is one of ANY, CO, or LNS. Let $c \notin \mathcal{A}$. The role of agent c will be to inform a of B and nothing else. Let s'' be the distribution reached by executing $bc; \sigma; ca$ in the initial distribution for agents $\mathcal{A} \cup \{c\}$. Observe that s is the restriction to $\mathcal{A}$ of s''. Also, call bc is ANY-, CO-, and LNS-permitted, as it is the first call. The last call ca is obviously ANY-permitted. It is CO-permitted because prefix $bc; \sigma$ does not contain a call between c and a. It is also LNS permitted, since c did not learn a in the first call and was not involved in σ. Therefore $bc; \sigma; ca$ is an ANY- CO- and LNS-permitted call sequence reaching s.

Now let P = SPI. Let in this case $c, d \notin \mathcal{A}$, and consider call sequence $bc; dc; \sigma; da$ for set of agents $\mathcal{A} \cup \{c, d\}$, resulting in distribution s''. In first call bc, b keeps its token, as in the initial distribution for $\mathcal{A}$, but c loses its token (so c can no longer inform a of B at the end, as in the previous case). In the second call dc, d keeps it token and learns B from c. Therefore, in the last call da, d can inform a of B, as desired. Also note that s is the restriction to $\mathcal{A}$ of s''. Therefore $bc; dc; \sigma; da$ is a SPI-permitted call sequence reaching s.

Finally, let P = TOK. This subcase is fairly similar to the subcase SPI. Again, as for SPI, let $c, d \notin \mathcal{A}$. However, now consider call sequence $cb; dc; \sigma; ca$. In the first call c hands its token to b. So b can still engage in σ as before. In the second call dc agent d hands back a token to agent c. Therefore, the final call ca (instead of da, for SPI) is TOK-permitted resulting in c again informing a of B. Therefore $cb; dc; \sigma; ca$ is a TOK-permitted call sequence reaching s. $\square$

Reachability in Unordered Distributions. To illustrate the difference between reachability in unordered and ordered distributions, let us consider the following example. In Theorem 3 we have shown that the ordered distribution $(ABCD, ABCD, ABC, ABD)$ is not CO-reachable. However, this holds only if we understand it as an ordered distribution. It is not difficult to see that the unordered distribution $\{ABCD, ABCD, ABC, ABD\}$ is CO-reachable by the call sequence $ab; ac; bd; cd$.

Theorem 7. *The protocols* ANY, TOK *and* SPI *reach the same unordered distributions.*

Proof. The fact that ANY and TOK reach the same set of ordered distributions (Theorem 2) implies that they also reach the same set of unordered ones. To show that TOK and SPI also reach the same set of unordered distributions we proceed as follows: assume that we have the unordered distribution $\{S_{a_1}, \ldots, S_{a_i}, S_{a_j}, \ldots S_{a_n}\}$ wherein (at least) the agent knowing S_{a_i} possesses a token. Both the TOK and the SPI-call between agents knowing S_{a_i} and S_{a_j} will lead to $\{S_{a_1}, \ldots, S_{a_i} \cup S_{a_j}, S_{a_i} \cup S_{a_j}, \ldots S_{a_n}\}$ where exactly one of the agents that know $S_{a_i} \cup S_{a_j}$ possesses a token. These two unordered distributions are the same, which proves the theorem.   □

5   Further Research: Parallel Gossip

We very succinctly describe some results for the setting wherein agents may make calls in parallel. Instead of *sequences of individual calls*, one now considers *sequences of rounds of calls*, where a round of calls consists of a set of calls made in parallel. Different semantics for parallel calls include the 'classical' 1970s setting of gossip [12] wherein calls made in parallel must be mutually disjoint, and the 'modern' 1990s setting of gossip [11] wherein agents, instead, may receive multiple calls. The latter leads to novel reachable distributions, for example, (AB, ABC, BC) is reachable by the simultaneous calls ab, ba, cb, wherein agent b simultaneously receives A from a and C from c. Let us call such a distribution *parallel reachable*, where the notion used so far is *sequential reachable*.

Although $(ABCD, ABCD, ABC, ABD)$ is not sequential CO-reachable (see Theorem 3), it is parallel CO-reachable by the sequence $\{ad, bc, ca, db\}; \{ab\}$ in two rounds. Similarly, $(ABCDEF, ABC, ABCDE, ABCDEF, DEF, ABDEF)$ is not sequential LNS-reachable (Theorem 4), but it is parallel LNS-reachable by the sequence $\{ab, cb, de, fe\}; \{ca, dc, fd, af\}; \{da\}$ in three rounds. Hence reachability in parallel gossip is very different and should be further investigated.

As we mentioned in the introduction the main motivation for studying reachability issues in gossip protocols is to provide an observer with some tools for understanding which protocol is currently being used by the agents. Some further research in this setting could also involve determining the reasoning power that such an observer should have or studying the design of a procedure/determining the resources needed for constructing such observers.

Beyond parallel calls and the observer construction, while in this paper we restricted our attention to only five protocols, our aim is to investigate reachability for protocols that have epistemic conditions. Examples are "call if you know that/consider it possible that an agent will learn a secret" and "don't call if you are an expert". In general, our results should be received as part of a bigger effort to compare the combinatorial properties of epistemic gossip protocols.

Acknowledgements. We would like to thank the anonymous reviewers for their helpful corrections and suggestions. Hans van Ditmarsch is also affiliated to IMSc, Chennai, as associate researcher.

References

1. Apt, K., Grossi, D., van der Hoek, W.: Epistemic protocols for distributed gossiping. In: Proceedings of 15th TARK (2015). https://doi.org/10.4204/EPTCS.215. 5

2. Attamah, M., van Ditmarsch, H., Grossi, D., van der Hoek, W.: Knowledge and gossip. In: Proceedings of 21st ECAI, pp. 21–26. IOS Press (2014). https://doi. org/10.3233/978-1-61499-419-0-21

3. Attamah, M., van Ditmarsch, H., Grossi, D., van der Hoek, W.: The pleasure of gossip. In: Başkent, C., Moss, L.S., Ramanujam, R. (eds.) Rohit Parikh on Logic, Language and Society. OCL, vol. 11, pp. 145–163. Springer, Cham (2017). https:// doi.org/10.1007/978-3-319-47843-2_9

4. van Ditmarsch, H., van Eijck, J., Pardo, P., Ramezanian, R., Schwarzentruber, F.: Epistemic protocols for dynamic gossip. J. Appl. Logic **20**, 1–31 (2017). https:// doi.org/10.1016/j.jal.2016.12.001

5. van Ditmarsch, H., van Eijck, J., Pardo, P., Ramezanian, R., Schwarzentruber, F.: Dynamic gossip. Bull. Iran. Math. Soc. **45**(3), 701–728 (2019). https://doi.org/10. 1007/s41980-018-0160-4

6. van Ditmarsch, H., Kokkinis, I., Stockmarr, A.: Reachability and expectation in gossiping. In: An, B., Bazzan, A., Leite, J., Villata, S., van der Torre, L. (eds.) PRIMA 2017. LNCS (LNAI), vol. 10621, pp. 93–109. Springer, Cham (2017). https://doi.org/10.1007/978-3-319-69131-2_6

7. van Ditmarsch, H., Kokkinis, I.: The expected duration of sequential gossiping. In: Belardinelli, F., Argente, E. (eds.) EUMAS/AT 2017. LNCS (LNAI), vol. 10767, pp. 131–146. Springer, Cham (2018). https://doi.org/10.1007/978-3-030-01713-2_10

8. Gattinger, M.: New directions in model checking dynamic epistemic logic. Ph.D. thesis, University of Amsterdam (2018). https://malv.in/phdthesis (ILLC Dissertation Series DS-2018-11)

9. Göbel, F., Cerdeira, J.O., Veldman, H.J.: Label-connected graphs and the gossip problem. Discrete Math. **87**(1), 29–40 (1991). https://doi.org/10.1016/0012-365X(91)90068-D

10. Hedetniemi, S., Hedetniemi, S., Liestman, A.: A survey of gossiping and broadcasting in communication networks. Networks **18**, 319–349 (1988). https://doi.org/10. 1002/net.3230180406

11. Kermarrec, A.M., van Steen, M.: Gossiping in distributed systems. SIGOPS Oper. Syst. Rev. **41**(5), 2–7 (2007). https://doi.org/10.1145/1317379.1317381

12. Knödel, W.: New gossips and telephones. Discrete Math. **13**, 95 (1975). https:// doi.org/10.1016/0012-365X(75)90090-4

13. Kokkinis, I.: Implementation for reachability and expectation in gossiping. https:// github.com/Jannis17/gossip_protocol_expectation

Author Index

Avni, Guy   1

Blažej, Václav   33
Brihaye, Thomas   48
Bruyère, Véronique   48, 63

Chonev, Ventsislav   79

Day, Joel D.   93
Delzanno, Giorgio   107
Dutta, Souradeep   22

Ehlers, Thorsten   93

Fatès, Nazim   121

Gattinger, Malvin   218
Goeminne, Aline   48

Hampson, Christopher   137
Henzinger, Thomas A.   1

Ibsen-Jensen, Rasmus   1

Köcher, Chris   149
Kokkinis, Ioannis   218
Křišt'an, Jan Matyáš   33
Kuijer, Louwe B.   218
Kulczynski, Mitja   93

Manea, Florin   93
Marcovici, Irène   121
Mover, Sergio   22

Novotný, Petr   1
Nowotka, Dirk   93

Pérez, Guillermo A.   63
Poulsen, Danny Bøgsted   93
Protasov, Vladimir Yu.   13

Raskin, Jean-François   63
Reynier, Pierre-Alain   164

Sandler, Andrei   178
Sankaranarayanan, Sriram   22
Schmitz, Sylvain   193
Servais, Frédéric   164

Taati, Siamak   121
Tamines, Clément   63
Thomasset, Nathan   48
Tredup, Ronny   202
Tveretina, Olga   178

Valla, Tomáš   33
van Ditmarsch, Hans   218

Zetzsche, Georg   193

Printed in the United States
By Bookmasters